# Lecture Notes in Control and Information Sciences

Edited by A.V. Balakrishnan and M. Thoma

## 68

T. Kaczorek

# Two-Dimensional Linear Systems

Springer-Verlag
Berlin Heidelberg New York Tokyo

**Series Editors**
A. V. Balakrishnan · M. Thoma

**Advisory Board**
L. D. Davisson · A. G. J. MacFarlane · H. Kwakernaak
J. L. Massey · Ya Z. Tsypkin · A. J. Viterbi

**Author**
Tadeusz Kaczorek
ul. Etiudy Rewolucyjnej 40 m. 33
02-643 Warszawa
Poland

ISBN 3-540-15086-2 Springer-Verlag Berlin Heidelberg New York Tokyo
ISBN 0-387-15086-2 Springer-Verlag New York Heidelberg Berlin Tokyo

This work is subject to copyright. All rights are reserved, whether the whole or part of the material is concerned, specifically those of translation, reprinting, re-use of illustrations, broadcasting, reproduction by photocopying machine or similar means, and storage in data banks. Under § 54 of the German Copyright Law where copies are made for other than private use, a fee is payable to "Verwertungsgesellschaft Wort", Munich.

© Springer-Verlag Berlin, Heidelberg 1985
Printed in Germany

Offsetprinting: Mercedes-Druck, Berlin
Binding: Lüderitz und Bauer, Berlin
2061/3020-543210

TO MY MOTHER

Preface

A growing interest has been developed over the past few years in problems involving signals and systems that depend on more than one variable. These multidimensional signals and systems have been studied in relation to several modern engineering fields such as multidimensional digital filtering, multivariable network realizability, multidimensional system synthesis digital picture processing, seismic data processing, X-ray image enhancement, the enhancement and analysis of aerial photographs for detection of forest fires or crop damage, the analysis of satellite weather photos, image deblurring, etc. Most of the major results concerning the multidimensional signals and systems are developed for two-dimensional /2-D/ cases.

These results may be grouped as follows.

1. <u>2-D systems and filters</u>. The 2-D linear shift invariant systems are described by a convolution of the input and the unit impulse response. Some of the problems already investigated refer to the questions of recursibility, stability and limit cycles.

2. <u>2-D state-space models</u>. Based on the state-space description several properties of 2-D systems such as controllability, observability, canonical forms, minimality, etc. have been investigated.

3. <u>2-D image processing, random fields and space-time processing.</u> These problems have drawn considerable attention and have shown great potential for practical applications such as X-ray image enhancement, image deblurring, weather prediction, seismic data analysis, radar and sonar array processing, etc.

4. <u>2-D feedback design techniques</u>. These problems refer to the general area of developing feedback design techniques so that the closed-loop system has pre-assigned desirable characteristics. The eigenvalue assignment exact model-matching, transfer function factorization, minimum energy control, observers have been considered

in many papers.

The main objective of this monograph is to present recent developments in 2-D linear system theory.

The monograph is organized as follows.

Chapter 1 presents Roesser's model, Attasi's model and two Fornasini-Marchesini's models. The transition matrices for the models are defined and the general response formulas are given.

The transfer function matrix, the realization problem and the separability of transfer function matrices are considered in Chapter 2.

Different notions of the controllabity, observability and reachability are described in Chapter 3. The minimum energy control of 2-D systems is also considered.

Chapter 4 gives definitions and stability tests for 2-D systems described by the transfer function matrices and the state equations.

The stabilization problems are also considered. Some new methods concerning eigenvalue assignment for 2-D and 3-D linear systems are given in Chapter 5. The asymptotic and deadbeat observers, the exact model matching and the decoupling are considered in Chapter 6.

Finally, Chapter 7 presents some new results concerning deadbeat control and deadbeat servo problems.

An Appendix of basic definitions, theorems and computational algorithms has been included for the sake of greater comprehensiveness.

The monograph is addressed to graduate students specializing in control, scientists and engineers engaged in control system research and development and mathematicians interested in control problems.

I wish to thank dr B.Eichsteadt and dr M.Kocięcki for their valuable remarks, suggestions and comments.

T.Kaczorek

CONTENTS

CHAPTER 1.  STATE-SPACE MODELS AND RESPONSE FORMULAE

1.1. State-Space models of two-dimensional linear systems ......... 1
1.2. Relations between the models .................................. 7
1.3. Transition matrix and general response formula for Roesser's model .............................................. 9
1.4. State-space models of three and N-dimensional linear systems ....................................................... 13
1.5. Transition matrix and general response formula for 3-D and N-D systems .......................................... 19
1.6. Cayley-Hamilton theorem ....................................... 23
1.7. 2-D division algorithm ........................................ 25
1.8. Computation of the transition matrix .......................... 28
1.9. Solution of Roesser's model as a function of 2-D eigenvalues .................................................. 33
1.10. General response formulae for Fornasini-Marchesini's models ....................................................... 35
1.11. Characteristic polynomial and Cayley-Hamilton theorem for the second Fornasini-Marchesini's model .................. 39

CHAPTER 2. TRANSFER FUNCTION MATRIX AND REALIZATION PROBLEM

2.1. Transfer function matrix of the Roesser's model ............... 44
2.2. Transfer function matrix of the Tzafestas-Pimenides' model ....................................................... 56
2.3. Transfer function matrix of the Fornasini-Marchesini's models ....................................................... 65
2.4. Matrix fraction description ................................... 67
2.5. Proper transfer function matrices ............................. 69
2.6. Realization problem ........................................... 78

## CHAPTER 3. CONTROLLABILITY AND OBSERVABILITY

3.1. Local controllability and observability of
Roesser's model .............................................. 107
3.2. Separate local controllability and observability
of Roesser's model .......................................... 112
3.3. Modal controllability and observability of
Roesser's model ............................................. 116
3.4. Separability of transfer function matrices ................... 127
3.5. Minimum energy control of Roesser's model ................... 134
3.6. Local controllability and observability of
Fornasini-Marchesini's models ............................... 138

## CHAPTER 4. STABILITY AND STABILIZATION

4.1. Stability of 2-D linear input-output systems ................. 148
4.2. Stability of Roesser's model ................................. 157
4.3. Asymptotic and exponental stability of
Fornasini-Marchesini's models ............................... 165
4.4. Margin of stability .......................................... 169
4.5. Stabilization of 2-D systems by state feedback or
output feedback ............................................. 171
4.6. The Lyapunov equation for 2-D systems ........................ 177

## CHAPTER 5. CHARACTERISTIC POLYNOMIAL AND EIGENVALUE ASSIGNMENT

5.1. Paraskevopoulos' method of coefficient assignment ............ 186
5.2. Characteristic polynomial assignment and determination
of the residual polynomial .................................. 197
5.3. Characteristic polynomial assignment by dynamic
output feedback ............................................. 201
5.4. Characteristic polynomial assignment using
PID controllers ............................................. 206
5.5. Eigenvalue assignment ........................................ 215

CHAPTER 6. OBSERVERS, EXACT MODEL MATCHING AND DECOUPLING

6.1. Asymptotic and deadbeat observers ............................. 243
6.2. Exact model matching via static state feedback ............... 259
6.3. Exact model matching via static output feedback .............. 269
6.4. Exact model matching via dynamic output feedback ............. 270
6.5. Sebek's method of exact model matching ....................... 274
6.6. Decoupling by state feedback ................................. 283

CHAPTER 7. DEADBEAT CONTROL AND DEADBEAT SERVO PROBLEM

7.1. Polynomial design of deadbeat control laws ................... 306
7.2. Output deadbeat control problem .............................. 310
7.3. Output deadbeat control of closed-loop systems ............... 317
7.4. Deadbeat control of open-loop system ......................... 320
7.5. Deadbeat servo problem for single-input
     single-output systems ........................................ 324
7.6. Deadbeat servo problem for multivariable linear system ....... 328

APPENDIX

1. Function of 2-D matrix ......................................... 338
2. Two-dimensional Z transformation ............................... 346
3. Euclidean algorithm, Hermite and Smith forms of 2-D
   polynomial matrices ............................................ 350
4. Factorization of 2-D polynomial matrices ....................... 357
5. Coprimeness of 2-D polynomials and polynomial matrices ......... 366
6. Matrix fraction description .................................... 379
7. 2-D polynomial matrix equations ................................ 386

# 1 STATE-SPACE MODELS AND RESPONSE FORMULAE

## 1.1 STATE-SPACE MODELS OF TWO-DIMENSIONAL LINEAR SYSTEMS

Roesser's model.

Roesser's model /RM/ is defined by the equations [13]

$$\begin{bmatrix} x^h/i+1,j/ \\ x^v/i,j+1/ \end{bmatrix} = \begin{bmatrix} A_{11} & A_{12} \\ A_{21} & A_{22} \end{bmatrix} \begin{bmatrix} x^h/i,j/ \\ x^v/i,j/ \end{bmatrix} + \begin{bmatrix} B_1 \\ B_2 \end{bmatrix} u/i,j/ \qquad /1.1/$$

$$y/i,j/ = \begin{bmatrix} C_1 & C_2 \end{bmatrix} \begin{bmatrix} x^h/i,j/ \\ x^v/i,j/ \end{bmatrix} + D\, u/i,j/ \qquad (i,j \geqslant 0) \qquad /1.2/$$

where  i is an integer-valued vertical coordinate,
j is an integer-valued horizontal coordinate,
$x^h/i,j/ \in R^{n_1}$ is the horizontal state vector,
$x^v/i,j/ \in R^{n_2}$ is the vertical state vector,
$u/i,j/ \in R^m$ is the input vector,
$y/i,j/ \in R^l$ is the output vector,
$A_{11}, A_{12}, A_{21}, A_{22}, B_1, B_2, C_1, C_2, D$ are real matrices of appropriate dimensions.

Boundary conditions for /1.1/ are given by

$$x^h/0,j/, \quad x^v/i,0/ \quad \text{for} \quad i,j = 0,1,2,\ldots \qquad /1.1a/$$

Introducing the matrices and vectors

$$A = \begin{bmatrix} A_{11} & A_{12} \\ A_{21} & A_{22} \end{bmatrix}, \quad B = \begin{bmatrix} B_1 \\ B_2 \end{bmatrix}, \quad C = \begin{bmatrix} C_1 & C_2 \end{bmatrix},$$

$$x' = \begin{bmatrix} x^h/i+1,j/ \\ x^v/i,j+1/ \end{bmatrix}, \quad x = \begin{bmatrix} x^h/i,j/ \\ x^v/i,j/ \end{bmatrix}, \quad u = u/i,j/, \quad y = y/i,j/$$

we can rewrite /1.1/ and /1.2/ in the form

$$x' = Ax + Bu \qquad /1.1^*/$$
$$y = Cx + Du \qquad /1.2^*/$$

### Example 1.1
Consider the equation [12]

$$\frac{\partial T/x,t/}{\partial x} = - \frac{\partial T/x,t/}{\partial t} - T/x,t/ + U/t/ \qquad /1.3/$$

with initial and boundary conditions

$$T/x,0/ = f_1/x/, \quad T/0,t/ = f_2/t/ \qquad /1.3a/$$

where $T/x,t/$ is an unknown function (usually the temperature) at $x$(space) $\in [0,x_f]$ and $t$(time) $\in [0,\infty]$, $U/t/$ is a given force function and $f_1/x/$, $f_2/t/$ are given functions.

The equation /1.3/ describes some thermal processes, for example in chemical reactors, heat exchangers and pipe furnaces /Fig. 1.1/.

Taking

$$T/i,j/ = T/i\Delta x, j\Delta t/, \quad U/j/ = U/j\Delta t/,$$

$$\frac{\partial T/x,t/}{\partial t} \simeq \frac{T/i,j+1/ - T/i,j/}{\Delta t} \qquad \frac{\partial T/x,t/}{\partial x} \simeq \frac{T/i,j/ - T/i-1,j/}{\Delta x}$$

we can write /1.3/ in the form

$$T/i,j+1/ = a_1 T/i,j/ + a_2 T/i-1,j/ + bU/j/ \qquad /1.4/$$

where

$$a_1 = 1 - \frac{\Delta t}{\Delta x} - \Delta t, \quad a_2 = \frac{\Delta t}{\Delta x}, \quad b = \Delta t.$$

If we define

$$x^h/i,j/ = T/i-1,j/ \quad \text{and} \quad x^v/i,j/ = T/i,j/$$

then from /1.4/ we obtain the Roesser's model

$$\begin{bmatrix} x^h/i+1,j/ \\ x^v/i,j+1/ \end{bmatrix} = \begin{bmatrix} 0 & 1 \\ a_2 & a_1 \end{bmatrix} \begin{bmatrix} x^h/i,j/ \\ x^v/i,j/ \end{bmatrix} + \begin{bmatrix} 0 \\ b \end{bmatrix} U/j/ \qquad /1.5/$$

### Example 1.2
Consider the equations

$$\frac{\partial u/x,t/}{\partial x} = L \frac{\partial i/x,t/}{\partial t}, \quad \frac{\partial i/x,t/}{\partial x} = C \frac{\partial u/x,t/}{\partial t} \qquad /1.6/$$

which describe voltage $u/x,t/$ and current $i/x,t/$ at $x$(space) $\in [0,1]$ and $t$(time) $\in [0,\infty]$ in a long transmission line (Fig. 1.2).
The initial and boundary conditions are given by

$$u/x,0/ = U/x/, \quad i/x,0/ = I/x/$$
$$u/0,t/ = U_1/t/, \quad u/1,t/ = U_2/t/ \qquad /1.6a/$$

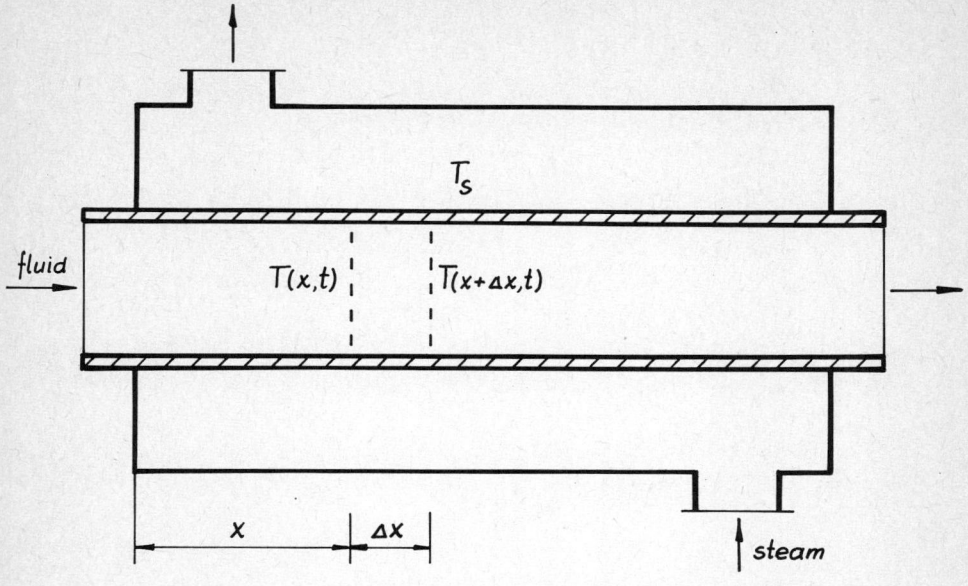

Fig. 1.1 Heat exchanger.

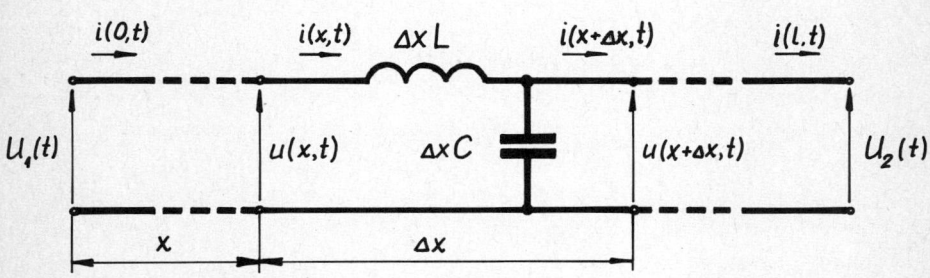

Fig. 1.2 Transmission line.

The equations /1.6/ can be rewritten in the form

$$\frac{\partial}{\partial t}\begin{bmatrix} u/x,t/ \\ i/x,t/ \end{bmatrix} = A \frac{\partial}{\partial x}\begin{bmatrix} u/x,t/ \\ i/x,t/ \end{bmatrix} \qquad /1.7/$$

where

$$A = \begin{bmatrix} 0 & \frac{1}{C} \\ \frac{1}{L} & 0 \end{bmatrix} \qquad /1.8/$$

Let us define

$$\begin{bmatrix} u/x,t/ \\ i/x,t/ \end{bmatrix} = T \begin{bmatrix} \bar{u}/x,t/ \\ \bar{i}/x,t/ \end{bmatrix} \qquad /1.9/$$

where

$$T = \begin{bmatrix} 1 & \frac{L}{\sqrt{LC}} \\ \frac{C}{\sqrt{LC}} & -1 \end{bmatrix}$$

is the matrix whose columns are the eigenvectors of /1.8/.
It is easy to check that

$$\frac{\partial}{\partial t}\begin{bmatrix} \bar{u}/x,t/ \\ \bar{i}/x,t/ \end{bmatrix} = \bar{A} \frac{\partial}{\partial x}\begin{bmatrix} \bar{u}/x,t/ \\ \bar{i}/x,t/ \end{bmatrix} \qquad /1.10/$$

where

$$\bar{A} = T^{-1}AT = \begin{bmatrix} \frac{1}{\sqrt{LC}} & 0 \\ 0 & -\frac{1}{\sqrt{LC}} \end{bmatrix}$$

To find the Roesser's model for /1.10/ we can apply the procedure used for /1.3/.

### Example 1.3
Consider the Darboux equation [12]

$$\frac{\partial^2 s/x,t/}{\partial x\, \partial t} = a_1 \frac{\partial s/x,t/}{\partial t} + a_2 \frac{\partial s/x,t/}{\partial x} + a_0\, s/x,t/ + b\, f/x,t/ \qquad /1.11/$$

with the initial and boundary conditions

$$s/x,0/ = S_1/x/, \quad s/0,t/ = S_2/t/ \qquad /1.11a/$$

where $s/x,t/$ is an unknown function at $x$(space) $\in [0, x_f]$ and $t$(time) $\in$ $[0,\infty]$, $a_0, a_1, a_2$ and $b$ are real coefficients, $f/x,t/$ is a given input function and $S_1/x/$, $S_2/t/$ are given functions.

The equation /1.11/ describes some linear processes of gas absorption, water stream heating and air drying.

Let us define

$$r/x,t/ = \frac{\partial s/x,t/}{\partial t} - a_2 \, s/x,t/ \qquad /1.12/$$

Using /1.12/ we can transform /1.11/ into an equivalent system of first order differential equations of the form

$$\begin{bmatrix} \frac{\partial r/x,t/}{\partial x} \\ \frac{\partial s/x,t/}{\partial t} \end{bmatrix} = \begin{bmatrix} a_1 & a_1 a_2 + a_0 \\ 1 & a_2 \end{bmatrix} \begin{bmatrix} r/x,t/ \\ s/x,t/ \end{bmatrix} + \begin{bmatrix} b \\ 0 \end{bmatrix} f/x,t/ \qquad /1.13/$$

From /1.12/ and /1.11a/ we have

$$r/0,t/ = \frac{\partial s/x,t/}{\partial t}\bigg|_{x=0} - a_2 \, s/0,t/ = \frac{d \, S_2/t/}{dt} - a_2 \, S_2/t/ =$$

$$= R/t/ \qquad /1.14/$$

Taking

$$r/i,j/ = r/i\Delta x, j\Delta t/$$

$$\frac{\partial r/x,t/}{\partial x} \simeq \frac{r/i+1,j/ - r/i,j/}{\Delta x} , \quad \frac{\partial s/x,t/}{\partial t} \simeq \frac{s/i,j+1/ - s/i,j/}{\Delta t}$$

we obtain from /1.13/ the following Roesser's model

$$\begin{bmatrix} r/i+1,j/ \\ s/i,j+1/ \end{bmatrix} = \begin{bmatrix} 1+a_1\Delta x & (a_1 a_2 + a_0)\Delta x \\ \Delta t & 1 + a_2\Delta t \end{bmatrix} \begin{bmatrix} r/i,j/ \\ s/i,j/ \end{bmatrix} + \begin{bmatrix} b \, \Delta x \\ 0 \end{bmatrix} f/i,j/ \qquad /1.15/$$

with boundary conditions

$$r/0,j/ = R/j \, \Delta t/$$
$$s/i,0/ = S_1/i \, \Delta x/ \qquad /1.15a/$$

## Attasi's model.

Attasi's model /AM/ is defined by the equations [1, 2]

$$\bar{x}/i+1,j+1/ = \bar{A}_1\bar{x}/i+1,j/ + \bar{A}_2\bar{x}/i,j+1/ - \bar{A}_1\bar{A}_2\bar{x}/i,j/ + \bar{B}u/i,j/ \quad /1.16/$$

$$y/i,j/ = \bar{C}\bar{x}/i,j/ \quad \text{with} \quad \bar{A}_1\bar{A}_2 = \bar{A}_2\bar{A}_1, \quad (i,j \geqslant 0) \quad /1.17/$$

where i, j are integer-valued vertical and horizontal coordinates, respectively,
$\bar{x}/i,j/ \in R^n$ is the local state vector at /i,j/,
$u/i,j/ \in R^m$ is the input vector,
$y/i,j/ \in R^l$ is the output vector,
$\bar{A}_1, \bar{A}_2, \bar{B}, \bar{C}$ are real matrices of appropriate dimensions.

Boundary conditions for /1.16/ are given by

$$\bar{x}/i,0/, \quad \bar{x}/0,j/ \quad \text{for } i,j = 0,1,2,\ldots \quad /1.17a/$$

## Fornasini - Marchesini's models.

The first Fornasini - Marchesini's model /F-MM I/ is defined by the equations [4]

$$\hat{x}/i+1,j+1/ = \hat{A}_0\hat{x}/i,j/ + \hat{A}_1\hat{x}/i+1,j/ + \hat{A}_2\hat{x}/i,j+1/ + \hat{B}u/i,j/ \quad /1.18/$$

$$y/i,j/ = \hat{C}\hat{x}/i,j/ \quad (i,j \geqslant 0) \quad /1.19/$$

where i, j are integer-valued vertical and horizontal coordinates, respectively,
$\hat{x}/i,j/ \in R^n$ is the local state vector at /i,j/,
$u/i,j/ \in R^m$ is the input vector,
$y/i,j/ \in R^l$ is the output vector,
$\hat{A}_0, \hat{A}_1, \hat{A}_2, \hat{B}, \hat{C}$ are real matrices of appropriate dimensions.

Boundary conditions for /1.18/ are given by

$$\hat{x}/i,0/, \quad \hat{x}/0,j/ \quad \text{for } i,j = 0,1,2,\ldots \quad /1.18a/$$

The second Fornasini - Marchesini's model /F - MM II / is defined by the equations [3]

$$x/i+1,j+1/ = A_1x/i,j+1/ + A_2x/i+1,j/ + B_{01}u/i+1,j/ + B_{10}u/i,j+1/$$

$$/1.20/$$

$$y/i,j/ = Cx/i,j/ + Du/i,j/ \qquad (i,j \geqslant 0) \qquad /1.21/$$

where i, j are integer-valued vertical and horizontal coordinates, respectively,

$x/i,j/ \in R^n$ is the local state vector at $/i,j/$,
$u/i,j/ \in R^m$ is the input vector,
$y/i,j/ \in R^l$ is the output vector,
$A_1$, $A_2$, $B_{10}$, $B_{01}$, C, D are real matrices of appropriate dimensions.

Boundary conditions for /1.20/ are given by

$$x/i,0/, \quad x/0,j/ \quad \text{for} \quad i,j = 1,2,\ldots \qquad /1.20a/$$

## 1.2 RELATIONS BETWEEN THE MODELS

From comparison of /1.16/ and /1.18/ it follows that AM is a special case of F-MM I for $\hat{A}_0 = -\bar{A}_1\bar{A}_2 = -\bar{A}_2\bar{A}_1$.
Let us define

$$x^h/i,j/ = \hat{x}/i,j+1/ - \hat{A}_1\hat{x}/i,j/ \quad \text{and} \quad x^v/i,j/ = \hat{x}/i,j/ .$$

Taking into account /1.18/ we can write

$$x^h/i+1,j/ = \hat{A}_0 x^v/i,j/ + \hat{A}_2\left[x^h/i,j/ + \hat{A}_1 x^v/i,j/\right] + \hat{B}u/i,j/ =$$
$$= \hat{A}_2 x^h/i,j/ + \left[\hat{A}_0 + \hat{A}_2\hat{A}_1\right]x^v/i,j/ + \hat{B}u/i,j/$$

and

$$x^v/i,j+1/ = x^h/i,j/ + \hat{A}_1 x^v/i,j/.$$

Hence

$$\begin{bmatrix} x^h/i+1,j/ \\ x^v/i,j+1/ \end{bmatrix} = \begin{bmatrix} \hat{A}_2 & \hat{A}_0+\hat{A}_2\hat{A}_1 \\ I_n & \hat{A}_1 \end{bmatrix} \begin{bmatrix} x^h/i,j/ \\ x^v/i,j/ \end{bmatrix} + \begin{bmatrix} \hat{B} \\ 0 \end{bmatrix} u/i,j/$$

and

$$y/i,j/ = \begin{bmatrix} 0 & \hat{C} \end{bmatrix} \begin{bmatrix} x^h/i,j/ \\ x^v/i,j/ \end{bmatrix}.$$

Thus, F-MM I can be recasted in RM with

$$\begin{bmatrix} A_{11} & A_{12} \\ A_{21} & A_{22} \end{bmatrix} = \begin{bmatrix} \hat{A}_2 & \hat{A}_0 + \hat{A}_2\hat{A}_1 \\ I_n & \hat{A}_1 \end{bmatrix}, \begin{bmatrix} B_1 \\ B_2 \end{bmatrix} = \begin{bmatrix} \hat{B} \\ 0 \end{bmatrix}, \begin{bmatrix} C_1 & C_2 \end{bmatrix} = \begin{bmatrix} 0 & \hat{C} \end{bmatrix}, D = 0 \quad /1.22/$$

It is easy to show that if $A_{21} = I_n$, $B_2 = 0$ and $C_1 = 0$ then RM can also be recasted in F-MM I with

$$\hat{A}_0 = A_{12} - A_{11}A_{22}, \quad \hat{A}_1 = A_{22}, \quad \hat{A}_2 = A_{11}, \quad \hat{B} = B_1, \quad \hat{C} = C_2.$$

In particular case AM can be recasted in RM with

$$\begin{bmatrix} A_{11} & A_{12} \\ A_{21} & A_{22} \end{bmatrix} = \begin{bmatrix} \bar{A}_2 & 0 \\ I_n & \bar{A}_1 \end{bmatrix}, \begin{bmatrix} B_1 \\ B_2 \end{bmatrix} = \begin{bmatrix} \bar{B} \\ 0 \end{bmatrix}, \begin{bmatrix} C_1 & C_2 \end{bmatrix} = \begin{bmatrix} 0 & \bar{C} \end{bmatrix}, D = 0$$

and if $A_{12} = 0$, $A_{21} = I_n$, $A_{11}A_{22} = A_{22}A_{11}$, $B_2 = 0$ and $C_1 = 0$ then RM can also be recasted in AM with

$$\bar{A}_1 = A_{22}, \quad \bar{A}_2 = A_{11}, \quad \bar{B} = B_1, \quad \bar{C} = C_2.$$

It will be shown that RM is a particular case of F-MM II.
Defining

$$x/i,j/ = \begin{bmatrix} x^h/i,j/ \\ x^v/i,j/ \end{bmatrix}$$

we can write /1.1/ and /1.2/ in the form

$$x/i+1,j+1/ = \begin{bmatrix} 0 & 0 \\ A_{21} & A_{22} \end{bmatrix} x/i+1,j/ + \begin{bmatrix} A_{11} & A_{12} \\ 0 & 0 \end{bmatrix} x/i,j+1/ + \begin{bmatrix} 0 \\ B_2 \end{bmatrix} u/i+1,j/ +$$

$$+ \begin{bmatrix} B_1 \\ 0 \end{bmatrix} u/i,j+1/$$

and

$$y/i,j/ = \begin{bmatrix} C_1 & C_2 \end{bmatrix} x/i,j/ + Du/i,j/.$$

Thus, RM is a particular case of F-MM II with

$$A_2 = \begin{bmatrix} 0 & 0 \\ A_{21} & A_{22} \end{bmatrix}, \quad A_1 = \begin{bmatrix} A_{11} & A_{12} \\ 0 & 0 \end{bmatrix}, \quad B_{01} = \begin{bmatrix} 0 \\ B_2 \end{bmatrix}, \quad B_{10} = \begin{bmatrix} B_1 \\ 0 \end{bmatrix}, \quad C = \begin{bmatrix} C_1 & C_2 \end{bmatrix}$$

/1.23/

Note that /1.18/ and /1.19/ can be written in the form

$$\begin{bmatrix} \hat{x}/i+1,j+1/ \\ \hat{x}/i+1,j/ \\ u/i+1,j/ \end{bmatrix} = \begin{bmatrix} \hat{A}_1 & 0 & 0 \\ I_n & 0 & 0 \\ 0 & 0 & 0 \end{bmatrix} \begin{bmatrix} \hat{x}/i+1,j/ \\ \hat{x}/i+1,j-1/ \\ u/i+1,j-1/ \end{bmatrix} + \begin{bmatrix} \hat{A}_2 & \hat{A}_0 & \hat{B} \\ 0 & 0 & 0 \\ 0 & 0 & 0 \end{bmatrix} \begin{bmatrix} \hat{x}/i,j+1/ \\ \hat{x}/i,j/ \\ u/i,j/ \end{bmatrix} +$$

$$+ \begin{bmatrix} 0 \\ 0 \\ I_m \end{bmatrix} u/i+1,j/ + \begin{bmatrix} 0 \\ 0 \\ 0 \end{bmatrix} u/i,j+1/$$

$$y/i,j/ = \begin{bmatrix} \hat{C} & 0 & 0 \end{bmatrix} \begin{bmatrix} \hat{x}/i,j/ \\ \hat{x}/i,j-1/ \\ u/i,j-1/ \end{bmatrix}.$$

Assuming the vector

$$x/i,j/ = \begin{bmatrix} \hat{x}/i,j/ \\ \hat{x}/i,j-1/ \\ u/i,j-1/ \end{bmatrix}$$

as local state vector of F-MM II it is easy to see that /1.18/ and /1.19/ can be rewritten in the form /1.20/ and /1.21/ with

$$A_2 = \begin{bmatrix} \hat{A}_1 & 0 & 0 \\ I_n & 0 & 0 \\ 0 & 0 & 0 \end{bmatrix}, \quad A_1 = \begin{bmatrix} \hat{A}_2 & \hat{A}_0 & \hat{B} \\ 0 & 0 & 0 \\ 0 & 0 & 0 \end{bmatrix}, \quad B_{01} = \begin{bmatrix} 0 \\ 0 \\ I_m \end{bmatrix}, \quad B_{10} = \begin{bmatrix} 0 \\ 0 \\ 0 \end{bmatrix},$$

$$C = \begin{bmatrix} \hat{C} & 0 & 0 \end{bmatrix}, \quad D = 0.$$

Therefore F-MM I can be embedded in F-MM II.

## 1.3 TRANSITION MATRIX AND GENERAL RESPONSE FORMULA FOR ROESSER´S MODEL

The following partial ordering is used for integer pairs

$/h,k/ \leqslant /i,j/$ if and only if $h \leqslant i$ and $k \leqslant j$ ;

$/h,k/ = /i,j/$ if and only if $h = i$ and $k = j$ ;

$/h,k/ < /i,j/$ if and only if $/h,k/ \leqslant /i,j/$ and $/h,k/ \neq /i,j/$ .

The transition /state – transition/ matrix $A^{i,j}$ for

$$A = \begin{bmatrix} A_{11} & A_{12} \\ A_{21} & A_{22} \end{bmatrix}, \quad A_{ii} \in R^{n_i \times n_i} \quad /i = 1,2/ \qquad /1.24/$$

is defined as follows [13]:

1° $A^{0,0} = I$ /the identity matrix/

2° $A^{1,0} = \begin{bmatrix} A_{11} & A_{12} \\ 0 & 0 \end{bmatrix}, \quad A^{0,1} = \begin{bmatrix} 0 & 0 \\ A_{21} & A_{22} \end{bmatrix}$ /1.25/

3° $A^{i,j} = A^{1,0} A^{i-1,j} + A^{0,1} A^{i,j-1}$ for $/i,j/ > /0,0/$

4° $A^{i,j} = 0$ /the zero matrix/ for $i < 0$ or $j < 0$

From /1.25/ for $j = 0$ we have

$$A^{i,0} = A^{1,0} A^{i-1,0} + A^{0,1} A^{i,-1} = A^{1,0} A^{i-1,0} \qquad /1.26/$$

because $A^{i,-1} = 0$. From /1.26/ it follows that

$$A^{i,0} = \left(A^{1,0}\right)^i \quad \text{for } i = 1,2,\ldots \qquad /1.27a/$$

In a similar way it can be proved that

$$A^{0,j} = \left(A^{0,1}\right)^j \quad \text{for } j = 1,2,\ldots \qquad /1.27b/$$

By induction on $/i,j/$ we shall show that

$$A^{1,0} A^{i-1,j} + A^{0,1} A^{i,j-1} = A^{i-1,j} A^{1,0} + A^{i,j-1} A^{0,1} \qquad /1.27c/$$

For $/i,j/ = /0,0/$ equation /1.25/ yields

$$A^{1,0} A^{-1,0} + A^{0,1} A^{0,-1} = A^{-1,0} A^{1,0} + A^{0,-1} A^{0,1}$$

and for $/i,j/ = /1,0/$, $/i,j/ = /0,1/$

$$A^{1,0} A^{0,0} + A^{0,1} A^{1,-1} = A^{0,0} A^{1,0} + A^{1,-1} A^{0,1}$$

$$A^{1,0} A^{-1,1} + A^{0,1} A^{0,0} = A^{-1,1} A^{1,0} + A^{0,0} A^{0,1}$$

Thus /1.27c/ is true for $/i,j/ = /0,0/$, $/i,j/ = /1,0/$ and $/i,j/ = /0,1/$. In a similar way it can be proven true for $/i,j/ = /i_o,1/$ and $/i,j/ = /1,j_o/$; $i_o, j_o > 1$. Assuming that the hypothesis /1.27c/ is true for all $/k,l/$ such that $/0,0/ \leqslant /k,l/ < /i,j/$ it will be shown that it is valid for $/i,j/$.

$$A^{1,0}\left(A^{i-2,j} A^{1,0} + A^{i-1,j-1} A^{0,1}\right) + A^{0,1}\left(A^{i-1,j-1} A^{1,0} + A^{i,j-2} A^{0,1}\right) =$$
$$= \left(A^{1,0} A^{i-2,j} + A^{0,1} A^{i-1,j-1}\right) A^{1,0} + \left(A^{1,0} A^{i-1,j-1} + A^{0,1} A^{i,j-2}\right) A^{0,1} =$$
$$= A^{i-1,j} A^{1,0} + A^{i,j-1} A^{0,1}.$$

This completes the proof of /1.27c/.

## Theorem 1.1 [13]

A solution to the equation /1.1/ with boundary conditions /1.1a/ is given by

$$\begin{bmatrix} x^h/i,j/ \\ x^v/i,j/ \end{bmatrix} = \sum_{k_1=0}^{i} A^{i-k_1,j} \begin{bmatrix} 0 \\ x^v/k_1,0/ \end{bmatrix} + \sum_{k_2=0}^{j} A^{i,j-k_2} \begin{bmatrix} x^h/0,k_2/ \\ 0 \end{bmatrix} +$$

$$+ \sum_{/0,0/\leqslant/k_1,k_2/</i,j/} \begin{bmatrix} A^{i-k_1-1,j-k_2} B^{1,0} + A^{i-k_1,j-k_2-1} B^{0,1} \end{bmatrix} u/k_1,k_2/$$
/1.28/

where

$$B^{1,0} = \begin{bmatrix} B_1 \\ 0 \end{bmatrix}, \quad B^{0,1} = \begin{bmatrix} 0 \\ B_2 \end{bmatrix}$$
/1.25a/

**Proof:** The proof is accomplished using induction on /i,j/.
For /i,j/ = /0,0/ equation /1.1/ yields

$$\begin{bmatrix} x^h/1,0/ \\ x^v/0,1/ \end{bmatrix} = A \begin{bmatrix} x^h/0,0/ \\ x^v/0,0/ \end{bmatrix} + \begin{bmatrix} B_1 \\ B_2 \end{bmatrix} u/0,0/$$

The same result follows from /1.28/. Thus the hypothesis is true for /i,j/=/0,0/. It is a matter of bookkeeping to show that the hypothesis holds true also for /i,j/ = /i$_o$,0/, /0,j$_o$/; i$_o$,j$_o$ > 0. Assuming that the hypothesis is true for all /k$_1$,k$_2$/ such that /0,0/ $\leqslant$ /k$_1$,k$_2$/ < /i,j/ it will be shown that the hypothesis is valid for /i,j/.
From /1.1/ it follows that

$$\begin{bmatrix} x^h/i,j/ \\ x^v/i,j/ \end{bmatrix} = A^{1,0} \begin{bmatrix} x^h/i-1,j/ \\ x^v/i-1,j/ \end{bmatrix} + A^{0,1} \begin{bmatrix} x^h/i,j-1/ \\ x^v/i,j-1/ \end{bmatrix} + B^{1,0} u/i-1,j/ +$$

$$+ B^{0,1} u/i,j-1/$$
/1.29/

Substituting the expressions for

$$\begin{bmatrix} x^h/i-1,j/ \\ x^v/i-1,j/ \end{bmatrix}, \begin{bmatrix} x^h/i,j-1/ \\ x^v/i,j-1/ \end{bmatrix} \quad \text{/which follow from /1.28//}$$

into /1.29/ we obtain

$$\begin{bmatrix} x^h/i,j/ \\ x^v/i,j/ \end{bmatrix} = A^{1,0} \left\{ \sum_{k_1=0}^{i-1} A^{i-k_1-1,j} \begin{bmatrix} 0 \\ x^v/k_1,0/ \end{bmatrix} + \sum_{k_2=0}^{j} A^{i-1,j-k_2} \begin{bmatrix} x^h/0,k_2/ \\ 0 \end{bmatrix} +$$

$$+ \sum_{/0,0/\leqslant/k_1,k_2/</i-1,j/} \begin{bmatrix} A^{i-k_1-2,j-k_2} B^{1,0} + A^{i-k_1-1,j-k_2-1} B^{0,1} \end{bmatrix} u/k_1,k_2/ \right\} +$$

$$+ A^{0,1}\left\{\sum_{k_1=0}^{i} A^{i-k_1,j-1}\begin{bmatrix}0\\x^v/k_1,0/\end{bmatrix} + \sum_{k_2=0}^{j-1} A^{i,j-k_2-1}\begin{bmatrix}x^h/0,k_2/\\0\end{bmatrix}\right. +$$

$$\left. + \sum_{/0,0/\leqslant/k_1,k_2/</i,j-1/} \left(A^{i-k_1-1,j-k_2-1}B^{1,0} + A^{i-k_1,j-k_2-2}B^{0,1}\right) u/k_1,k_2/\right\} +$$

$$+ B^{1,0} u/i-1,j/ + B^{0,1} u/i,j-1/.$$

Rearranging the summations and taking into account /1.25/ and /1.27/ we obtain

$$\begin{bmatrix}x^h/i,j/\\x^v/i,j/\end{bmatrix} = \sum_{k_1=0}^{i-1} \left(A^{1,0}A^{i-k_1-1,j} + A^{0,1}A^{i-k_1,j-1}\right)\begin{bmatrix}0\\x^v/k_1,0/\end{bmatrix} +$$

$$+ A^{0,1}A^{0,j-1}\begin{bmatrix}0\\x^v/k_1,0/\end{bmatrix} + \sum_{k_2=0}^{j-1}\left(A^{1,0}A^{i-1,j-k_2} + A^{0,1}A^{i,j-k_2-1}\right)\begin{bmatrix}x^h/0,k_2/\\0\end{bmatrix} +$$

$$+ A^{1,0}A^{i-1,0}\begin{bmatrix}x^h/0,k_2/\\0\end{bmatrix} + \sum_{/0,0/\leqslant/k_1,k_2/</i-1,j-1/}\left(A^{1,0}A^{i-k_1-2,j-k_2} + \right.$$

$$\left. + A^{0,1}A^{i-k_1-1,j-k_2-1}\right)B^{1,0} u/k_1,k_2/ + \sum_{/0,0/\leqslant/k_1,k_2/</i-1,j-1/}\left(A^{1,0}\cdot\right.$$

$$\left. A^{i-k_1-1,j-k_2-1} + A^{0,1}A^{i-k_1,j-k_2-2}\right)B^{0,1} u/k_1,k_2/ +$$

$$+ A^{1,0}\sum_{k_1=0}^{i-2}\left(A^{i-k_1-2,0}B^{1,0} + A^{i-k_1-1,-1}B^{0,1}\right) u/k_1,k_2/ +$$

$$+ A^{0,1}\sum_{k_2=0}^{j-2}\left(A^{-1,j-k_2-1}B^{1,0} + A^{0,j-k_2-2}B^{0,1}\right) u/k_1,k_2/ +$$

$$+ B^{1,0} u/i-1,j/ + B^{0,1} u/i,j-1/ =$$

$$= \sum_{k_1=0}^{i-1} A^{i-k_1,j}\begin{bmatrix}0\\x^v/k_1,0/\end{bmatrix} + A^{0,j}\begin{bmatrix}0\\x^v/k_1,0/\end{bmatrix} + \sum_{k_2=0}^{j-1} A^{i,j-k_2}\begin{bmatrix}x^h/0,k_2/\\0\end{bmatrix} +$$

$$+ A^{i,0}\begin{bmatrix}x^h/0,k_2/\\0\end{bmatrix} + \sum_{/0,0/\leqslant/k_1,k_2/</i-1,j-1/}\left(A^{i-k_1-1,j-k_2}B^{1,0} + \right.$$

$$+ A^{i-k_1,j-k_2-1} \begin{pmatrix} 0,1 \\ B \end{pmatrix} u/k_1,k_2/ + \left( \sum_{k_1=0}^{i-2} A^{i-k_1-1} B^{1,0} + \right.$$

$$+ \sum_{k_2=0}^{j-2} A^{0,j-k_2-1} B^{0,1} \Biggr) u/k_1,k_2/ + B^{1,0} u/i-1,j/ + B^{0,1} u/i,j-1/ =$$

$$= \sum_{k_1=0}^{i} A^{i-k_1,j} \begin{bmatrix} 0 \\ x^v/k_1,0/ \end{bmatrix} + \sum_{k_2=0}^{j} A^{i,j-k_2} \begin{bmatrix} x^h/0,k_2/ \\ 0 \end{bmatrix} +$$

$$+ \sum_{/0,0/\leqslant/k_1,k_2/</i,j/} \left( A^{i-k_1-1,j-k_2} B^{1,0} + A^{i-k_1,j-k_2-1} B^{0,1} \right) u/k_1,k_2/$$

A different proof of the theorem based on the superposition property of the system is given in [13].
From Theorem 1.1 for $u/i,j/ = 0$ for all $/i,j/$ and $x^h/0,j/ = 0$, $x^v/i,0/ = 0$ for $i,j = 1,2,...$ it follows that

$$\begin{bmatrix} x^h/i,j/ \\ x^v/i,j/ \end{bmatrix} = A^{i,j} \begin{bmatrix} x^h/0,0/ \\ x^v/0,0/ \end{bmatrix} \qquad /1.30/$$

Substitution of /1.28/ into /1.2/ yields the general response formula

$$y/i,j/ = \begin{bmatrix} C_1 & C_2 \end{bmatrix} \left\{ \sum_{k_1=0}^{i} A^{i-k_1,j} \begin{bmatrix} 0 \\ x^v/k_1,0/ \end{bmatrix} + \sum_{k_2=0}^{j} A^{i,j-k_2} \begin{bmatrix} x^h/0,k_2/ \\ 0 \end{bmatrix} + \right.$$

$$+ \sum_{/0,0/\leqslant/k_1,k_2/</i,j/} \left( A^{i-k_1-1,j-k_2} B^{1,0} + A^{i-k_1,j-k_2-1} B^{0,1} \right) u/k_1,k_2/ \Biggr\} +$$

$$+ Du/i,j/ \qquad /1.31/$$

## 1.4 STATE - SPACE MODELS OF THREE AND N - DIMENSIONAL LINEAR SYSTEMS

Roesser's model was extended for three-dimensional /3-D/ systems by Tzafestas and Pimenides in [17]. The Tzafestas-Pimenides' model /T-PM/ is defined by the equations

$$\begin{bmatrix} x^h/i+1,j,k/ \\ x^v/i,j+1,k/ \\ x^d/i,j,k+1/ \end{bmatrix} = \begin{bmatrix} A_{11} & A_{12} & A_{13} \\ A_{21} & A_{22} & A_{23} \\ A_{31} & A_{32} & A_{33} \end{bmatrix} \begin{bmatrix} x^h/i,j,k/ \\ x^v/i,j,k/ \\ x^d/i,j,k/ \end{bmatrix} + \begin{bmatrix} B_1 \\ B_2 \\ B_3 \end{bmatrix} u/i,j,k/ \qquad /1.32/$$

$$(i,j,k \geq 0)$$

$$y/i,j,k/ = \begin{bmatrix} C_1 & C_2 & C_3 \end{bmatrix} \begin{bmatrix} x^h/i,j,k/ \\ x^v/i,j,k/ \\ x^d/i,j,k/ \end{bmatrix} + Du/i,j,k/ \qquad /1.33/$$

where i,j,k are integer-valued vertical, horizontal and depth coordinates, respectively,

$x^h/i,j,k/ \in R^{n_1}$ is the horizontal state vector,

$x^v/i,j,k/ \in R^{n_2}$ is the vertical state vector,

$x^d/i,j,k/ \in R^{n_3}$ is the depth state vector,

$u/i,j,k/ \in R^m$ is the input vector,

$y/i,j,k/ \in R^l$ is the output vector,

$A_{ij}$, $B_i$, $C_i$ /i,j = 1,2,3/ and D are real matrices of appropriate dimensions.

Boundary conditions for /1.32/ are given by

$x^h/0,j,k/$, $x^v/i,0,k/$, $x^d/i,j,0/$ for i,j,k = 0,1,2,... /1.32a/

Introducing the matrices and vectors

$$A = \begin{bmatrix} A_{11} & A_{12} & A_{13} \\ A_{21} & A_{22} & A_{23} \\ A_{31} & A_{32} & A_{33} \end{bmatrix}, \quad B = \begin{bmatrix} B_1 \\ B_2 \\ B_3 \end{bmatrix}, \quad C = \begin{bmatrix} C_1 & C_2 & C_3 \end{bmatrix}$$

$$x' = \begin{bmatrix} x^h/i+1,j,k/ \\ x^v/i,j+1,k/ \\ x^d/i,j,k+1/ \end{bmatrix}, \quad x = \begin{bmatrix} x^h/i,j,k/ \\ x^v/i,j,k/ \\ x^d/i,j,k/ \end{bmatrix}, \quad u = u/i,j,k/, \quad y = y/i,j,k/$$

we can rewrite /1.32/ and /1.33/ in the form

$x' = Ax + Bu$ /1.32'/

$y = Cx + Du$ /1.33'/

Roesser's model for N-dimensional /N-D/ linear systems is defined by the equations [9]

$$\begin{bmatrix} x_1/i_1+1,i_2,\ldots,i_N/ \\ x_2/i_1,i_2+1,\ldots,i_N/ \\ \cdots\cdots\cdots\cdots \\ x_N/i_1,i_2,\ldots,i_N+1/ \end{bmatrix} = \begin{bmatrix} A_{11} & A_{12} & \cdots & A_{1N} \\ A_{21} & A_{22} & \cdots & A_{2N} \\ \cdots & \cdots & \cdots & \cdots \\ A_{N1} & A_{N2} & \cdots & A_{NN} \end{bmatrix} \begin{bmatrix} x_1/i_1,i_2,\ldots,i_N/ \\ x_2/i_1,i_2,\ldots,i_N/ \\ \cdots\cdots\cdots\cdots \\ x_N/i_1,i_2,\ldots,i_N/ \end{bmatrix} +$$

$$+ \begin{bmatrix} B_1 \\ B_2 \\ \cdots \\ B_N \end{bmatrix} u/i_1,i_2,\ldots,i_N/ \qquad\qquad /1.34/$$

$$(i_1,i_2,\ldots,i_N \geqslant 0)$$

$$y/i_1,i_2,\ldots,i_N/ = \begin{bmatrix} C_1 & C_2 & \cdots & C_N \end{bmatrix} \begin{bmatrix} x_1/i_1,i_2,\ldots,i_N/ \\ x_2/i_1,i_2,\ldots,i_N/ \\ \cdots\cdots\cdots\cdots \\ x_N/i_1,i_2,\ldots,i_N/ \end{bmatrix} +$$

$$+ Du/i_1,i_2,\ldots,i_N/ \qquad\qquad /1.35/$$

where $i_1,i_2,\ldots,i_N$ are integer-valued coordinates,

$x_j/i_1,i_2,\ldots,i_N/ \in R^{n_j}$ is the j-th /j = 1,2,...,N/ state vector,

$u/i_1,i_2,\ldots,i_N/ \in R^m$ is the input vector,

$y/i_1,i_2,\ldots,i_N/ \in R^l$ is the output vector,

$A_{ij}$, $B_i$, $C_i$ /i,j = 1,2,...,N/ and D are real matrices of appropriate dimensions.

Boundary conditions for /1.34/ are given by

$$x_j/i_1,i_2,\ldots,0_j,\ldots,i_N/ \text{ for } i_1,i_2,\ldots,i_{j-1},i_{j+1},\ldots,i_N \in Z^+ \qquad /1.34a/$$

where $/i_1,i_2,\ldots,0_j,\ldots,i_N/$ denotes the N-tuple with zero on the j-th position and $Z^+$ is the set of nonnegative integer numbers.

Introducing the matrices and vectors

$$A = \begin{bmatrix} A_{11} & A_{12} & \cdots & A_{1N} \\ A_{21} & A_{22} & \cdots & A_{2N} \\ \cdots & \cdots & \cdots & \cdots \\ A_{N1} & A_{N2} & \cdots & A_{NN} \end{bmatrix}, \quad B = \begin{bmatrix} B_1 \\ B_2 \\ \cdots \\ B_N \end{bmatrix}, \quad C = \begin{bmatrix} C_1 & C_2 & \cdots & C_N \end{bmatrix}$$

$$x' = \begin{bmatrix} x_1/i_1+1, i_2, \ldots, i_N/ \\ x_2/i_1, i_2+1, \ldots, i_N/ \\ \cdots \cdots \cdots \\ x_N/i_1, i_2, \ldots, i_N+1/ \end{bmatrix}, \quad x = \begin{bmatrix} x_1/i_1, i_2, \ldots, i_N/ \\ x_2/i_1, i_2, \ldots, i_N/ \\ \cdots \cdots \cdots \\ x_N/i_1, i_2, \ldots, i_N/ \end{bmatrix}$$

$$u = u/i_1, i_2, \ldots, i_N/, \quad y = y/i_1, i_2, \ldots, i_N/$$

we can write /1.34/ and /1.35/ in the form /1.32´/ and /1.33´/.

## Example 1.4

Consider two-dimensional diffusion problem described by the equation [14]

$$\frac{\partial s/x,y,t/}{\partial t} = a_1 \frac{\partial^2 s/x,y,t/}{\partial x^2} + a_2 \frac{\partial^2 s/x,y,t/}{\partial y^2} + a_0 u/x,y,t/ \qquad /1.36/$$

where $s/x,y,t/$ is an unknown function at $/x,y,t/$, $u/x,y,t/$ is a given force function and $a_0$, $a_1$, $a_2$ are given real coefficients.
Using

$$s/i,j,k/ = s/i\Delta x, j\Delta y, k\Delta t/, \quad u/i,j,k/ = u/i\Delta x, j\Delta y, k\Delta t/$$

$$\frac{\partial s/x,y,t/}{\partial t} \cong \frac{s/i,j,k/ - s/i,j,k-1/}{\Delta t}$$

$$\frac{\partial^2 s/x,y,t/}{\partial x^2} \cong \frac{s/i,j,k/ - 2s/i-1,j,k/ + s/i-2,j,k/}{(\Delta x)^2}$$

$$\frac{\partial^2 s/x,y,t/}{\partial y^2} \cong \frac{s/i,j,k/ - 2s/i,j-1,k/ + s/i,j-2,k/}{(\Delta y)^2}$$

we can write /1.36/ in the form

$$\left[\frac{1}{\Delta t} - \frac{a_1}{(\Delta x)^2} - \frac{a_2}{(\Delta y)^2}\right] s/i,j,k/ = \frac{1}{\Delta t} s/i,j,k-1/ - \frac{2a_1}{(\Delta x)^2} s/i-1,j,k/ +$$

$$+ \frac{a_1}{(\Delta x)^2} s/i-2,j,k/ - \frac{2a_2}{(\Delta y)^2} s/i,j-1,k/ + \frac{a_2}{(\Delta y)^2} s/i,j-2,k/ +$$

$$+ a_0 u/i,j,k/ \qquad /1.37/$$

Defining

$$x_1^h/i,j,k/ = -\frac{2a_1}{(\Delta x)^2} s/i-1,j,k/ + \frac{a_1}{(\Delta x)^2} s/i-2,j,k/$$

$$x_2^h/i,j,k/ = s/i-2,j,k/ \qquad /1.38/$$

$$x_1^v/i,j,k/ = -\frac{2a_2}{(\Delta y)^2} s/i,j-1,k/ + \frac{a_2}{(\Delta y)^2} s/i,j-2,k/$$

$$x_2^v/i,j,k/ = s/i,j-2,k/$$

$$x^d/i,j,k/ = s/i,j,k-1/$$

from /1.37/ we obtain

$$s/i,j,k/ = \frac{1}{a\Delta t} x^d/i,j,k/ + \frac{1}{a} x_1^h/i,j,k/ + \frac{1}{a} x_1^v/i,j,k/ +$$

$$+ \frac{a_0}{a} u/i,j,k/ \qquad /1.39/$$

where

$$a = \frac{1}{\Delta t} - \frac{a_1}{(\Delta x)^2} - \frac{a_2}{(\Delta y)^2}$$

From /1.38/ and /1.39/ we have

$$x_1^h/i+1,j,k/ = -\frac{2a_1}{(\Delta x)^2}\left[\frac{1}{a\Delta t} x^d/i,j,k/ + \frac{1}{a} x_1^h/i,j,k/ + \frac{1}{a} x_1^v/i,j,k/ + \right.$$

$$\left. + \frac{a_0}{a} u/i,j,k/\right] + \frac{a_1}{(\Delta x)^2}\left[-\frac{(\Delta x)^2}{2a_1} x_1^h/i,j,k/ + \frac{1}{2} x_2^h/i,j,k/\right] =$$

$$= -\left[\frac{2a_1}{a(\Delta x)^2} - \frac{1}{2}\right] x_1^h/i,j,k/ + \frac{a_1}{2(\Delta x)^2} x_2^h/i,j,k/ - \frac{2a_1}{a(\Delta x)^2} x_1^v/i,j,k/ +$$

$$- \frac{2a_1}{a(\Delta x)^2 \Delta t} x^d/i,j,k/ - \frac{2a_0 a_1}{a(\Delta x)^2} u/i,j,k/ \qquad /1.40/$$

$$x_2^h/i+1,j,k/ = s/i-1,j,k/ = -\frac{(\Delta x)^2}{2a_1} x_1^h/i,j,k/ + \frac{1}{2} x_2^h/i,j,k/ \qquad /1.41/$$

$$x_1^v/i,j+1,k/ = -\frac{2a_2}{(\Delta y)^2}\left[\frac{1}{a\Delta t} x^d/i,j,k/ + \frac{1}{a} x_1^h/i,j,k/ + \frac{1}{a} x_1^v/i,j,k/ + \right.$$

$$\left. + \frac{a_0}{a} u/i,j,k/\right] + \frac{a_2}{(\Delta y)^2}\left[-\frac{(\Delta y)^2}{2a_2} x_1^v/i,j,k/ + \frac{1}{2} x_2^v/i,j,k/\right] =$$

$$= -\frac{2a_2}{a(\Delta y)^2} x_1^h/i,j,k/ - \left[\frac{2a_2}{a(\Delta y)^2} + \frac{1}{2}\right] x_1^v/i,j,k/ + \frac{a_2}{2(\Delta y)^2} x_2^v/i,j,k/ +$$

$$- \frac{2a_2}{a(\Delta y)^2 \Delta t} x^d/i,j,k/ - \frac{2a_0 a_2}{a(\Delta y)^2} u/i,j,k/ \qquad /1.42/$$

$$x_2^v/i,j+1,k/ = -\frac{(\Delta y)^2}{2a_2} x_1^v/i,j,k/ + \frac{1}{2} x_2^v/i,j,k/ \qquad /1.43/$$

$$x^d/i,j,k+1/ = s/i,j,k/ = \frac{1}{a} x_1^h/i,j,k/ + \frac{1}{a} x_1^v/i,j,k/ +$$

$$+ \frac{1}{a\Delta t} x^d/i,j,k/ + \frac{a_0}{a} u/i,j,k/ \qquad /1.44/$$

Setting

$$x^h/i,j,k/ = \begin{bmatrix} x_1^h/i,j,k/ \\ x_2^h/i,j,k/ \end{bmatrix}, \quad x^v/i,j,k/ = \begin{bmatrix} x_1^v/i,j,k/ \\ x_2^v/i,j,k/ \end{bmatrix}$$

from /1.40/ - /1.44/ we obtain

$$\begin{bmatrix} x^h/i+1,j,k/ \\ x^v/i,j+1,k/ \\ x^d/i,j,k+1/ \end{bmatrix} = \begin{bmatrix} A_{11} & A_{12} & A_{13} \\ A_{21} & A_{22} & A_{23} \\ A_{31} & A_{32} & A_{33} \end{bmatrix} \begin{bmatrix} x^h/i,j,k/ \\ x^v/i,j,k/ \\ x^d/i,j,k/ \end{bmatrix} + \begin{bmatrix} B_1 \\ B_2 \\ B_3 \end{bmatrix} u/i,j,k/$$

where

$$A_{11} = \begin{bmatrix} -\frac{2a_1}{a(\Delta x)^2} - \frac{1}{2} & \frac{a_1}{2(\Delta x)^2} \\ -\frac{(\Delta x)^2}{2a_1} & \frac{1}{2} \end{bmatrix}, \quad A_{12} = \begin{bmatrix} -\frac{2a_1}{a(\Delta x)^2} & 0 \\ 0 & 0 \end{bmatrix},$$

$$A_{13} = \begin{bmatrix} -\frac{2a_1}{a(\Delta x)^2 \Delta t} \\ 0 \end{bmatrix}, \quad A_{21} = \begin{bmatrix} -\frac{2a_2}{a(\Delta y)^2} & 0 \\ 0 & 0 \end{bmatrix},$$

$$A_{22} = \begin{bmatrix} -\frac{2a_2}{a(\Delta y)^2} - \frac{1}{2} & \frac{a_2}{2(\Delta y)^2} \\ -\frac{(\Delta y)^2}{2a_2} & \frac{1}{2} \end{bmatrix}, \quad A_{23} = \begin{bmatrix} -\frac{2a_2}{a(\Delta y)^2 \Delta t} \\ 0 \end{bmatrix},$$

$$A_{31} = \begin{bmatrix} \frac{1}{a} & 0 \end{bmatrix}, \quad A_{32} = \begin{bmatrix} \frac{1}{a} & 0 \end{bmatrix}, \quad A_{33} = \begin{bmatrix} \frac{1}{a\Delta t} \end{bmatrix},$$

$$B_1 = \begin{bmatrix} -\frac{2a_0 a_1}{a(\Delta x)^2} \\ 0 \end{bmatrix}, \quad B_2 = \begin{bmatrix} -\frac{2a_0 a_2}{a(\Delta y)^2} \\ 0 \end{bmatrix}, \quad B_3 = \begin{bmatrix} \frac{a_0}{a} \end{bmatrix}.$$

Different examples of derivation of the Roesser's model for 3-D systems are given in [17, 18].

## 1.5 TRANSITION MATRIX AND GENERAL RESPONSE FORMULA FOR 3-D AND N-D SYSTEMS

The following partial ordering is used for integer N-tuples
$/i_1,\ldots,i_j,\ldots,i_N/ \leqslant /k_1,\ldots,k_j,\ldots,k_N/$ if and only if $i_j \leqslant k_j$ for $j = 1,2,\ldots,N$;
$/i_1,\ldots,i_j,\ldots,i_N/ = /k_1,\ldots,k_j,\ldots,k_N/$ if and only if $i_j = k_j$ for $j = 1,2,\ldots,N$;
$/i_1,\ldots,i_j,\ldots,i_N/ < /k_1,\ldots,k_j,\ldots,k_N/$ if and only if $/i_1,\ldots,i_j,\ldots,i_N/ \leqslant$
$\leqslant /i_1,\ldots,i_j,\ldots,i_N/$ and $/i_1,\ldots,i_j,\ldots,i_N/ \neq /i_1,\ldots,i_j,\ldots,i_N/$.

The transition /state transition/ matrix $A^{i_1,\ldots,i_j,\ldots,i_N}$ for

$$A = \begin{bmatrix} A_{11} & A_{12} & \cdots & A_{1N} \\ A_{21} & A_{22} & \cdots & A_{2N} \\ \cdot & \cdot & & \cdot \\ A_{N1} & A_{N2} & \cdots & A_{NN} \end{bmatrix}, \quad A_{ii} \in R^{n_i \times n_i} \quad /i = 1,2,\ldots,N/$$

is defined as follows $|9|$:

$1^0 \quad A^{0,\ldots,0,\ldots,0} = I$ /the identity matrix/,

$$2^0 \quad A^{0,\ldots,1_j,\ldots,0} = \begin{bmatrix} 0 & 0 & \cdots & 0 \\ \cdot & \cdot & & \cdot \\ 0 & 0 & \cdots & 0 \\ A_{j1} & A_{j2} & \cdots & A_{jN} \\ 0 & 0 & \cdots & 0 \\ \cdot & \cdot & & \cdot \\ 0 & 0 & \cdots & 0 \end{bmatrix} \qquad /1.45/$$

where $/0,\ldots,1_j,\ldots,0/$ denotes the N-tuple with 1 on the j-th position and 0 elsewhere,

$3^0 \quad A^{i_1,\ldots,i_j,\ldots,i_N} = \sum_{j=1}^{N} A^{0,\ldots,1_j,\ldots,0} A^{i_1,\ldots,i_j-1,\ldots,i_N}$,

$4^0 \quad A^{i_1,\ldots,i_j,\ldots,i_N} = 0$ /the zero matrix/ for $i_1 < 0$ or $i_2 < 0 \ldots$
$\ldots$ or $i_N < 0$.

In a similar way as in 2-D case it can be proved that

$$A^{0,\ldots,i_j,\ldots,0} = \left(A^{0,\ldots,1_j,\ldots,0}\right)^{i_j} \quad \text{for} \quad j=1,2,\ldots,N \qquad /1.46/$$

## Theorem 1.2

A solution to the equation /1.34/ with boundary conditions /1.34a/ is given by

$$x/i_1,i_2,\ldots,i_N/ =$$

$$= \sum_{j=1}^{N} \sum_{\substack{/i_1,\ldots,i_j,\ldots,i_N/\geqslant \\ \geqslant /k_1,\ldots,0_j,\ldots,k_N/\geqslant \\ \geqslant /0,\ldots,0,\ldots,0/}} A^{i_1-k_1,\ldots,i_j,\ldots,i_N-k_N} \begin{bmatrix} 0 \\ \vdots \\ 0 \\ x_j/k_1,\ldots,0_j,\ldots,k_N/ \\ 0 \\ \vdots \\ 0 \end{bmatrix} +$$

$$+ \sum_{\substack{/i_1,\ldots,i_j,\ldots,i_N/\geqslant \\ \geqslant /k_1,\ldots,k_j,\ldots,k_N/\geqslant \\ \geqslant /0,\ldots,0,\ldots,0/}} \left( \sum_{j=1}^{N} A^{i_1-k_1,\ldots,i_j-k_j-1,\ldots,i_N-k_N} B^{0,\ldots,1_j,\ldots,0} \right) u/k_1,\ldots,k_j,\ldots,k_N/$$

where /1.47/

$$B^{0,\ldots,1_j,\ldots,0} = \begin{bmatrix} 0 \\ \vdots \\ 0 \\ B_j \\ 0 \\ \vdots \\ 0 \end{bmatrix} \quad /j=1,2,\ldots,N/$$

Theorem 1.2 can be proved in a similar way as Theorem 1.1.
A proof of Theorem 1.2 based on the superposition property is given in [9]
From Theorem 1.2 for $u/k_1,\ldots,k_N/ = 0$ for all $/k_1,\ldots,k_N/$
and $x_j/i_1,\ldots,0_j,\ldots,i_N/ = 0$ for $i_1,\ldots,i_{j-1},i_{j+1},\ldots,i_N = 1,2,\ldots;\quad j=1,2,\ldots,N$

it follows that

$$\begin{bmatrix} x_1/i_1,i_2,\ldots,i_N/ \\ x_2/i_1,i_2,\ldots,i_N/ \\ \vdots \\ x_N/i_1,i_2,\ldots,i_N/ \end{bmatrix} = A^{i_1,i_2,\ldots,i_N} \begin{bmatrix} x_1/0,0,\ldots,0/ \\ x_2/0,0,\ldots,0/ \\ \vdots \\ x_N/0,0,\ldots,0/ \end{bmatrix} \qquad /1.48/$$

Substitution of /1.47/ into /1.35/ yields the general response formula

$y/i_1,i_2,\ldots,i_N/ =$

$$= \begin{bmatrix} c_1 & c_2 & \cdots & c_N \end{bmatrix} \left\{ \sum_{j=1}^{N} \sum_{\substack{/i_1,\ldots,i_j,\ldots,i_N/\geqslant \\ \geqslant /k_1,\ldots,0_j,\ldots,k_N/\geqslant \\ \geqslant /0,\ldots,0,\ldots,0/}} A^{i_1-k_1,\ldots,i_j,\ldots,i_N-k_N} \begin{bmatrix} 0 \\ \vdots \\ 0 \\ x_j/k_1,\ldots,0_j,\ldots,k_N/ \\ 0 \\ \vdots \\ 0 \end{bmatrix} + \right.$$

$$\left. + \sum_{\substack{/i_1,\ldots,i_j,\ldots,i_N/\geqslant \\ \geqslant /k_1,\ldots,k_j,\ldots,k_N/\geqslant \\ \geqslant /0,\ldots,0,\ldots,0/}} \left( \sum_{j=1}^{N} A^{i_1-k_1,\ldots,i_j-k_j-1,\ldots,i_N-k_N} B^{0,\ldots,1_j,\ldots,0} \right) u/k_1,\ldots,k_j,\ldots,k_N/ \right\} +$$

/1.49/

$+ Du/i_1,i_2,\ldots,i_N/$

In particular case for N=3 from /1.47/ and /1.49/ we obtain

$$\begin{bmatrix} x^h/i,j,k/ \\ x^v/i,j,k/ \\ x^d/i,j,k/ \end{bmatrix} = \sum_{k_2=0}^{j} \sum_{k_3=0}^{k} A^{i,j-k_2,k-k_3} \begin{bmatrix} x^h/0,k_2,k_3/ \\ 0 \\ 0 \end{bmatrix} +$$

$$+ \sum_{k_1=0}^{i} \sum_{k_2=0}^{j} A^{i-k_1,j-k_2,k} \begin{bmatrix} 0 \\ 0 \\ x^d/k_1,k_2,0/ \end{bmatrix} +$$

$$+ \sum_{k_1=0}^{i} \sum_{k_3=0}^{k} A^{i-k_1,j,k-k_3} \begin{bmatrix} 0 \\ x^v/k_1,0,k_3/ \\ 0 \end{bmatrix} + \qquad /1.50/$$

$$+ \sum_{/0,0,0/ \leqslant /k_1,k_2,k_3/ \leqslant /i,j,k/} \left( A^{i-k_1-1,j-k_2,k-k_3} B^{1,0,0} + \right.$$

$$\left. + A^{i-k_1,j-k_2-1,k-k_3} B^{0,1,0} + \right.$$

and

$$y/i,j,k/ = \begin{bmatrix} c_1 & c_2 & c_3 \end{bmatrix} \Biggl\{ \sum_{k_2=0}^{j} \sum_{k_3=0}^{k} A^{i,j-k_2,k-k_3} \begin{bmatrix} x^h/0,k_2,k_3/ \\ 0 \\ 0 \end{bmatrix} +$$

$$+ \sum_{k_1=0}^{i} \sum_{k_3=0}^{k} A^{i-k_1,j,k-k_3} \begin{bmatrix} 0 \\ x^v/k_1,0,k_3/ \\ 0 \end{bmatrix} +$$

$$+ \sum_{k_1=0}^{i} \sum_{k_2=0}^{j} A^{i-k_1,j-k_2,k} \begin{bmatrix} 0 \\ 0 \\ x^d/k_1,k_2,0/ \end{bmatrix} + \qquad /1.51/$$

$$+ \sum_{/0,0,0/ \leqslant /k_1,k_2,k_3/ < /i,j,k/} \Bigl( A^{i-k_1-1,j-k_2,k-k_3} B^{1,0,0} +$$

$$+ A^{i-k_1,j-k_2-1,k-k_3} B^{0,1,0} +$$

$$+ A^{i-k_1,j-k_2,k-k_3-1} B^{0,0,1} \Bigr) u/k_1,k_2,k_3/ \Biggr\} + Du/i,j,k/$$

where

$$B^{1,0,0} = \begin{bmatrix} B_1 \\ 0 \\ 0 \end{bmatrix}, \quad B^{0,1,0} = \begin{bmatrix} 0 \\ B_2 \\ 0 \end{bmatrix}, \quad B^{0,0,1} = \begin{bmatrix} 0 \\ 0 \\ B_3 \end{bmatrix}$$

The formulae given above were also derived in [8, 18].

## 1.6 CAYLEY – HAMILTON THEOREM

**Definition 1.1** [13]

The determinant

$$p(z_1,z_2) = \det\begin{bmatrix} I_{n_1}z_1 - A_{11} & -A_{12} \\ -A_{21} & I_{n_2}z_2 - A_{22} \end{bmatrix} = \sum_{i=0}^{n_1}\sum_{j=0}^{n_2} a_{ij}z_1^i z_2^j \quad /1.52/$$

$$(a_{n_1 n_2} = 1)$$

is called the 2-D characteristic polynomial /function/ of the matrix

$$A = \begin{bmatrix} A_{11} & A_{12} \\ A_{21} & A_{22} \end{bmatrix} \quad /1.53/$$

The equation $p(z_1,z_2) = 0$ is called the 2-D characteristic equation of the matrix /1.53/.

For example, the 2-D characteristic polynomial of the matrix

$$A = \begin{bmatrix} 1 & -1 \\ -1 & 2 \end{bmatrix}$$

has the form

$$p(z_1,z_2) = \begin{vmatrix} z_1-1 & 1 \\ 1 & z_2-2 \end{vmatrix} = z_1 z_2 - 2z_1 - z_2 + 1$$

**Theorem 1.3**

$$\sum_{i=0}^{n_1}\sum_{j=0}^{n_2} a_{ij} A^{i+h, j+k} = 0 \quad \text{for} \quad h \geqslant 0 \quad \text{or} \quad k \geqslant 0 \quad /1.54/$$

**Proof:** In Appendix it is shown that

$$Z[A^{i,j}] = \sum_{i=0}^{\infty}\sum_{j=0}^{\infty} A^{i,j} z_1^{-i} z_2^{-j} = \begin{bmatrix} I_{n_1} - A_{11}z_1^{-1} & -A_{12}z_1^{-1} \\ -A_{21}z_2^{-1} & I_{n_2} - A_{22}z_2^{-1} \end{bmatrix}^{-1} \quad /1.55/$$

Taking into account that

$$\begin{bmatrix} I_{n_1}z_1 - A_{11} & -A_{12} \\ -A_{21} & I_{n_2}z_2 - A_{22} \end{bmatrix} = \begin{bmatrix} I_{n_1}z_1 & 0 \\ 0 & I_{n_2}z_2 \end{bmatrix}\begin{bmatrix} I_{n_1} - A_{11}z_1^{-1} & -A_{12}z_1^{-1} \\ -A_{21}z_2^{-1} & I_{n_2} - A_{22}z_2^{-1} \end{bmatrix} \quad /1.56/$$

we can write

$$\sum_{i=0}^{\infty}\sum_{j=0}^{\infty} A^{i,j} z_1^{-i} z_2^{-j} = \begin{bmatrix} I_{n_1} z_1 - A_{11} & -A_{12} \\ -A_{21} & I_{n_2} z_2 - A_{22} \end{bmatrix}^{-1} \begin{bmatrix} I_{n_1} z_1 & 0 \\ 0 & I_{n_2} z_2 \end{bmatrix} \quad /1.57/$$

and

$$\sum_{i=0}^{\infty}\sum_{j=0}^{\infty} A^{i,j} z_1^{-i} z_2^{-j} \left[ \sum_{h=0}^{n_1}\sum_{k=0}^{n_2} a_{hk} z_1^h z_2^k \right] =$$

$$= \sum_{h=-i}^{\infty}\sum_{k=-j}^{\infty} \left[ \sum_{i=0}^{n_1}\sum_{j=0}^{n_2} a_{ij} A^{i+h,j+k} \right] z_1^{-h} z_2^{-k} =$$

$$= \begin{bmatrix} I_{n_1} z_1 - A_{11} & -A_{12} \\ -A_{21} & I_{n_2} z_2 - A_{22} \end{bmatrix}_{ad} \begin{bmatrix} I_{n_1} z_1 & 0 \\ 0 & I_{n_2} z_2 \end{bmatrix} = \sum_{i=0}^{n_1}\sum_{j=0}^{n_2} \hat{B}_{ij} z_1^i z_2^j \quad /1.58/$$

with $\hat{B}_{00} = 0$.

The relation /1.54/ follows from /1.58/, since $\hat{B}_{ij} = 0$ for $i < 0$ or $j < 0$. □

In particular case for $h = k = 0$ from /1.54/ we obtain the following

## 2-D Cayley - Hamilton Theorem

Every matrix /1.53/ satisfies its own characteristic equation, i.e.

$$\sum_{i=0}^{n_1}\sum_{j=0}^{n_2} a_{ij} A^{i,j} = 0 \quad /1.59/$$

An alternate proof of that theorem is given in [14].

## 1.7  2-D DIVISION ALGORITHM

**Definition 1.2** [13]

The 2-D eigenvalues of the matrix /1.53/ are the pairs $/z_1,z_2/$ that simultaneously solve the following set of equations

$$P_{i,-1}/z_1,z_2/ = 0 \quad \text{for} \quad 0 < i \leq n_1 \qquad /1.60a/$$
$$P_{-1,j}/z_1,z_2/ = 0 \quad \text{for} \quad 0 < j \leq n_2 \qquad /1.60b/$$
$$p/z_1,z_2/ = 0 \qquad /1.60c/$$

where

$$P_{i,j}/z_1,z_2/ = \sum_{k=i}^{n_1} \sum_{l=j}^{n_2} a_{kl} z_1^{k-i} z_2^{l-j} \qquad /1.61/$$

for any integer pair $/i,j/$.

**Theorem 1.3**  /2-D Division Algorithm/ [13]

Any polynomial $w/z_1,z_2/$ may be expressed as follows

$$w/z_1,z_2/ = \sum_{i=1}^{n_1} q_{1i}/z_2/\, P_{i,-1}/z_1,z_2/ + \sum_{j=1}^{n_2} q_{2j}/z_1/\, P_{-1,j}/z_1,z_2/ +$$

$$+ \; m/z_1,z_2/\, p/z_1,z_2/ \; + \; r/z_1,z_2/ \qquad /1.62/$$

where $\deg r/z_1,z_2/ < \deg p/z_1,z_2/$.

**Proof:** Let

$$w/z_1,z_2/ = \sum_{i=0}^{N_1} \sum_{j=0}^{N_2} b_{ij} z_1^i z_2^j \qquad /1.62'/$$

where $N_1 \geq n_1$ and $N_2 \geq n_2$.

Each term in $w/z_1,z_2/$ of degree $/i,j/ \geq /n_1,n_2/$ may be reduced by subtracting $b_{ij} z_1^{i-n_1} z_2^{j-n_2} p/z_1,z_2/$ from it. Repeating this until there are no terms of degree greater than (or equal to) $/n_1,n_2/$ we obtain

$$w/z_1,z_2/ - m/z_1,z_2/ \, p/z_1,z_2/ = \sum_{i=0}^{N_1} \sum_{j=0}^{N_2} c_{ij} z_1^i z_2^j \qquad /1.63/$$

with $c_{ij} = 0$ for $i \geqslant n_1$ and $j \geqslant n_2$ and $c_{ij} \neq 0$ for some $0 \leqslant i < n_1$, $0 \leqslant j \leqslant N_2$ or $0 \leqslant i \leqslant N_1$, $0 \leqslant j < n_2$.

Any term in /1.63/ of degree /k,l/, with $k > n_1$ and $l = n_2 - j$, may be deleted by subtracting $c_{kl} z_1^{k-n_1-1} p_{-1,j}/z_1,z_2/$ from it. Let the reduction process be performed for $j = 1, 2, \ldots, n_2$ leaving no terms of degree greater than / equal to $/n_1+1, n_2-j/$ after each step. A similar process is repeated for polynomials $p_{i,-1}/z_1,z_2/$, $i = 1, 2, \ldots, n_1$. The result is a remainder of the form

$$r/z_1,z_2/ = w/z_1,z_2/ - m/z_1,z_2/ \, p/z_1,z_2/ +$$

$$- \sum_{i=1}^{n_1} q_{1i}/z_2/ \, p_{i,-1}/z_1,z_2/ - \sum_{j=1}^{n_2} q_{2j}/z_1/ \, p_{-1,j}/z_1,z_2/$$

with $\deg r/z_1,z_2/ < /n_1, n_2/$.

$\square$

Taking into account

$$q_{1i}/z_2/ = \sum_{k=0}^{N_2-n_2-1} (q_{1i})_k \, z_2^k , \qquad q_{2j}/z_1/ = \sum_{l=0}^{N_1-n_1-1} (q_{2j})_l \, z_1^l ,$$

$$m/z_1,z_2/ = \sum_{i=0}^{N_1-n_1} \sum_{j=0}^{N_2-n_2} m_{ij} z_1^i z_2^j$$

/1.62/ may be written as

$$w/z_1,z_2/ = \sum_{i=1}^{n_1} \sum_{j=1}^{N_2-n_2} q_{-i,j} \, p_{i,-j}/z_1,z_2/ + \sum_{j=1}^{n_2} \sum_{i=1}^{N_1-n_1} q_{i,-j} \, p_{-i,j}/z_1,z_2/ +$$

$$+ \sum_{i=0}^{N_1-n_1} \sum_{j=0}^{N_2-n_2} m_{ij} \, p_{-i,-j}/z_1,z_2/ + r/z_1,z_2/ \qquad /1.62''/$$

where $q_{-i,j} = (q_{1i})_{j-1}$, $q_{i,-j} = (q_{2j})_{i-1}$.

From Theorem 1.3 and /1.60/ it follows that if $/z_1^o, z_2^o/$ is an eigenvalue of A, then

$$w/z_1^o, z_2^o/ = r/z_1^o, z_2^o/. \qquad /1.64/$$

In general case the determination of 2-D eigenvalues of a matrix A is not a trivial problem. In particular case when a characteristic polynomial $p/z_1, z_2/$ is separable /factorable/, i.e. when it can be written in the form

$$p/z_1, z_2/ = p_1/z_1/ \cdot p_2/z_2/ \qquad /1.65/$$

then 2-D eigenvalues can be easily found.


Lemma 1.1

If the 2-D characteristic polynomial $p/z_1, z_2/$ is separable /1.65/ and

$$p_i/z_i/ = (z_i - z_{i1})(z_i - z_{i2}) \cdots (z_i - z_{in_i}) \quad , \quad i = 1,2 \qquad /1.66/$$

then the set of 2-D eigenvalues is

$$\left\{ /z_{1i}, z_{2j}/ : /0,0/ < /i,j/ \leq /n_1, n_2/ \right\} \qquad /1.67/$$

where $z_{10} = z_{20} = 0$.

Proof: From assumption /1.65/ it follows that

$$p_{i,-1}/z_1, z_2/ = \hat{p}_{1i}/z_1/ \cdot z_2 \cdot p_2/z_2/$$

and

$$p_{-1,j}/z_1, z_2/ = z_1 \cdot p_1/z_1/ \cdot \hat{p}_{2j}/z_2/$$

for some polynomials $\hat{p}_{1i}/z_1/$, $\hat{p}_{2j}/z_2/$.

Note that $p_2/z_{2j}/ = 0$ implies $p_{i,-1}/z_1, z_{2j}/ = 0$ and $p_1/z_{1i}/ = 0$ implies $p_{-1,j}/z_{1i}, z_2/ = 0$. Thus, for $/z_{1i}, z_{2j}/$ all polynomials $p/z_1, z_2/$, $p_{i,-1}/z_1, z_2/$ and $p_{-1,j}/z_1, z_2/$ are equal zero.

□

The number $(n_1 + 1)(n_2 + 1) - 1$ of distinct 2-D eigenvalues of a matrix A with separable $p/z_1, z_2/$ is the same as the number of coefficients in

$$r/z_1,z_2/ = \sum_{i=0}^{n_1} \sum_{j=0}^{n_2} r_{ij} z_1^i z_2^j \quad \text{with} \quad r_{n_1 n_2} = 0.$$

Thus, for a given polynomial $w/z_1,z_2/$ solving the set of $(n_1+1)(n_2+1)-1$ equations

$$w/z_{1i},z_{2j}/ = r/z_{1i},z_{2j}/ \quad \text{for} \quad /0,0/ \leqslant /i,j/ \leqslant /n_1,n_2/ \qquad /1.68/$$

we can find the coefficients of $r/z_1,z_2/$.


## 1.8 COMPUTATION OF THE TRANSITION MATRIX

Computation of the transition matrix $A^{i,j}$ using /1.25/ becomes quite tedious as i and j become large. An algorithm based on Cayley-Hamilton Theorem will be presented for finding $A^{i,j}$.

For /1.62'/ let us define the matrix

$$w/A/ = \sum_{i=0}^{N_1} \sum_{j=0}^{N_2} b_{ij} A^{i,j}. \qquad /1.69/$$

From /1.62"/ it follows that

$$w/A/ = \sum_{i=1}^{n_1} \sum_{j=1}^{N_2-n_2} q_{-i,j}\, P_{i,-j}/A/ + \sum_{j=1}^{n_2} \sum_{i=1}^{N_1-n_1} q_{i,-j}\, P_{-i,j}/A/ +$$

$$+ \sum_{i=0}^{N_1-n_1} \sum_{j=0}^{N_2-n_2} m_{ij}\, P_{-i,-j}/A/ + r/A/.$$

From Theorem 1.3 and /1.25/ we have

$$P_{i,j}/A/ = 0 \quad \text{for} \quad i<0 \text{ or } j<0 \text{ and for } i=j=0.$$

Hence
$$w/A/ = r/A/. \qquad /1.70/$$

If $p/z_1,z_2/$ is separable and 2-D eigenvalues of A are distinct, then solving the set of equations /1.68/ we can find the coefficients of $r/z_1,z_2/$. Using /1.70/ we can evaluate the desired matrix $w/A/$.

From the above considerations the following algorithm for finding $A^{i,j}$ $/i,j/ \geqslant /n_1,n_2/$ follows:

## Algorithm 1.1

**Step 1.** For given matrix A find the 2-D characteristic polynomial $p/z_1,z_2/$ and evaluate 2-D eigenvalues $/z_{1k},z_{21}/$ of the matrix A.

**Step 2.** If the 2-D eigenvalues $/z_{1k},z_{21}/$ are distinct solving the set of equations

$$z_{1k}^i z_{21}^j = r/z_{1k},z_{21}/ \quad \text{for} \quad /0,0/ < /k,l/ \leqslant /n_1,n_2/ \qquad /1.71/$$

find the coefficients $r_{kl}$ of

$$r/z_1,z_2/ = \sum_{k=0}^{n_1} \sum_{l=0}^{n_2} r_{kl} z_1^k z_2^l \; ; \quad r_{n_1 n_2} = 0 \, .$$

**Step 3.** Using the formula

$$A^{i,j} = \sum_{k=0}^{n_1} \sum_{l=0}^{n_2} r_{kl} A^{k,l} \qquad /1.72/$$

find $A^{i,j}$ for $/i,j/ \geqslant /n_1,n_2/$.

## Example 1.5

Find $A^{i,j}$, $/i,j/ \geqslant /n_1,n_2/$, for the matrix

$$A = \begin{bmatrix} 2 & 1 \\ 0 & -3 \end{bmatrix} \qquad /1.73/$$

**Step 1.** The 2-D characteristic polynomial of the matrix /1.73/ has the form

$$p/z_1,z_2/ = \begin{vmatrix} z_1 - 2 & -1 \\ 0 & z_2 + 3 \end{vmatrix} = (z_1 - 2)(z_2 + 3)$$

The 2-D eigenvalues of the matrix /1.73/ are /2, 0/, /0, -3/ and /2, -3/.

**Step 2.** In this case $r/z_1,z_2/$ has the form

$$r/z_1,z_2/ = r_{10} z_1 + r_{01} z_2 + r_{00}$$

Substituting the 2-D eigenvalues into the equation

$$r_{10} z_1 + r_{01} z_2 + r_{00} = z_1^i z_2^j$$

we obtain

$$2r_{10} + r_{00} = 0, \quad -3r_{01} + r_{00} = 0 \quad \text{and} \quad 2r_{10} - 3r_{01} + r_{00} = 2^i(-3)^j.$$

The solution to the equation is

$$r_{10} = 2^{i-1}(-3)^j, \quad r_{01} = 2^i(-3)^{j-1}, \quad r_{00} = -2^i(-3)^j$$

and $r/z_1, z_2/ = 2^{i-1}(-3)^{j-1}(-3z_1 + 2z_2 + 6)$.

Step 3. Using /1.72/ we obtain

$$A^{i,j} = 2^{i-1}(-3)^{j-1}\left(-3A^{1,0} + 2A^{0,1} + 6I\right) =$$

$$= 2^{i-1}(-3)^{j-1}\left(-3\begin{bmatrix} 2 & 1 \\ 0 & 0 \end{bmatrix} + 2\begin{bmatrix} 0 & 0 \\ 0 & -3 \end{bmatrix} + 6\begin{bmatrix} 1 & 0 \\ 0 & 1 \end{bmatrix}\right) =$$

$$= 2^{i-1}(-3)^j \begin{bmatrix} 0 & 1 \\ 0 & 0 \end{bmatrix} \quad \text{for } /i,j/ \geqslant /1,1/.$$

For finding $A^{i,j}$ the algorithm based on Sylvester's theorem can also be used [7]. If the characteristic polynomial $p/z_1, z_2/$ is separable and $p_i/z_i/$ has the form /1.66/, then from Corollary A.1 the following algorithm for finding $A^{i,j}$ follows:

Algorithm 1.2

Step 1. Find the 2-D characteristic polynomial $p/z_1,z_2/$ and the set of 2-D eigenvalues /1.67/ of the matrix A.

Step 2. Using the formula

$$v_{i,j}/z_1,z_2/ = \prod_{\substack{k=0 \\ k \neq i}}^{n_1} \frac{z_1 - z_{1k}}{z_{1i} - z_{1k}} \prod_{\substack{l=0 \\ l \neq j}}^{n_2} \frac{z_2 - z_{21}}{z_{2j} - z_{21}} \qquad \begin{array}{l} /1.74/ \\ /0,0/ \leqslant /i,j/ \leqslant \\ \leqslant /n_1,n_2/ \end{array}$$

find $v_{i,j}$ for $/0,0/ \leqslant /i,j/ \leqslant /n_1,n_2/$.

Step 3. Evaluate $Z_{i,j} = v_{i,j}/A/$ for $/0,0/ \leqslant /i,j/ \leqslant /n_1,n_2/$.

Step 4. Using the formula

$$f/A/ = \sum_{/0,0/ \leqslant /i,j/ \leqslant /n_1,n_2/} Z_{i,j} \, f/z_{1i}, z_{2j}/ \qquad /1.75/$$

find $f/A/$.

## Example 1.6

Find $A^{i,j}$, $/i,j/ \geqslant /n_1,n_2/$, for the matrix $/1.73/$.

Step 1. The 2-D characteristic polynomial of the matrix $/1.73/$ has the form $p/z_1,z_2/ = (z_1 - 2)(z_2 + 3)$ and the set of 2-D eigenvalues is
$$\left\{/z_{1i},z_{2j}/:\ /0,0/ \leqslant /i,j/ \leqslant /n_1,n_2/\right\} = \left\{/2,0/,\ /0,-3/,\ /2,-3/\right\}$$

Step 2. Taking into account that $z_{10} = z_{20} = 0$, $z_{11} = 2$, $z_{21} = -3$ and using $/1.74/$ we obtain

$$v_{1,1}/z_1,z_2/ = \frac{z_1 - z_{10}}{z_{11} - z_{10}} \frac{z_2 - z_{20}}{z_{21} - z_{20}} = -\frac{1}{6} z_1 z_2$$

Note that we do not need to find $v_{1,0}$ and $v_{0,1}$ since $z_{1k}^i z_{21}^j = 0$ for $k = l = 0$.

Step 3. We have

$$Z_{1,1} = v_{1,1}/A/ = -\frac{1}{6} A^{1,1} = -\frac{1}{6}\left[A^{1,0} A^{0,1} + A^{0,1} A^{1,0}\right] =$$

$$= -\frac{1}{6}\left\{\begin{bmatrix}2 & 1\\0 & 0\end{bmatrix}\begin{bmatrix}0 & 0\\0 & -3\end{bmatrix} + \begin{bmatrix}0 & 0\\0 & -3\end{bmatrix}\begin{bmatrix}2 & 1\\0 & 0\end{bmatrix}\right\} = \frac{1}{2}\begin{bmatrix}0 & 1\\0 & 0\end{bmatrix}$$

Step 4. Using $/1.75/$ for $f/z_1,z_2/ = z_1^i z_2^j$ we obtain

$$A^{i,j} = \sum_{/0,0/ \leqslant /k,l/ \leqslant /1,1/} Z_{i,j}\ z_{1k}^i z_{2l}^j = Z_{1,1}\ z_{11}^i z_{21}^j =$$

$$= 2^{i-1}(-3)^j \begin{bmatrix}0 & 1\\0 & 0\end{bmatrix} \quad \text{for } /i,j/ \geqslant /1,1/,$$

i.e. the same result as in the Example 1.5.

The Algorithms 1.1 and 1.2 with slight modifications can be extended for matrices A with separable 2-D characteristic polynomials and multiple 2-D eigenvalues /see the Appendix/ [10].

From formula $Z_{i,j} = v_{i,j}/A/$ and $/1.74/$ it follows that $Z_{i,j}$ are completely determined by the matrix A and are independent of the function $f/z_1,z_2/$. Thus, to find $Z_{i,j}$ the following procedure can also be used:

1° Choose suitable polynomials

$p_{kl}/z_1,z_2/$ for $/0,0/ \leqslant /k,l/ \leqslant /n_1,n_2/$

with least possible degrees.

2° Solving the set of equations

$$p_{kl}/A/ = \sum_{/0,0/ \leq /i,j/ \leq /n_1,n_2/} z_{i,j}^{i,j} p_{ij}/z_{1i}, z_{2j}/ \qquad /1.76/$$

find the matrices $Z_{i,j}$ for $/0,0/ \leq /i,j/ \leq /n_1,n_2/$.

## Example 1.7

Find the matrices $Z_{i,j}$ for the matrix /1.73/.

The 2-D eigenvalues of the matrix /1.73/ are /2,0/, /0,-3/, /2,-3/ (see Example 1.5) and $n_1 = n_2 = 1$. Choosing $p_{10}/z_1,z_2/ = 1$, $p_{01}/z_1,z_2/ = z_1^o z_2$, $p_{11}/z_1,z_2/ = z_1 z_2^o$ and using /1.76/ we obtain

$$Z_{1,0} + Z_{0,1} + Z_{1,1} = I$$
$$-3(Z_{0,1} + Z_{1,1}) = A^{0,1}$$
$$2(Z_{1,0} + Z_{1,1}) = A^{1,0}$$

Hence

$$Z_{1,0} = \frac{1}{3} A^{0,1} + I = \begin{bmatrix} 1 & 0 \\ 0 & 0 \end{bmatrix}, \quad Z_{0,1} = I - \frac{1}{2} A^{1,0} = \begin{bmatrix} 0 & -\frac{1}{2} \\ 0 & 1 \end{bmatrix}$$

and

$$Z_{1,1} = I - Z_{1,0} - Z_{0,1} = \begin{bmatrix} 0 & \frac{1}{2} \\ 0 & 0 \end{bmatrix}.$$

From /1.25/ it follows that $A^{i,j}$ may be expressed as a function of the matrices $A^{1,0}$ and $A^{0,1}$. Let us define a new operator $P^{i,j}$ as follows:

$P^{i,j} A = $ the sum of products of all non-equivalent permutations of the matrices $\underbrace{A^{1,0} A^{1,0} \dots A^{1,0}}_{i}$ and $\underbrace{A^{0,1} A^{0,1} \dots A^{0,1}}_{j}$

It is well-known that the number of products of all non-equivalent permutations of the matrices $\underbrace{A^{1,0} A^{1,0} \dots A^{1,0}}_{i}$ and $\underbrace{A^{0,1} A^{0,1} \dots A^{0,1}}_{j}$ is equal to $\frac{(i+j)!}{i! \cdot j!}$.

For example

$$P^{2,1} A = A^{1,0} A^{1,0} A^{0,1} + A^{1,0} A^{0,1} A^{1,0} + A^{0,1} A^{1,0} A^{1,0} =$$
$$= A^{2,0} A^{0,1} + A^{1,0} A^{0,1} A^{1,0} + A^{0,1} A^{2,0}$$

and

$$P^{2,2} A = A^{1,0} A^{1,0} A^{0,1} A^{0,1} + A^{1,0} A^{0,1} A^{1,0} A^{0,1} + A^{1,0} A^{0,1} A^{0,1} A^{1,0} +$$
$$+ A^{0,1} A^{1,0} A^{1,0} A^{0,1} + A^{0,1} A^{1,0} A^{0,1} A^{1,0} + A^{0,1} A^{0,1} A^{1,0} A^{1,0} =$$

$$= A^{2,0}A^{0,2} + A^{1,0}A^{0,1}A^{1,0}A^{0,1} + A^{1,0}A^{0,2}A^{1,0} +$$
$$+ A^{0,1}A^{2,0}A^{0,1} + A^{0,1}A^{1,0}A^{0,1}A^{1,0} + A^{0,2}A^{2,0}.$$

**Lemma 1.2**

$P^{i,j}A = A^{i,j}$  for  $/i,j/ > /0,0/$

Proof: The proof is accomplished by induction on $/i,j/$.
Using $/1.25/$ it is easy to check that

$P^{i,0}A = [A^{1,0}]^i = A^{i,0}$  for  $i = 1,2,...$
$P^{0,j}A = [A^{0,1}]^j = A^{0,j}$  for  $j = 1,2,...$
$P^{1,1}A = A^{1,0}A^{0,1} + A^{0,1}A^{1,0} = A^{1,1}$

Thus our hypothesis is true for $/i,0/$, $/0,j/$ $(i,j = 1,2,...)$ and $/1,1/$.
Assuming that the hypothesis is true for $/i-1,j/$ and $/i,j-1/$ it will
be shown that the hypothesis is valid for $/i,j/$.
From $/1.25/$ and the definition of $P^{i,j}$ we have

$$A^{i,j} = A^{1,0}A^{i-1,j} + A^{0,1}A^{i,j-1} = A^{1,0}P^{i-1,j}A + A^{0,1}P^{i,j-1}A =$$
$$= P^{i,j}A.$$

□

## 1.9 SOLUTION OF ROESSER'S MODEL AS A FUNCTION OF 2-D EIGENVALUES

**Theorem 1.5**

If the 2-D characteristic polynomial $p/z_1,z_2/$ is separable and the set
of distinct 2-D eigenvalues of the matrix A has the form $/1.67/$, then
a solution to equation $/1.1/$ with boundary conditions $/1.1a/$ is given
by

$$\begin{bmatrix} x^h/i,j/ \\ x^v/i,j/ \end{bmatrix} = \sum_{\substack{/n_1,n_2/ \\ > /k,l/ \\ > /0,0/}} Z_{k,l} \left\{ \sum_{k_1=0}^{i} z_{1k}^{i-k_1} z_{21}^{j} \begin{bmatrix} 0 \\ x^v/k_1,0/ \end{bmatrix} + \right.$$

$$\left. + \sum_{k_2=0}^{j} z_{1k}^{i} z_{21}^{j-k_2} \begin{bmatrix} x^h/0,k_2/ \\ 0 \end{bmatrix} + \sum_{\substack{/i,j/ \\ > /k_1,k_2/ \\ \geq /0,0/}} \left( z_{1k}^{i-k_1-1} z_{21}^{j-k_2} B^{1,0} + \right. \right.$$

$$+ z_{1k}^{i-k_1} z_{21}^{j-k_2-1} B^{0,1}\Big) u/k_1,k_2/ \Bigg\} \quad \text{for } /i,j/ > /0,0/ \qquad /1.77/$$

where $Z_{k,1} = v_{k,1}/A/$ and $v_{k,1}/z_1,z_2/$ is defined by /1.74/.

<u>Remark</u>: All terms involving powers of $z_1$ or $z_2$ with negative exponents are deleted in /1.77/.

<u>Proof</u>: Consider the function

$$f/z_1,z_2/ = \sum_{k_1=0}^{i} z_1^{i-k_1} z_2^j \begin{bmatrix} 0 \\ x^v/k_1,0/ \end{bmatrix} + \sum_{k_2=0}^{j} z_1^i z_2^{j-k_2} \begin{bmatrix} x^h/0,k_2/ \\ 0 \end{bmatrix} +$$

$$+ \sum_{\substack{/i,j/ > \\ > /k_1,k_2/ \geq \\ \geq /0,0/}} \left( z_1^{i-k_1-1} z_2^{j-k_2} B^{1,0} + z_1^{i-k_1} z_2^{j-k_2-1} B^{0,1} \right) u/k_1,k_2/ \qquad /1.78/$$

which is analytic in an open set containing the 2-D eigenvalues $/z_{1i},z_{2j}/$ for $/0,0/ < /i,j/ \leq /n_1,n_2/$ of the matrix A.
Using the formula /1.75/ for /1.78/ we obtain /1.77/. $\square$

The solution to the equation /1.1/ with boundary conditions /1.1a/ can be found by the use of the following:

<u>Algorithm 1.3</u>

<u>Step 1</u>. Find the 2-D characteristic polynomial $p/z_1,z_2/$ and the set of 2-D eigenvalues /1.67/ of the matrix A.

<u>Step 2</u>. Find the matrices $Z_{k,1}$ for $/0,0/ < /k,1/ \leq /n_1,n_2/$.

<u>Step 3</u>. Using formula /1.77/ find the desired solution.

<u>Example 1.8</u>

Find a solution to the equation

$$\begin{bmatrix} x^h/i+1,j/ \\ x^v/i,j+1/ \end{bmatrix} = \begin{bmatrix} 2 & 1 \\ 0 & -3 \end{bmatrix} \begin{bmatrix} x^h/i,j/ \\ x^v/i,j/ \end{bmatrix} + \begin{bmatrix} 1 \\ 2 \end{bmatrix} e^{-i} 2^{-j} \qquad /1.79/$$

with zero boundary conditions.

Using the results of Examples 1.5 and 1.7 together with formula /1.77/ we obtain

$$\begin{bmatrix} x^h/i,j/ \\ x^v/i,j/ \end{bmatrix} = \sum_{\substack{/1,1/ \geqslant \\ /k,1/ > \\ /0,0/ \geqslant}} \sum_{\substack{/i,j/ > \\ /k_1,k_2/ \geqslant \\ /0,0/ \geqslant}} Z_{k,1} \left( z_{1k}^{i-k_1-1} z_{21}^{j-k_2} B^{1,0} + \right.$$

$$\left. + z_{1k}^{i-k_1} z_{21}^{j-k_2-1} B^{0,1} \right) u/k_1,k_2/ =$$

$$= \sum_{\substack{/i,j/ > \\ /k_1,k_2/ \geqslant \\ /0,0/ \geqslant}} \left\{ \left( z_{11}^{i-k_1-1} z_{20}^{j-k_2} + z_{11}^{i-k_1} z_{21}^{j-k_2-1} \right) \begin{bmatrix} 1 \\ 0 \end{bmatrix} + \right.$$

$$\left. + z_{10}^{i-k_1} z_{21}^{j-k_2-1} \begin{bmatrix} -1 \\ 2 \end{bmatrix} \right\} e^{-k_1} 2^{-k_2} =$$

$$= \sum_{k_1=0}^{i-1} 2^{i-k_1-j-1} e^{-k_1} \begin{bmatrix} 1 \\ 0 \end{bmatrix} + \sum_{k_2=0}^{j-1} (-3)^{j-k_2-1} 2^{-k_2} e^{-i} \begin{bmatrix} -1 \\ 2 \end{bmatrix} +$$

$$+ \sum_{\substack{/i,j/ > \\ /k_1,k_2/ \geqslant \\ /0,0/ \geqslant}} 2^{i-k_1-k_2} (-3)^{j-k_2-1} e^{-k_1} \begin{bmatrix} 1 \\ 0 \end{bmatrix} \quad \text{for } /i,j/ > /0,0/.$$

The above considerations with slight modifications can be extended for matrices A with separable 2-D characteristic polynomials and multiple 2-D eigenvalues.


## 1.10 GENERAL RESPONSE FORMULAE FOR FORNASINI-MARCHESINI'S MODELS


Let us consider the first Fornasini-Marchesini's model /F-MM I/ defined by the equations /1.18/ and /1.19/.

### Theorem 1.6

A solution to the equations /1.18/ with boundary conditions /1.18a/ has the form

$$x/i,j/ = \sum_{k=0}^{i} A_{22}^{i-k,j} x/k,0/ + \sum_{l=0}^{j} A_{21}^{i,j-1} \left[ x/0,l+1/ - \hat{A}_1 x/0,l/ \right] +$$

$$+ \sum_{/0,0/ \leq /k,l/ < /i,j/} A_{21}^{i-k-1,j-1} \hat{B} u/k,l/ \, , \quad /i,j/ \geq /0,0/ \qquad /1.80/$$

where

$$\begin{bmatrix} A_{11}^{k,l} & A_{12}^{k,l} \\ A_{21}^{k,l} & A_{22}^{k,l} \end{bmatrix} = \begin{bmatrix} \hat{A}_2 & \hat{A}_0 + \hat{A}_2 \hat{A}_1 \\ I_n & \hat{A}_1 \end{bmatrix}^{k,l} = A^{k,l} \qquad /1.81/$$

**Proof:** From /1.22/ and /1.28/ it follows that

$$\begin{bmatrix} x/i,j+1/ - \hat{A}_1 x/i,j/ \\ x/i,j/ \end{bmatrix} = \sum_{k=0}^{i} A^{i-k,j} \begin{bmatrix} 0 \\ x/k,0/ \end{bmatrix} +$$

$$+ \sum_{l=0}^{j} A^{i,j-1} \begin{bmatrix} x/0,l+1/ - \hat{A}_1 x/0,l/ \\ 0 \end{bmatrix} + \sum_{\substack{/i,j/> \\ >/k,l/ \geq \\ \geq /0,0/}} A^{i-k-1,j-1} \begin{bmatrix} \hat{B} \\ 0 \end{bmatrix} u/k,l/ \qquad /1.82/$$

since $B^{0,1} = 0$.

Taking into account /1.81/ we obtain the desired formula /1.80/. □

Substitution of /1.80/ into /1.19/ yields the general response formula

$$y/i,j/ = \sum_{k=0}^{i} \hat{C} A_{22}^{i-k,j} x/k,0/ + \sum_{l=0}^{j} \hat{C} A_{21}^{i,j-1} \left[ x/0,l+1/ - \hat{A}_1 x/0,l/ \right] +$$

$$+ \sum_{/0,0/ \leq /k,l/ \leq /i,j/} \hat{C} A_{21}^{i-k-1,j-1} \hat{B} u/k,l/ \qquad /1.83/$$

for $/i,j/ \geq /0,0/$.

Now let us consider the second Fornasini-Marchesini's model /F-MM II/ defined by the equations /1.20/ and /1.21/.

The transition /state-transition/ matrix $A^{i,j}$ for FM-M II is defined as follows:

1° $A^{0,0} = I$ /the identity matrix/

2° $A^{i,j} = A_1 A^{i-1,j} + A_2 A^{i,j-1}$ for /i,j/ > /0,0/ /1.84/

3° $A^{-i,j} = A^{i,-j} = 0$ /the zero matrix/ for $i<0$ or $j<0$

In a similar way as for RM it can be shown that

$$A^{0,j} = [A_2]^j \; ; \; A^{i,0} = [A_1]^i \quad \text{for } i>0, \; j>0 \quad /1.85a/$$

$$A_1 A^{i-1,j} + A_2 A^{i,j-1} = A^{i-1,j} A_1 + A^{i,j-1} A_2 \quad \text{for } /i,j/ \geqslant /0,0/ \quad /1.85b/$$

### Theorem 1.7

A solution to the equation /1.20/ with boundary conditions /1.20a/ has the form

$$x/i,j/ = \sum_{k=1}^{i} A^{i-k,j-1} \left[ A_2 x/k,0/ + B_{01} u/k,0/ \right] +$$

$$+ \sum_{l=1}^{j} A^{i-1,j-l} \left[ A_1 x/0,l/ + B_{10} u/0,l/ \right] + \qquad /1.86/$$

$$\sum_{k=1}^{i} \sum_{l=1}^{j} \left[ A^{i-k-1,j-l} B_{10} + A^{i-k,j-l-1} B_{01} \right] u/k,l/; \quad /i,j/ > /0,0/.$$

Proof: Let the input u/i,j/ for all /i,j/ as well as the boundary conditions /1.20a/ be zero except x/k,0/. We shall show that

$$x/i,j/ = A^{i-k,j-1} A_2 x/k,0/ \quad (k>0) \qquad /1.87/$$

The proof will be accomplished by induction.
From /1.20/ we have

$$x/k,j/ = [A_2]^j x/k,0/ = [A_2]^{j-1} A_2 x/k,0/ =$$

$$= A^{0,j-1} A_2 x/k,0/ \quad \text{for } j = 1,2,\ldots \;;$$

$$x/i,1/ = [A_1]^{i-k} A_2 x/k,0/ = A^{i-k,0} A_2 x/k,0/ \quad \text{for } i = k, \; k+1, \; k+2, \ldots$$

because of /1.85a/, for k = 1,2,....
Thus our assertion is true for any /i,0/ and /0,j/, i,j = 1,2,....
Similarly we can prove that if the input u/i,j/ for all /i,j/ and the

boundary conditions /1.20a/ are zero except x/0,1/, then

$$x/i,j/ = A^{i-1,j-1} A_1 x/0,1/ \quad (1>0) \quad /1.88/$$

Assuming the hypothesis true for all /k,l/ such that /0,0/ $<$ /k,l/ $<$ /i,j/ we shall show that it is true for /i,j/. From /1.2o/ and /1.84/ we obtain

$$x/i,j/ = A_1 x/i-1,j/ + A_2 x/i,j-1/ =$$
$$= A_1 A^{i-k-1,j-1} A_2 x/k,0/ + A_2 A^{i-k,j-2} A_2 x/k,0/ =$$
$$= A^{i-k,j-1} A_2 x/k,0/ \ .$$

Assume u/k,l/ for some /1,1/ $\leqslant$ /k,l/ $<$ /i,j/ be the only nonzero input and all boundary conditions /1.20a/ be zero. Then from /1.20/ we have

$$x/k,l+1/ = B_{01} u/k,l/ \quad \text{and} \quad x/k+1,l/ = B_{10} u/k,l/ \ .$$

Therefore

$$x/i,j/ = A^{i-k-1,j-1} x/k+1,l/ + A^{i-k,j-l-1} x/k,l+1/ =$$
$$= \left( A^{i-k-1,j-1} B_{10} + A^{i-k,j-l-1} B_{01} \right) u/k,l/ \quad /1.89/$$

If $u/k,0/$ $\left( u/0,l/ \right)$ is the only nonzero input and all boundary conditions are zero, then

$$x/i,j/ = A^{i-k,j-1} B_{01} u/k,0/ \quad \text{for} \quad i = k, k+1, k+2, \ldots \quad /1.89'/$$

$$\left( x/i,j/ = A^{i-1,j-l} B_{10} u/0,l/ \quad \text{for} \quad j = l, l+1, l+2, \ldots \right) \quad /1.89''/$$

for any $k,l = 1,2,\ldots$.

By superposition of the effects of all boundary conditions and inputs from /1.87/, /1.88/, /1.89/, /1.89'/ and /1.89''/ we obtain /1.86/. □

Substitution of /1.86/ into /1.21/ yields the general response formula

$$y/i,j/ = \sum_{k=1}^{i} C A^{i-k,j-1} \left[ A_2 x/k,0/ + B_{01} u/k,0/ \right] +$$
$$+ \sum_{l=1}^{j} C A^{i-1,j-l} \left[ A_1 x/0,l/ + B_{10} u/0,l/ \right] +$$

$$+ \sum_{k=1}^{i} \sum_{l=1}^{j} C\left[A^{i-k-1,j-1} B_{10} + A^{i-k,j-1-1} B_{01}\right] u/k,l/ + Du/i,j/ \qquad /1.90/$$

for $/i,j/ > /0,0/$.

## 1.11 CHARACTERISTIC POLYNOMIAL AND CAYLEY-HAMILTON THEOREM FOR THE SECOND FORNASINI-MARCHESINI´S MODEL.

**Definition 1.3**

The determinant

$$p/z_1,z_2/ = \det\left[I_n z_1 z_2 - A_1 z_2 - A_2 z_1\right] = \sum_{i=0}^{n} \sum_{j=0}^{n} a_{ij} z_1^i z_2^j \qquad /1.91/$$

is called the 2-D characteristic polynomial /function/ of F-MM II. The equation $p/z_1,z_2/ = 0$ is called the 2-D characteristic equation of F-MM II.

For example, the 2-D characteristic polynomial of F-MM II with

$$A_1 = \begin{bmatrix} 1 & 0 \\ 2 & 1 \end{bmatrix}, \quad A_2 = \begin{bmatrix} 1 & -1 \\ 0 & 2 \end{bmatrix} \qquad /1.92a/$$

has the form

$$p/z_1,z_2/ = \begin{vmatrix} z_1 z_2 - z_2 - z_1 & z_1 \\ -2z_2 & z_1 z_2 - z_2 - 2z_1 \end{vmatrix} =$$

$$= z_1^2 z_2^2 - 3z_1^2 z_2 - 2z_1 z_2^2 + 2z_1^2 + z_2^2 + 5z_1 z_2 \qquad /1.92b/$$

**Theorem 1.8** /2-D Cayley-Hamilton theorem/

The F-MM II transition matrix satisfies its own characteristic equation, i.e.

$$\sum_{i=0}^{n} \sum_{j=0}^{n} a_{ij} A^{i,j} = 0 \qquad /1.93/$$

**Proof:** Substituting the adjoint matrix

$$\text{adj}\left[I_n z_1 z_2 - A_1 z_2 - A_2 z_1\right] = \sum_{i=0}^{n-1} \sum_{j=0}^{n-1} B_{ij} z_1^i z_2^j$$

and /1.91/ into the equation

$$\text{adj}\left[I_n z_1 z_2 - A_1 z_2 - A_2 z_1\right]\left[I_n z_1 z_2 - A_1 z_2 - A_2 z_1\right] = I_n p/z_1, z_2/$$

we obtain

$$\sum_{i=0}^{n-1} \sum_{j=0}^{n-1} B_{ij}\left[I_n z_1^{i+1} z_2^{j+1} - A_1 z_1^i z_2^{j+1} - A_2 z_1^{i+1} z_2^j\right] =$$

$$= \sum_{i=0}^{n} \sum_{j=0}^{n} I_n a_{ij} z_1^i z_2^j \qquad /1.94/$$

From /1.84/ it follows that

$$A^{i+1,j+1} = A_1 A^{i,j+1} + A_2 A^{i+1,j} \qquad /1.95/$$

□

For example, taking into account that for /1.92a/

$$A^{1,1} = \begin{bmatrix} 0 & -2 \\ 6 & 2 \end{bmatrix},\ A^{1,2} = \begin{bmatrix} -5 & -7 \\ 14 & 2 \end{bmatrix},\ A^{2,1} = \begin{bmatrix} -3 & -3 \\ 14 & 0 \end{bmatrix},$$

$$A^{2,0} = \begin{bmatrix} 1 & 0 \\ 4 & 1 \end{bmatrix},\ A^{0,2} = \begin{bmatrix} 1 & -3 \\ 0 & 4 \end{bmatrix},\ A^{2,2} = \begin{bmatrix} -22 & -10 \\ 32 & -12 \end{bmatrix}$$

we get

$$A^{2,2} - 3A^{2,1} - 3A^{1,2} + 2A^{2,0} + A^{0,2} + 5A^{1,1} = 0$$

Therefore, the transition matrix of F-MM II with /1.92a/ satisfies its own characteristic equation $p/z_1, z_2/ = 0$ with $p/z_1, z_2/$ given by /1.92b/.

## PROBLEMS.

1. Show that
$$E^{0,1} A^{i,0} = 0 \quad /i = 1,2,.../$$

for any

$$E = \begin{bmatrix} E_1 & 0 \\ 0 & E_2 \end{bmatrix} \quad \text{and} \quad A = \begin{bmatrix} A_1 & A_2 \\ A_3 & A_4 \end{bmatrix}$$

where $E_1 \subset R^{n \times n}$, $E_2 \subset R^{m \times m}$, $A_1 \subset R^{n \times n}$, $A_4 \subset R^{m \times m}$.

**Hint**: Make use of the fact that $A^{i,0}$ has the same form as $A^{1,0}$.

2. Using the superposition property of the system prove formula /1.47/.
**Hint**: Superpose the effects of all inputs and boundary conditions.

3. Show that
$$A^{i,j} = 0, \quad /i,j/ \geqslant /1,1/$$

for any RM with

$$A = \begin{bmatrix} A_1 & 0 \\ 0 & A_2 \end{bmatrix}, \quad A_i \in R^{n_i \times n_i} \quad /i=1,2/.$$

**Hint**: Use equation /1.25/.

4. Show that for any matrix A and $/i,j/ \geqslant /0,0/$

$$A^{i,j} = \begin{cases} \sum_{k=0}^{i} A^{k,0} A^{0,1} A^{i-k,j-1} & \text{for } j > 0 \\ \sum_{k=0}^{j} A^{0,k} A^{1,0} A^{i-1,j-k} & \text{for } i > 0 \end{cases}$$

**Hint**: Use the induction on /i,j/.

5. Evaluate $A^{i,j}$, $/i,j/ > /0,0/$ for the matrix

$$A = \begin{bmatrix} 1 & 3 & 5 \\ 0 & 2 & 4 \\ 0 & 0 & 2 \end{bmatrix}$$

**Hint**: Use the formula /A.18/.

**Solution**:

$$A^{i,j} = 2^j \begin{bmatrix} 0 & 3 & 5+6j \\ 0 & 0 & 0 \\ 0 & 0 & 0 \end{bmatrix} \quad \text{for} \quad /i,j/ > /0,0/.$$

6. Show that if 2-D eigenvalues of a matrix A are distinct and the 2-D characteristic polynomial is separable, then

   $1^o$ the matrices $Z_{ij}$ (defined by $Z_{ij} = v_{ij}/A/$ and /1.74/) for $/0,0/ < /i,j/ \leqslant /n_1,n_2/$ are linearly independent,

   $2^o \quad \sum_{/0,0/ < /i,j/ \leqslant /n_1,n_2/} Z_{ij} = I,$

   $3^o \quad Z_{ij} Z_{kl} = \begin{cases} Z_{kl} & \text{for } i = k \text{ and } j = l \quad /k \neq l/ \\ 0 & \text{for } i \neq k \text{ and } j \neq l \end{cases}$

   **Hints:** 1/ note that $\sum_{/0,0/ < /i,j/ \leqslant /n_1,n_2/} c_{ij} Z_{ij} = 0$ implies

   $\sum_{/0,0/ < /i,j/ \leqslant /n_1,n_2/} c_{ij} v_{ij}/z_1,z_2/ = 0$,

   2/ use /1.75/ for $f/z_1,z_2/ = 1$,

   3/ use /1.75/ for $f/z_1,z_2/ = v_{ij}/z_1,z_2/ \, v_{kl}/z_1,z_2/$.

7. Show that for any square, partitioned matrix A

   $$AA^{i,j} = I^{1,0} A^{i+1,j} + I^{0,1} A^{i,j+1} \quad \text{for} \quad /i,j/ > /0,0/$$

   where

   $$I^{1,0} = \begin{bmatrix} I_{n_1} & 0 \\ 0 & 0 \end{bmatrix}, \quad I^{0,1} = \begin{bmatrix} 0 & 0 \\ 0 & I_{n_2} \end{bmatrix}.$$

   **Hint:** Use /1.25/ and note that $I^{1,0} A^{0,1} = 0$, $I^{0,1} A^{1,0} = 0$.

REFERENCES.

[1] Attasi S.: Modélisation et traitement des suites à deux indices. IRIA Rapport Laboria, Sept. 1975.

[2] Attasi S.: Systèmes lineaires homogènes à deux indices. IRIA Rapport Laboria No.31, Sept. 1973.

[3] Fornasini E., Marchesini G.: Doubly indexed dynamical systems; state space models and structural properties. Mathematical Systems Theory, vol.12, no.1, 1978.

[4] Fornasini E., Marchesini G.: State-space realization theory of two-dimensional filters. IEEE Trans. Autom. Contr. vol. AC-21, no.4, Aug. 1976, pp. 484-491.

[5] Givone D.D., Roesser R.P.: Minimization of multidimensional linear iterative circuits. IEEE Trans. on Computers vol. C-22, no.7, 1973, pp.673-678.

[6] Givone D.D., Roesser R.P.: Multidimensional linear iterative circuits - general properties. IEEE Trans on Computers vol. C-21, no.10, 1972, pp.1067-1073.

[7] Kaczorek T.: Extension of Sylvester's theorem to two-dimensional systems. Bull. Acad. Polon. Sci. Ser. Sci. Techn. vol. XXX, no. 1-2, 1982, pp.53-58.

[8] Kaczorek T.: Minimum energy control. Control and Cybernetics, vol. 12, no.3-4, 1983, pp.121-131.

[9] Klamka J.: Controllability of M-dimensional systems. Foundations of Control Engineering, vol.8, no.2, 1983, pp.65-74

[10] Klamka J.: Function of 2-D matrices. Foundations of Control Engineering vol.9, 1984 /in print/

[11] Kung S., Lèvy B., Morf M., Kailath T.: New results in 2-D systems theory, part I : 2-D polynomial matrices, factorization and coprimeness, part II : 2-D state-space models realization and the notions of controllability, observability and minimality. Proc. of IEEE, vol.65, no.6, 1977.

[12] Marszałek W.: Two-dimensional state-space discrete models for hyperbolic partial differential equations. Applied Mathematical Modelling, vol.8, Febr. 1984, pp.11-14.

[13] Roesser R.P.: A discrete state-space model for linear image processing. IEEE Trans Autom. Contr. vol. AC-20, no.1, Febr. 1975, pp.1-10.

[14] Stavroulakis P., Tzafestas S.G.: State reconstruction in low-sensitive design of 3-dimensional systems. IEE Proceedings, vol.130, Pt.D., no.6, Nov. 1983, pp.333-340.

[15] Trikha C.H., Chopra Y.C., Bajwa J.S.: Three-dimensional unilateral linear iterative circuits. Int. J. Systems Sci. vol.14, no.2, 1983, pp.117-139.

[16] Turhan Çiftçibaşi, Önder Yüksel: On the Cayley-Hamilton theorem for two-dimensional systems. IEEE Trans Autom. Contr. vol. AC-27, no.1, 1982, pp.193-194.

[17] Tzafestas S.G., Pimenides T.G.: Exact model matching control of three-dimensional systems using state and output feedback. Int. J. Systems Sci. vol.13, no.11, 1982, pp.1171-1187.

[18] Tzafestas S.G., Pimenides T.G.: Transfer function computation and factorization of 3-dimensional systems in state-space. IEE Proceedings, vol.130, Pt.D., no.5, Sept. 1983, pp.231-242.

# 2 TRANSFER FUNCTION MATRIX AND REALIZATION PROBLEM

## 2.1 TRANSFER FUNCTION MATRIX OF THE ROESSER'S MODEL

The transfer function matrix for 2-D systems is defined in a similar way as for 1-D systems.

Consider a 2-D linear system with inputs $u_1/i,j/$, $u_2/i,j/$, ..., $u_m/i,j/$ and outputs $y_1/i,j/$, $y_2/i,j/$, ..., $y_l/i,j/$. The transfer function $G_{hk}/z_1,z_2/$ between the k-th input and the h-th output of the system is defined as the ratio of the 2-D Z transform $Y_h/z_1,z_2/$ of the h-th output $y_h/i,j/$ to the 2-D Z transform $U_k/z_1,z_2/$ of the k-th input $u_k/i,j/$ for zero boundary conditions and other inputs also equal zero, i.e.

$$G_{hk}/z_1,z_2/ = \frac{Y_h/z_1,z_2/}{U_k/z_1,z_2/} \quad \begin{vmatrix} x^h/0,j/=0, & j \geqslant 0 \\ x^v/i,0/=0, & i \geqslant 0 \end{vmatrix} \qquad /2.1/$$

$$\underline{u_i}/i,j/ \begin{cases} \neq 0 & \text{for } i = k \\ = 0 & \text{for } i \neq k \end{cases}$$

for $h = 1,2, ..., l$ ; $k = 1,2, ..., m$ .

The matrix

$$G/z_1,z_2/ = \begin{bmatrix} G_{11}/z_1,z_2/ & \cdots & G_{1m}/z_1,z_2/ \\ \cdots & \cdots & \cdots \\ G_{11}/z_1,z_2/ & \cdots & G_{1m}/z_1,z_2/ \end{bmatrix} \qquad /2.2/$$

is called the transfer function matrix of the system.

The transfer function matrix gives an external description of the system.

Let us consider the Roesser's model /RM/ defined by the equations /1.1/ and /1.2/. Taking the 2-D Z transform of /1.1/, /1.2/ and making use of /A.2.4/ we obtain

$$\begin{bmatrix} z_1 X^h/z_1,z_2/ \\ z_2 X^v/z_1,z_2/ \end{bmatrix} = A \begin{bmatrix} X^h/z_1,z_2/ \\ X^v/z_1,z_2/ \end{bmatrix} + B U/z_1,z_2/ + \begin{bmatrix} z_1 X^h/0,z_2/ \\ z_2 X^v/z_1,0/ \end{bmatrix} \qquad /2.3/$$

and

$$Y/z_1,z_2/ = C \begin{bmatrix} X^h/z_1,z_2/ \\ X^v/z_1,z_2/ \end{bmatrix} + D U/z_1,z_2/ \qquad /2.4/$$

where $X^h/z_1,z_2/$, $X^v/z_1,z_2/$, $X^h/0,z_2/$, $X^v/z_1,0/$, $U/z_1,z_2/$ and $Y/z_1,z_2/$ are the 2-D Z transforms of $x^h/i,j/$, $x^v/i,j/$, $x^h/0,j/$, $x^v/i,0/$, $u/i,j/$ and $y/i,j/$, respectively.

From /2.3/ we have

$$\begin{bmatrix} X^h/z_1,z_2/ \\ X^v/z_1,z_2/ \end{bmatrix} = [Z-A]^{-1} B U/z_1,z_2/ + \begin{bmatrix} z_1 X^h/0,z_2/ \\ z_2 X^v/z_1,0/ \end{bmatrix} \qquad /2.5/$$

where

$$Z = \begin{bmatrix} I_{n_1} z_1 & 0 \\ 0 & I_{n_2} z_2 \end{bmatrix} \qquad /2.6/$$

and $I_k$ is the identity matrix of order k.

Substitution of /2.5/ into /2.4/ yields

$$Y/z_1,z_2/ = \left\{ C[Z-A]^{-1} B + D \right\} U/z_1,z_2/ + C[Z-A]^{-1} \begin{bmatrix} z_1 X^h/0,z_2/ \\ z_2 X^v/z_1,0/ \end{bmatrix} \qquad /2.7/$$

For zero boundary conditions ($x^h/0,j/ = 0$, $x^v/i,0/ = 0$ for $/i,j/ \geqslant /0,0/$) we have

$$Y/z_1,z_2/ = G/z_1,z_2/ \, U/z_1,z_2/ \qquad /2.7'/$$

where

$$G/z_1,z_2/ = C[Z-A]^{-1} B + D \qquad /2.8/$$

is the transfer function matrix of RM.

## 2.1.1 Mertzios-Paraskevopoulos' Formula.

Let

$$[Z-A]^{-1} = \frac{B/z_1,z_2/}{p/z_1,z_2/} \qquad /2.9/$$

where

$$B/z_1,z_2/ = \mathrm{Adj}[Z-A] = \sum_{i=0}^{n_1} \sum_{j=0}^{n_2} B_{ij} z_1^i z_2^j \qquad /2.10/$$

$$p/z_1,z_2/ = \det[Z-A] = \sum_{i=0}^{n_1} \sum_{j=0}^{n_2} a_{ij} z_1^i z_2^j \qquad /2.11/$$

$$B_{n_1 n_2} = 0, \quad B_{n_1-1, n_2} = \begin{bmatrix} I_{n_1} & 0 \\ 0 & 0 \end{bmatrix}, \quad B_{n_1, n_2-1} = \begin{bmatrix} 0 & 0 \\ 0 & I_{n_2} \end{bmatrix}, \quad a_{n_1 n_2} = 1 \qquad /2.12/$$

### Theorem 2.1 [10]

The transfer function matrix $G/z_1, z_2/$ may be expressed directly in terms of $A^{i,j}$, $B^{1,0}$, $B^{0,1}$, $C$ and $p/z_1, z_2/$ as follows:

$$G/z_1, z_2/ = \frac{1}{p/z_1, z_2/} \left[ \sum_{i=0}^{n_1} \sum_{j=0}^{n_2} \sum_{k=0}^{n_1-i} \sum_{l=0}^{n_2-j} a_{i+k, j+l} C \left( A^{k-1, l} B^{1,0} + A^{k, l-1} B^{0,1} \right) z_1^i z_2^j \right] + D \qquad /2.13/$$

where $A^{i,j}$, $B^{1,0}$ and $B^{0,1}$ are defined by /1.25/ and /1.25a/, respectively.

**Proof:** Substitution of /2.6/, /2.10/ and /2.11/ into the relation

$$[Z - A] \operatorname{Adj}[Z - A] = I_n \det[Z - A]$$

yields

$$\left( \begin{bmatrix} I_{n_1} z_1 & 0 \\ 0 & I_{n_2} z_2 \end{bmatrix} - A \right) \sum_{i=0}^{n_1} \sum_{j=0}^{n_2} B_{ij} z_1^i z_2^j = \sum_{i=0}^{n_1} \sum_{j=0}^{n_2} I_n a_{ij} z_1^i z_2^j \qquad /2.14/$$

Equating the coefficients of each term $z_1^i z_2^j$ on both sides of /2.14/ we obtain

$$B_{i-1, j}^{1,0} + B_{i, j-1}^{0,1} - A B_{ij} = I_n a_{ij} \quad \text{for} \quad /1,1/ \leq /i,j/ \leq /n_1, n_2/ \qquad /2.15/$$

$$B_{n_1 j}^{1,0} = 0 \quad \text{for} \quad 0 \leq j \leq n_2, \quad B_{i n_2}^{0,1} = 0 \quad \text{for} \quad 0 \leq i \leq n_1 \qquad /2.16a/$$

$$B_{0, j-1}^{0,1} - A B_{0j} = I_n a_{0j} \quad \text{for} \quad 1 \leq j \leq n_2 \qquad /2.16b/$$

$$B_{i-1, 0}^{1,0} - A B_{i0} = I_n a_{i0} \quad \text{for} \quad 1 \leq i \leq n_1 \qquad /2.16c/$$

$$-A B_{00} = I_n a_{00} \qquad /2.16d/$$

From /1.25/ and the relations

$$I^{1,0} A^{0,1} = 0, \quad I^{0,1} A^{1,0} = 0$$

it follows that
$$A A^{i,j} = I^{1,0} A^{i+1,j} + I^{0,1} A^{i,j+1} \qquad /2.17/$$

Let
$$A^{i,j} = \begin{bmatrix} A_{11}^{i,j} & A_{12}^{i,j} \\ A_{21}^{i,j} & A_{22}^{i,j} \end{bmatrix} \quad \text{for} \quad /0,0/ \leqslant /i,j/ \leqslant /n_1,n_2/ \qquad /2.18/$$

We shall show that the matrix

$$B_{ij} = \sum_{k=0}^{n_1-i} \sum_{l=0}^{n_2-j} a_{i+k,j+l} \begin{bmatrix} A_{11}^{k-1,l} & A_{12}^{k,l-1} \\ A_{21}^{k-1,l} & A_{22}^{k,l-1} \end{bmatrix} \qquad /2.19/$$

satisfies /2.15/ and /2.16/.
Using /2.17/ we obtain

$$AB_{ij} + I_n a_{ij} = \sum_{k=0}^{n_1-i} \sum_{l=0}^{n_2-j} a_{i+k,j+l} A \begin{bmatrix} A_{11}^{k-1,l} & A_{12}^{k,l-1} \\ A_{21}^{k-1,l} & A_{22}^{k,l-1} \end{bmatrix} + I_n a_{ij} =$$

$$= \sum_{k=0}^{n_1-i} \sum_{l=0}^{n_2-j} a_{i+k,j+l} \left[ I_n^{1,0} \begin{bmatrix} A_{11}^{k,l} \\ A_{21}^{k,l} \end{bmatrix} + I_n^{0,1} \begin{bmatrix} A_{11}^{k-1,l+1} \\ A_{21}^{k-1,l+1} \end{bmatrix} \; \middle| \; I^{1,0} \begin{bmatrix} A_{12}^{k+1,l-1} \\ A_{22}^{k+1,l-1} \end{bmatrix} + \right.$$

$$\left. + I_n^{0,1} \begin{bmatrix} A_{12}^{k,l} \\ A_{22}^{k,l} \end{bmatrix} \right] + a_{ij} \left[ I_n^{1,0} + I_n^{0,1} \right] =$$

$$= \sum_{k=0}^{n_1-i} \sum_{l=0}^{n_2-j+1} a_{i+k,j+l-1} I_n^{0,1} \begin{bmatrix} A_{11}^{k-1,l} & A_{12}^{k,l-1} \\ A_{21}^{k-1,l} & A_{22}^{k,l-1} \end{bmatrix} +$$

$$+ \sum_{k=0}^{n_1-i+1} \sum_{l=0}^{n_2-j} a_{i+k-1,j+l} I_n^{1,0} \begin{bmatrix} A_{11}^{k-1,l} & A_{12}^{k,l-1} \\ A_{21}^{k-1,l} & A_{22}^{k,l-1} \end{bmatrix} =$$

$$= B_{i,j-1}^{0,1} + B_{i-1,j}^{1,0}.$$

In a similar way we can show that the matrix /2.19/ satisfies /2.16/. Note that for $/0,0/ \leqslant /k,1/ \leqslant /n_1,n_2/$ we have

$$\begin{bmatrix} A_{11}^{k-1,1} & A_{12}^{k,1-1} \\ A_{21}^{k-1,1} & A_{22}^{k,1-1} \end{bmatrix} B = A B^{k-1,1} \; 1,0 + A B^{k,1-1} \; 0,1 \quad \text{for } /0,0/ \; /k,1/ \; /n_1,n_2/$$

and

$$B_{ij} B = \sum_{k=0}^{n_1-i} \sum_{l=0}^{n_2-j} a_{i+k,j+l} \left[ A B^{k-1,1 \; 1,0} + A B^{k,1-1 \; 0,1} \right] \qquad /2.20/$$

Substitution of /2.9/ and /2.10/ into /2.8/ yields

$$G/z_1,z_2/ = \frac{C B/z_1,z_2/ B}{p/z_1,z_2/} + D =$$

$$= \frac{1}{p/z_1,z_2/} \left[ \sum_{i=0}^{n_1} \sum_{j=0}^{n_2} C B_{ij} B z_1^i z_2^j \right] + D \qquad /2.21/$$

Substituting /2.20/ into /2.21/ we obtain the desired formula /2.13/.
□

From /2.13/ it follows that to find $G/z_1,z_2/$ we have to evaluate $A^{i,j}$ for $/0,0/ \leqslant /i,j/ \leqslant /n_1,n_2/$ using the algorithms given in Section 1.8.

Example 2.1

Find $G/z_1,z_2/$ for

$$A = \begin{bmatrix} A_{11} & A_{12} \\ A_{21} & A_{22} \end{bmatrix} = \begin{bmatrix} 1 & 2 & | & 0 & -1 \\ -1 & 0 & | & 1 & 0 \\ \hline 1 & 0 & | & 2 & 0 \\ 0 & 1 & | & 1 & 1 \end{bmatrix}, \quad B = \begin{bmatrix} B_1 \\ B_2 \end{bmatrix} = \begin{bmatrix} 1 & -1 \\ 1 & 0 \\ \hline 1 & 2 \\ 0 & 1 \end{bmatrix},$$

$$C = \begin{bmatrix} C_1 & C_2 \end{bmatrix} = \begin{bmatrix} 1 & 0 & | & 0 & 1 \\ 1 & -1 & | & 1 & -1 \end{bmatrix}, \quad \text{and} \quad D = \begin{bmatrix} 1 & 2 \\ 1 & -1 \end{bmatrix}. \qquad /2.22/$$

The characteristic polynomial has the form

$$p/z_1,z_2/ = \det[Z - A] = \begin{vmatrix} z_1-1 & -2 & 0 & 1 \\ 1 & z_1 & -1 & 0 \\ -1 & 0 & z_2-2 & 0 \\ 0 & -1 & -1 & z_2-1 \end{vmatrix} =$$

$$= \left(z_1^2 - z_1 + 2\right)\left(z_2^2 - 3z_2 + 2\right) + z_1 - 3z_2 + 5 =$$

$$= z_1^2 z_2^2 - 3z_1^2 z_2 - z_1 z_2^2 + 2z_1^2 + 2z_2^2 + 3z_1 z_2 - z_1 - 9z_2 + 9$$

and

$a_{00}=9$, $a_{10}=-1$, $a_{01}=-9$, $a_{11}=3$, $a_{20}=2$, $a_{02}=2$, $a_{21}=-3$, $a_{12}=-1$, $a_{22}=1$.

We have

$$A^{1,0} = \begin{bmatrix} 1 & 2 & 0 & -1 \\ -1 & 0 & 1 & 0 \\ 0 & 0 & 0 & 0 \\ 0 & 0 & 0 & 0 \end{bmatrix}, \quad A^{0,1} = \begin{bmatrix} 0 & 0 & 0 & 0 \\ 0 & 0 & 0 & 0 \\ 1 & 0 & 2 & 0 \\ 0 & 1 & 1 & 1 \end{bmatrix},$$

$$A^{1,1} = \begin{bmatrix} 0 & -1 & -1 & -1 \\ 1 & 0 & 2 & 0 \\ 1 & 2 & 0 & -1 \\ -1 & 0 & 1 & 0 \end{bmatrix}, \quad A^{2,0} = \begin{bmatrix} -1 & 2 & 2 & -1 \\ -1 & -2 & 0 & 1 \\ 0 & 0 & 0 & 0 \\ 0 & 0 & 0 & 0 \end{bmatrix},$$

$$A^{0,2} = \begin{bmatrix} 0 & 0 & 0 & 0 \\ 0 & 0 & 0 & 0 \\ 2 & 0 & 4 & 0 \\ 1 & 1 & 3 & 1 \end{bmatrix}, \quad A^{2,1} = \begin{bmatrix} 3 & -1 & 2 & -1 \\ 1 & 3 & 1 & 0 \\ -1 & 2 & 2 & -1 \\ -1 & -2 & 0 & 1 \end{bmatrix},$$

$$A^{1,2} = \begin{bmatrix} -1 & -1 & -3 & -1 \\ 2 & 0 & 4 & 0 \\ 2 & 3 & -1 & -3 \\ 1 & 2 & 3 & -1 \end{bmatrix}, \quad B^{1,0} = \begin{bmatrix} 1 & -1 \\ 1 & 0 \\ 0 & 0 \\ 0 & 0 \end{bmatrix}, \quad B^{0,1} = \begin{bmatrix} 0 & 0 \\ 0 & 0 \\ 1 & 2 \\ 0 & 1 \end{bmatrix}.$$

Using /2.13/ we obtain

$$G/z_1,z_2/ = \frac{1}{p/z_1,z_2/} \left[ \sum_{i=0}^{2} \sum_{j=0}^{2} \sum_{k=0}^{2-i} \sum_{l=0}^{2-j} a_{i+k,j+l} \, C\left(A^{k-1,l} B^{1,0}\right) + \right.$$

$$+ A^{k,1-1} B^{0,1}\bigg) z_1^i z_2^j \bigg] + D =$$

$$= \frac{1}{p/z_1,z_2/} \bigg[ C \bigg( \sum_{i=0}^{1} \sum_{j=0}^{2} a_{i+1,j} z_1^i z_2^j B^{1,0} + \sum_{i=0}^{2} \sum_{j=0}^{1} a_{i,j+1} z_1^i z_2^j B^{0,1} \bigg) +$$

$$+ C A^{1,0} \bigg( \sum_{j=0}^{2} a_{2,j} z_2^j B^{1,0} + \sum_{i=0}^{1} \sum_{j=0}^{1} a_{i+1,j+1} z_1^i z_2^j B^{0,1} \bigg) +$$

$$+ C A^{0,1} \bigg( \sum_{i=0}^{1} \sum_{j=0}^{1} a_{i+1,j+1} z_1^i z_2^j B^{1,0} + \sum_{i=0}^{2} a_{i,2} z_1^i B^{0,1} \bigg) +$$

$$+ C A^{1,1} \bigg( \sum_{j=0}^{1} a_{2,j+1} z_2^j B^{1,0} + \sum_{i=0}^{1} a_{i+1,2} z_1^i B^{0,1} \bigg) +$$

$$+ C A^{2,0} \bigg( \sum_{j=0}^{1} a_{2,j+1} z_2^j B^{0,1} \bigg) +$$

$$+ C A^{0,2} \bigg( \sum_{i=0}^{1} a_{i+1,2} z_1^i B^{1,0} \bigg) +$$

$$+ C A^{1,2} B^{1,0} + C A^{2,1} B^{0,1} \bigg] + D =$$

$$= \frac{1}{p/z_1,z_2/} \begin{bmatrix} G_{11}/z_1,z_2/ & G_{12}/z_1,z_2/ \\ G_{21}/z_1,z_2/ & G_{22}/z_1,z_2/ \end{bmatrix}$$

where

$$G_{11}/z_1,z_2/ = z_1 z_2^2 + z_1^2 + 2z_2^2 - 2z_1 z_2 - 7z_2 + 11 \; ,$$

$$G_{12}/z_1,z_2/ = z_1^2 z_2 - z_1 z_2^2 + 3z_1 z_2 + 4z_2^2 - 2z_1^2 - 3z_1 - 13z_2 + 11 \; ,$$

$$G_{21}/z_1,z_2/ = z_1^2 z_2 + 4z_2^2 - 3z_1^2 - 2z_1 z_2 - 6z_2 - z_1 - 9 \; ,$$

$$G_{22}/z_1,z_2/ = z_1^2 z_2 - z_1 z_2^2 - z_2^2 - 2z_1^2 - 2z_1 z_2 + 10z_2 + z_1 - 9 \; .$$

## 2.1.2 Koo-Chen's Algorithm.

Koo-Chen's algorithm [9] is an extension of well-known Fadeeva-Leverrier algorithm for 1-D systems [16].

**Theorem 2.2**

$$a_{ij} = -\frac{1}{n-i-j} \, \text{tr}\left[A\,B_{ij}\right] \qquad /2.23/$$

where $n = n_1 + n_2$ and $\text{tr}\left[A\,B_{ij}\right]$ is the trace of the matrix $A\,B_{ij}$, i.e. the sum of the diagonal elements of $A\,B_{ij}$.

**Proof:** It can be shown /see Problem 2.2/ that

$$\frac{\partial p/z_1,z_2/}{\partial z_1} = \text{tr}\, B^{1,0}/z_1,z_2/ \qquad /2.24a/$$

and

$$\frac{\partial p/z_1,z_2/}{\partial z_2} = \text{tr}\, B^{0,1}/z_1,z_2/ \qquad /2.24b/$$

where $p/z_1,z_2/$ and $B/z_1,z_2/$ are defined by /2.11/ and /2.10/, respectively.

Taking into account that

$$\frac{\partial p/z_1,z_2/}{\partial z_1} = \sum_{i=0}^{n_1} \sum_{j=0}^{n_2} i\,a_{ij}\,z_1^{i-1}\,z_2^{j}$$

$$\frac{\partial p/z_1,z_2/}{\partial z_2} = \sum_{i=0}^{n_1} \sum_{j=0}^{n_2} j\,a_{ij}\,z_1^{i}\,z_2^{j-1}$$

and equating the coefficients of like power of terms $z_1^i z_2^j$ from /2.24/ we obtain

$$i\,a_{ij} = \text{tr}\left[B_{i-1,j}^{1,0}\right] \quad \text{and} \quad j\,a_{ij} = \text{tr}\left[B_{i,j-1}^{0,1}\right] \qquad /2.25/$$

From /2.15/ we have

$$\text{tr}\left[B_{i-1,j}^{1,0}\right] + \text{tr}\left[B_{i,j-1}^{0,1}\right] = \text{tr}\left[A\,B_{ij}\right] + \text{tr}\left[a_{ij}\,I_n\right] \qquad /2.26/$$

Substitution of /2.25/ into /2.26/ yields

$$(i+j)\,a_{ij} = \text{tr}\left[A\,B_{ij}\right] + n\,a_{ij}$$

from which /2.23/ follows immediately. □

Using /2.15/ and /2.23/ we can find $G(z_1,z_2)$ for the given matrices A, B, C and D. The matrices $B_{ij}$ and the coefficients $a_{ij}$ can be evaluated by the use of the following [9]

## Algorithm 2.1

Step 1. Set

$$a_{n_1 n_2} = 1$$

$$B_{i n_2}^{0,1} = 0 \quad \text{for} \quad i = n_1, n_1-1, \ldots, 0$$

$$B_{n_1 j}^{1,0} = 0 \quad \text{for} \quad j = n_2, n_2-1, \ldots, 0$$

Step 2. For $i = n_1-1, n_1-2, \ldots, 0$ evaluate

$$B_{i n_2} = \left[ I_n a_{i+1,n_2} + A B_{i+1,n_2} \right]^{1,0}$$

$$a_{i n_2} = -\frac{1}{n_1 - i} \text{tr}\left[ A B_{i n_2} \right]$$

/2.27a/

Step 3. For $j = n_2-1, n_2-2, \ldots, 0$ evaluate

$$B_{n_1 j} = \left[ I_n a_{n_1,j+1} + A B_{n_1,j+1} \right]^{0,1}$$

$$a_{n_1 j} = -\frac{1}{n_2 - j} \text{tr}\left[ A B_{n_1 j} \right]$$

/2.27b/

Step 4. For $i = n_1-k$; $j = n_2-1, n_2-2, \ldots, 0$ evaluate

$$B_{ij} = \left[ I_n a_{i,j+1} + A B_{i,j+1} \right]^{0,1} + \left[ I_n a_{i+1,j} + A B_{i+1,j} \right]^{1,0}$$

$$a_{ij} = -\frac{1}{n-i-j} \text{tr}\left[ A B_{ij} \right]$$

/2.28/

taking $k = 1, 2, \ldots, n_1$.

## Example 2.2

Evaluate $G(z_1, z_2)$ for the matrices A, B, C, D given by /2.22/.
Since $a_{22} = 1$ and $B_{22} = 0$ we get

$$M_{22} = AB_{22} + I_4 a_{22} = \begin{bmatrix} 1 & 0 & 0 & 0 \\ 0 & 1 & 0 & 0 \\ 0 & 0 & 1 & 0 \\ 0 & 0 & 0 & 1 \end{bmatrix}$$

Taking into account that

$$B_{12}^{1,0} = M_{22}^{1,0}, \quad B_{21}^{0,1} = M_{22}^{0,1}, \quad \text{and} \quad B_{12}^{0,1} = 0, \quad B_{22}^{1,0} = 0$$

we obtain

$$B_{12} = B_{12}^{1,0} + B_{12}^{0,1} = \begin{bmatrix} 1 & 0 & 0 & 0 \\ 0 & 1 & 0 & 0 \\ 0 & 0 & 0 & 0 \\ 0 & 0 & 0 & 0 \end{bmatrix}, \quad B_{21} = B_{21}^{1,0} + B_{21}^{0,1} = \begin{bmatrix} 0 & 0 & 0 & 0 \\ 0 & 0 & 0 & 0 \\ 0 & 0 & 1 & 0 \\ 0 & 0 & 0 & 1 \end{bmatrix}.$$

From /2.23/ we have

$$a_{12} = -\operatorname{tr} AB_{12} = -\operatorname{tr} \begin{bmatrix} 1 & 2 & 0 & 0 \\ -1 & 0 & 0 & 0 \\ 1 & 0 & 0 & 0 \\ 0 & 1 & 0 & 0 \end{bmatrix} = -1$$

and

$$a_{21} = -\operatorname{tr} AB_{21} = -\operatorname{tr} \begin{bmatrix} 0 & 0 & 0 & -1 \\ 0 & 0 & 1 & 0 \\ 0 & 0 & 2 & 0 \\ 0 & 0 & 1 & 1 \end{bmatrix} = -3.$$

Then we evaluate

$$M_{12} = AB_{12} + I_4 a_{12} = \begin{bmatrix} 0 & 2 & 0 & 0 \\ -1 & -1 & 0 & 0 \\ 1 & 0 & -1 & 0 \\ 0 & 1 & 0 & -1 \end{bmatrix}, \quad M_{21} = AB_{21} + I_4 a_{21} = \begin{bmatrix} -3 & 0 & 0 & -1 \\ 0 & -3 & 1 & 0 \\ 0 & 0 & -1 & 0 \\ 0 & 0 & 1 & -2 \end{bmatrix};$$

$$B_{02}^{1,0} = M_{12}^{1,0}, \quad B_{20}^{0,1} = M_{21}^{0,1}, \quad B_{02}^{0,1} = 0, \quad B_{20}^{1,0} = 0;$$

$$B_{02} = B_{02}^{1,0} + B_{02}^{0,1} = \begin{bmatrix} 0 & 2 & 0 & 0 \\ -1 & -1 & 0 & 0 \\ 0 & 0 & 0 & 0 \\ 0 & 0 & 0 & 0 \end{bmatrix}, \quad B_{20} = B_{20}^{1,0} + B_{20}^{0,1} = \begin{bmatrix} 0 & 0 & 0 & 0 \\ 0 & 0 & 0 & 0 \\ 0 & 0 & -1 & 0 \\ 0 & 0 & 1 & -2 \end{bmatrix}$$

$$B_{11}^{1,0} = M_{21}^{1,0}, \quad B_{11}^{0,1} = M_{12}^{0,1};$$

$$B_{11} = B_{11}^{1,0} + B_{11}^{0,1} = \begin{bmatrix} -3 & 0 & 0 & -1 \\ 0 & -3 & 1 & 0 \\ 1 & 0 & -1 & 0 \\ 0 & 1 & 0 & -1 \end{bmatrix};$$

$$a_{02} = -\frac{1}{2} \operatorname{tr} AB_{02} = -\frac{1}{2} \operatorname{tr} \begin{bmatrix} -2 & 0 & 0 & 0 \\ 0 & -2 & 0 & 0 \\ 0 & 2 & 0 & 0 \\ -1 & -1 & 0 & 0 \end{bmatrix} = 2;$$

$$a_{20} = -\frac{1}{2} \operatorname{tr} AB_{20} = -\frac{1}{2} \operatorname{tr} \begin{bmatrix} 0 & 0 & -1 & 2 \\ 0 & 0 & -1 & 0 \\ 0 & 0 & -2 & 0 \\ 0 & 0 & 0 & -2 \end{bmatrix} = 2;$$

$$a_{11} = -\frac{1}{2} \operatorname{tr} AB_{11} = -\frac{1}{2} \operatorname{tr} \begin{bmatrix} -3 & -7 & 2 & 0 \\ 4 & 0 & -1 & 1 \\ -1 & 0 & -2 & -1 \\ 1 & -2 & 0 & -1 \end{bmatrix} = 3;$$

$$M_{02} = AB_{02} + I_4 a_{02} = \begin{bmatrix} 0 & 0 & 0 & 0 \\ 0 & 0 & 0 & 0 \\ 0 & 2 & 2 & 0 \\ -1 & -1 & 0 & 2 \end{bmatrix}, \quad M_{20} = AB_{20} + I_4 a_{20} = \begin{bmatrix} 2 & 0 & -1 & 2 \\ 0 & 2 & -1 & 0 \\ 0 & 0 & 0 & 0 \\ 0 & 0 & 0 & 0 \end{bmatrix},$$

$$M_{11} = AB_{11} + I_4 a_{11} = \begin{bmatrix} 0 & -7 & 2 & 0 \\ 4 & 3 & -1 & 1 \\ -1 & 0 & 1 & -1 \\ 1 & -2 & 0 & 2 \end{bmatrix};$$

$$B_{10}^{1,0} = M_{20}^{1,0}, \quad B_{10}^{0,1} = M_{11}^{0,1}, \quad B_{01}^{1,0} = M_{11}^{1,0}, \quad B_{01}^{0,1} = M_{02}^{0,1};$$

$$B_{10} = B_{10}^{1,0} + B_{10}^{0,1} = \begin{bmatrix} 2 & 0 & -1 & 2 \\ 0 & 2 & -1 & 0 \\ -1 & 0 & 1 & -1 \\ 1 & -2 & 0 & 2 \end{bmatrix}, \quad B_{01} = B_{01}^{1,0} + B_{01}^{0,1} = \begin{bmatrix} 0 & -7 & 2 & 0 \\ 4 & 3 & -1 & 1 \\ 0 & 2 & 2 & 0 \\ -1 & -1 & 0 & 2 \end{bmatrix};$$

$$a_{10} = -\frac{1}{3} \operatorname{tr} AB_{10} = -\frac{1}{3} \operatorname{tr} \begin{bmatrix} 1 & 6 & -3 & 0 \\ -3 & 0 & 2 & -3 \\ 0 & 0 & 1 & 0 \\ 0 & 0 & 0 & 1 \end{bmatrix} = -1;$$

$$a_{01} = -\frac{1}{3} \text{tr } AB_{01} = -\frac{1}{3} \text{tr} \begin{bmatrix} 9 & 0 & 0 & 0 \\ 0 & 9 & 0 & 0 \\ 0 & -3 & 6 & 0 \\ 3 & 4 & 1 & 3 \end{bmatrix} = -9 \; ;$$

$$M_{10} = AB_{10} + I_4 a_{10} = \begin{bmatrix} 0 & 6 & -3 & 0 \\ -3 & -1 & 2 & -3 \\ 0 & 0 & 0 & 0 \\ 0 & 0 & 0 & 0 \end{bmatrix}, \quad M_{01} = AB_{01} + I_4 a_{01} = \begin{bmatrix} 0 & 0 & 0 & 0 \\ 0 & 0 & 0 & 0 \\ 0 & -3 & -3 & 0 \\ 3 & 4 & 1 & -6 \end{bmatrix};$$

$$B_{00}^{1,0} = M_{10}^{1,0}, \quad B_{00}^{0,1} = M_{01}^{0,1},$$

$$B_{00} = B^{1,0} + B^{0,1} = \begin{bmatrix} 0 & 6 & -3 & 0 \\ -3 & -1 & 2 & -3 \\ 0 & -3 & -3 & 0 \\ 3 & 4 & 1 & -6 \end{bmatrix};$$

$$a_{00} = -\frac{1}{4} \text{tr } AB_{00} = -\frac{1}{4} \text{tr} \begin{bmatrix} -9 & 0 & 0 & 0 \\ 0 & -9 & 0 & 0 \\ 0 & 0 & -9 & 0 \\ 0 & 0 & 0 & -9 \end{bmatrix} = 9 \; .$$

Using /2.21/ we obtain

$$G/z_1, z_2/ = \frac{1}{p/z_1, z_2/} \left[ \sum_{i=0}^{n_1} \sum_{j=0}^{n_2} CB_{ij} B z_1^i z_2^j \right] + D =$$

$$= \frac{1}{p/z_1, z_2/} \begin{bmatrix} 1 & 0 & 0 & 1 \\ 1 & -1 & 1 & -1 \end{bmatrix} \begin{bmatrix} z_1 z_2^2 - 3z_1 z_2 + 2z_1 & 2z_2^2 - 7z_2 + 6 \\ -z_2^2 + 4z_2 - 3 & z_1 z_2^2 - 3z_1 z_2 - z_2^2 + 2z_1 + 3z_2 - 1 \\ z_1 z_2 - z_1 & z_2 - 3 \\ z_1 - z_2 + 3 & z_1 z_2 - 2z_1 - z_2 + 4 \end{bmatrix}$$

$$\begin{bmatrix} -z_1 + 2z_2 - 3 & -z_1 z_2 + 2z_1 \\ z_1 z_2 - z_1 - z_2 + 2 & z_2 - 3 \\ z_1^2 z_2 - z_1 z_2 - z_1^2 + 2z_2 - 3 & -z_1 \\ z_1^2 + 1 & z_1^2 z_2 - z_1 z_2 - 2z_1^2 + 2z_1 + 2z_2 - 6 \end{bmatrix} \begin{bmatrix} 1 & -1 \\ 1 & 0 \\ 1 & 2 \\ 0 & 1 \end{bmatrix} +$$

$$+ \begin{bmatrix} 1 & 2 \\ 1 & -1 \end{bmatrix} = \begin{bmatrix} G_{11}/z_1,z_2/ & G_{12}/z_1,z_2/ \\ G_{21}/z_1,z_2/ & G_{22}/z_1,z_2/ \end{bmatrix}$$

where

$$p/z_1,z_2/ = z_1^2 z_2^2 - 3z_1^2 z_2 - z_1 z_2^2 + 2z_1^2 + 2z_2^2 + 3z_1 z_2 - z_1 - 9z_2 + 9$$

$$G_{11}/z_1,z_2/ = z_1 z_2^2 + z_1^2 + 2z_2^2 - 2z_1 z_2 - 7z_2^2 + 11$$

$$G_{12}/z_1,z_2/ = z_1^2 z_2 - z_1 z_2^2 + 3z_1 z_2 + 4z_2^2 - 2z_1^2 - 3z_1 - 13z_2 + 11$$

$$G_{21}/z_1,z_2/ = z_1^2 z_2 + 4z_2^2 - 3z_1^2 - 2z_1 z_2 - 6z_2 - z_1 - 9$$

$$G_{22}/z_1,z_2/ = z_1^2 z_2 - z_1 z_2^2 - z_2^2 - 2z_1^2 - 2z_1 z_2 + 10z_2 + z_1 - 9 \ .$$

## 2.2 TRANSFER FUNCTION MATRIX OF THE TZAFESTAS-PIMENIDES' MODEL

Consider Tzafestas-Pimenides' model /T-PM/ defined by the equations /1.32/ and /1.33/. Taking the 3-D Z transform of /1.32/ and /1.33/ with zero boundary conditions we obtain

$$\begin{bmatrix} z_1 X^h/z_1,z_2,z_3/ \\ z_2 X^v/z_1,z_2,z_3/ \\ z_3 X^d/z_1,z_2,z_3/ \end{bmatrix} = A \begin{bmatrix} X^h/z_1,z_2,z_3/ \\ X^v/z_1,z_2,z_3/ \\ X^d/z_1,z_2,z_3/ \end{bmatrix} + B\,U/z_1,z_2,z_3/ \qquad /2.29/$$

$$Y/z_1,z_2,z_3/ = C \begin{bmatrix} X^h/z_1,z_2,z_3/ \\ X^v/z_1,z_2,z_3/ \\ X^d/z_1,z_2,z_3/ \end{bmatrix} + D\,U/z_1,z_2,z_3/ \qquad /2.30/$$

where $X^h/z_1,z_2,z_3/$, $X^v/z_1,z_2,z_3/$, $X^d/z_1,z_2,z_3/$, $U/z_1,z_2,z_3/$ and $Y/z_1,z_2,z_3/$ are the 3-D Z transforms of $x^h/i,j,k/$, $x^v/i,j,k/$, $x^d/i,j,k/$, $u/i,j,k/$ and $y/i,j,k/$, respectively.

From /2.29/ we have

$$\begin{bmatrix} X^h/z_1,z_2,z_3/ \\ X^v/z_1,z_2,z_3/ \\ X^d/z_1,z_2,z_3/ \end{bmatrix} = [Z - A]^{-1} B\,U/z_1,z_2,z_3/ \qquad /2.31/$$

where

$$Z = \begin{bmatrix} I_{n_1} z_1 & 0 & 0 \\ 0 & I_{n_2} z_2 & 0 \\ 0 & 0 & I_{n_3} z_3 \end{bmatrix} \qquad /2.32/$$

Substitution of /2.31/ into /2.30/ yields

$$Y/z_1,z_2,z_3/ = G/z_1,z_2,z_3/\; U/z_1,z_2,z_3/$$

where

$$G/z_1,z_2,z_3/ = C[Z - A]^{-1} B + D \qquad /2.33/$$

is the transfer function matrix of T-PM.


## 2.2.1 Extension of Koo-Chen's algorithm for 3-D systems.

The Koo-Chen algorithm was extended for 3-D systems by Tzafestas and Pimenides in [14].
Let

$$[Z - A]^{-1} = \frac{B/z_1,z_2,z_3/}{p/z_1,z_2,z_3/} \qquad /2.34/$$

where

$$B/z_1,z_2,z_3/ = \text{Adj}[Z - A] = \sum_{i=0}^{n_1} \sum_{j=0}^{n_2} \sum_{k=0}^{n_3} B_{ijk}\, z_1^i\, z_2^j\, z_3^k \qquad /2.35/$$

$$p/z_1,z_2,z_3/ = \det[Z - A] = \sum_{i=0}^{n_1} \sum_{j=0}^{n_2} \sum_{k=0}^{n_3} a_{ijk}\, z_1^i\, z_2^j\, z_3^k \qquad /2.36/$$

$$B_{n_1 n_2 n_3} = 0, \quad B_{n_1-1,n_2,n_3} = \begin{bmatrix} I_{n_1} & 0 & 0 \\ 0 & 0 & 0 \\ 0 & 0 & 0 \end{bmatrix}, \quad B_{n_1,n_2-1,n_3} = \begin{bmatrix} 0 & 0 & 0 \\ 0 & I_{n_2} & 0 \\ 0 & 0 & 0 \end{bmatrix},$$

$$B_{n_1,n_2,n_3-1} = \begin{bmatrix} 0 & 0 & 0 \\ 0 & 0 & 0 \\ 0 & 0 & I_{n_3} \end{bmatrix}, \quad a_{n_1 n_2 n_3} = 1.$$

Substitution of /2.32/, /2.35/ and /2.36/ into the equation

$$[Z - A] \text{Adj}[Z - A] = I_n \det[Z - A]$$

yields

$$\left( \begin{bmatrix} I_{n_1} z_1 & 0 & 0 \\ 0 & I_{n_2} z_2 & 0 \\ 0 & 0 & I_{n_3} z_3 \end{bmatrix} - A \right) \sum_{i=0}^{n_1} \sum_{j=0}^{n_2} \sum_{k=0}^{n_3} B_{ijk} z_1^i z_2^j z_3^k =$$

$$= \sum_{i=0}^{n_1} \sum_{j=0}^{n_2} \sum_{k=0}^{n_3} I_n a_{ijk} z_1^i z_2^j z_3^k . \qquad /2.37/$$

Equating the coefficients of each term $z_1^i z_2^j z_3^k$ on both sides of /2.37/ we obtain

$$B_{i-1\ j\ k}^{1,0,0} + B_{i\ j-1\ k}^{0,1,0} + B_{i\ j\ k-1}^{0,0,1} - A B_{ijk} = I_n a_{ijk} \qquad /2.38/$$

for $/1,1,1/ \leqslant /i,j,k/ \leqslant /n_1,n_2,n_3/$,

$$B_{in_2n_3}^{0,1,1} = 0, \quad B_{n_1 j n_3}^{1,0,1} = 0, \quad B_{n_1 n_2 k}^{1,1,0} = 0 \quad \text{for } 0 \leqslant i \leqslant n_1, \ 0 \leqslant j \leqslant n_2,$$

$$0 \leqslant k \leqslant n_3 \qquad /2.39a/$$

$$B_{i-1\ j-1\ 0}^{1,1,0} - A B_{ij0} = I_n a_{ij0} \quad \text{for } /1,1/ \leqslant /i,j/ \leqslant /n_1,n_2/ \qquad /2.39b/$$

$$B_{0\ j-1\ k-1}^{0,1,1} - A B_{0jk} = I_n a_{0jk} \quad \text{for } /1,1/ \leqslant /j,k/ \leqslant /n_2,n_3/ \qquad /2.39c/$$

$$B_{i-1\ 0\ k-1}^{1,0,1} - A B_{i0k} = I_n a_{i0k} \quad \text{for } /1,1/ \leqslant /i,k/ \leqslant /n_1,n_3/ \qquad /2.39d/$$

and

$$A B_{000} + I_n a_{000} = 0 \qquad /2.39e/$$

where

$$B_{ijk}^{1,0,0} = \begin{bmatrix} B_{ijk\ 11} & B_{ijk\ 12} & B_{ijk\ 13} \\ 0 & 0 & 0 \\ 0 & 0 & 0 \end{bmatrix}$$

$$B_{ijk}^{0,1,0} = \begin{bmatrix} 0 & 0 & 0 \\ B_{ijk\ 21} & B_{ijk\ 22} & B_{ijk\ 23} \\ 0 & 0 & 0 \end{bmatrix}$$

$$B_{ijk}^{0,0,1} = \begin{bmatrix} 0 & 0 & 0 \\ 0 & 0 & 0 \\ B_{ijk\,31} & B_{ijk\,32} & B_{ijk\,33} \end{bmatrix}$$

$$B_{ijk}^{1,1,0} = \begin{bmatrix} B_{ijk\,11} & B_{ijk\,12} & B_{ijk\,13} \\ B_{ijk\,21} & B_{ijk\,22} & B_{ijk\,23} \\ 0 & 0 & 0 \end{bmatrix}$$

$$B_{ijk}^{1,0,1} = \begin{bmatrix} B_{ijk\,11} & B_{ijk\,12} & B_{ijk\,13} \\ 0 & 0 & 0 \\ B_{ijk\,31} & B_{ijk\,32} & B_{ijk\,33} \end{bmatrix}$$

$$B_{ijk}^{0,1,1} = \begin{bmatrix} 0 & 0 & 0 \\ B_{ijk\,21} & B_{ijk\,22} & B_{ijk\,23} \\ B_{ijk\,31} & B_{ijk\,32} & B_{ijk\,33} \end{bmatrix}$$

From the well-known relation

$$\frac{d}{dx} \begin{vmatrix} a_{11}/x/ & \cdots & a_{1n}/x/ \\ \cdots & \cdots & \cdots \\ a_{n1}/x/ & \cdots & a_{nn}/x/ \end{vmatrix} = \sum_{i=1}^{n} \begin{vmatrix} a_{11}/x/ & \cdots & a_{1n}/x/ \\ \cdots & \cdots & \cdots \\ a_{i-1,1}/x/ & \cdots & a_{i-1,n}/x/ \\ \frac{d}{dx}a_{i1}/x/ & \cdots & \frac{d}{dx}a_{in}/x/ \\ a_{i+1,1}/x/ & \cdots & a_{i+1,n}/x/ \\ \cdots & \cdots & \cdots \\ a_{n1}/x/ & & a_{nn}/x/ \end{vmatrix} \qquad /2.40/$$

it follows that

$$\frac{\partial p/z_1,z_2,z_3/}{\partial z_1} = \frac{\partial}{\partial z_1} \begin{vmatrix} I_{n_1}z_1 - A_{11} & -A_{12} & -A_{13} \\ -A_{21} & I_{n_2}z_2 - A_{22} & -A_{23} \\ -A_{31} & -A_{32} & I_{n_3}z_3 - A_{33} \end{vmatrix} =$$

$$= \operatorname{tr} B/z_1,z_2,z_3/^{1,0,0}$$

$$\frac{\partial p/z_1,z_2,z_3/}{\partial z_2} = \frac{\partial}{\partial z_2} \begin{vmatrix} I_{n_1}z_1 - A_{11} & -A_{12} & -A_{13} \\ -A_{21} & I_{n_2}z_2 - A_{22} & -A_{23} \\ -A_{31} & -A_{32} & I_{n_3}z_3 - A_{33} \end{vmatrix} =$$

$$= \mathrm{tr}\, B/z_1,z_2,z_3/ \overset{0,1,0}{} \qquad \qquad /2.41/$$

$$\frac{\partial p/z_1,z_2,z_3/}{\partial z_3} = \frac{\partial}{\partial z_3} \begin{vmatrix} I_{n_1}z_1 - A_{11} & -A_{12} & -A_{13} \\ -A_{21} & I_{n_2}z_2 - A_{22} & -A_{23} \\ -A_{31} & -A_{32} & I_{n_3}z_3 - A_{33} \end{vmatrix} =$$

$$= \mathrm{tr}\, B/z_1,z_2,z_3/ \overset{0,0,1}{}$$

where $\mathrm{tr}\, B/z_1,z_2,z_3/$ is the trace of $B/z_1,z_2,z_3/$, i.e. the sum of the diagonal elements of the matrix $B/z_1,z_2,z_3/$ defined by /2.35/.

Taking into account that

$$\frac{\partial p/z_1,z_2,z_3/}{\partial z_1} = \sum_{i=0}^{n_1}\sum_{j=0}^{n_2}\sum_{k=0}^{n_3} i\, a_{ijk}\, z_1^{i-1} z_2^j z_3^k$$

$$\frac{\partial p/z_1,z_2,z_3/}{\partial z_2} = \sum_{i=0}^{n_1}\sum_{j=0}^{n_2}\sum_{k=0}^{n_3} j\, a_{ijk}\, z_1^i z_2^{j-1} z_3^k \qquad /2.42/$$

$$\frac{\partial p/z_1,z_2,z_3/}{\partial z_3} = \sum_{i=0}^{n_1}\sum_{j=0}^{n_2}\sum_{k=0}^{n_3} k\, a_{ijk}\, z_1^i z_2^j z_3^{k-1}$$

and equating the coefficients of like power of terms $z_1^i z_2^j z_3^k$ from /2.41/ we obtain

$$i\, a_{ijk} = \mathrm{tr}\, B_{i-1,j,k}^{1,0,0}, \qquad j\, a_{ijk} = \mathrm{tr}\, B_{i,j-1,k}^{0,1,0},$$

$$k\, a_{ijk} = \mathrm{tr}\, B_{i,j,k-1}^{0,0,1}. \qquad \qquad /2.43/$$

From /2.38/ we have

$$\mathrm{tr}\, B_{i-1,j,k}^{1,0,0} + \mathrm{tr}\, B_{i,j-1,k}^{0,1,0} + \mathrm{tr}\, B_{i,j,k-1}^{0,0,1} =$$

$$= \mathrm{tr}\!\left[A\, B_{ijk}\right] + \mathrm{tr}\!\left[I_n\, a_{ijk}\right]. \qquad /2.44/$$

Substitution of /2.43/ into /2.44/ yields

$$(i+j+k)a_{ijk} = \mathrm{tr}\!\left[A\, B_{ijk}\right] + n\, a_{ijk}$$

and

$$a_{ijk} = -\frac{1}{n-i-j-k}\, \mathrm{tr}\!\left[A\, B_{ijk}\right] \qquad /i+j+k \neq n/ \qquad /2.45/$$

Hence the matrices $B_{ijk}$ and the coefficients $a_{ijk}$ can be found by the use of the algorithm which is a straightforward extension for 3-D systems of the Algorithm 2.1.

The desired $G/z_1,z_2,z_3/$ can be evaluated from the formula

$$G/z_1,z_2,z_3/ = \frac{1}{p/z_1,z_2,z_3/} \sum_{i=0}^{n_1} \sum_{j=0}^{n_2} \sum_{k=0}^{n_3} C\, B_{ijk}\, B\, z_1^i z_2^j z_3^k + D. \qquad /2.46/$$

### 2.2.2 Tzafestas-Pimenides´ Formula.

The Mertzios-Paraskevopoulos´ formula was extended for 3-D case by Tzafestas and Pimenides in [14].

#### Theorem 2.3

The transfer function matrix $G/z_1,z_2,z_3/$ of the 3-D system /T-PM/ is given by

$$G/z_1,z_2,z_3/ =$$

$$= \frac{1}{p/z_1,z_2,z_3/} \Bigg[ \sum_{i=0}^{n_1} \sum_{j=0}^{n_2} \sum_{k=0}^{n_3} \sum_{l=0}^{n_1-i} \sum_{q=0}^{n_2-j} \sum_{r=0}^{n_3-k} a_{i+l,j+q,k+r}\, C\!\left(A^{l-1,q,r} B^{1,0,0} \right. +$$

$$\left. + A^{l,q-1,r} B^{0,1,0} + A^{l,q,r-1} B^{0,0,1}\right) z_1^i z_2^j z_3^k \Bigg] + D \qquad /2.47/$$

where the transition /state transition/ matrix $A^{i,j,k}$ for the T-PM is defined as follows:

1° $A^{0,0,0} = I_n$ /the identity matrix/,

2° $A^{1,0,0} = \begin{bmatrix} A_{11} & A_{12} & A_{13} \\ 0 & 0 & 0 \\ 0 & 0 & 0 \end{bmatrix}$, $A^{0,1,0} = \begin{bmatrix} 0 & 0 & 0 \\ A_{21} & A_{22} & A_{23} \\ 0 & 0 & 0 \end{bmatrix}$,

$A^{0,0,1} = \begin{bmatrix} 0 & 0 & 0 \\ 0 & 0 & 0 \\ A_{31} & A_{32} & A_{33} \end{bmatrix}$ ; /2.47a/

3° $A^{i,j,k} = A^{1,0,0} A^{i-1,j,k} + A^{0,1,0} A^{i,j-1,k} + A^{0,0,1} A^{i,j,k-1}$

for $/i,j,k/ > /0,0,0/$ ;

4° $A^{i,j,k} = 0$ /the zero matrix/ for $i<0$ or $j<0$ or $k<0$

and

$B^{1,0,0} = \begin{bmatrix} B_1 \\ 0 \\ 0 \end{bmatrix}$, $B^{0,1,0} = \begin{bmatrix} 0 \\ B_2 \\ 0 \end{bmatrix}$, $B^{0,0,1} = \begin{bmatrix} 0 \\ 0 \\ B_3 \end{bmatrix}$.

**Proof:** Let

$A^{i,j,k} = \begin{bmatrix} A_{11}^{i,j,k} & A_{12}^{i,j,k} & A_{13}^{i,j,k} \\ A_{21}^{i,j,k} & A_{22}^{i,j,k} & A_{23}^{i,j,k} \\ A_{31}^{i,j,k} & A_{32}^{i,j,k} & A_{33}^{i,j,k} \end{bmatrix}$

In a similar way as for 2-D case it can be proved that

$B_{ijk} = \sum_{l=0}^{n_1-i} \sum_{q=0}^{n_2-j} \sum_{r=0}^{n_3-k} a_{i+l,j+q,k+r} \begin{bmatrix} A_{11}^{l-1,q,r} & A_{12}^{l,q-1,r} & A_{13}^{l,q,r-1} \\ A_{21}^{l-1,q,r} & A_{22}^{l,q-1,r} & A_{23}^{l,q,r-1} \\ A_{31}^{l-1,q,r} & A_{32}^{l,q-1,r} & A_{33}^{l,q,r-1} \end{bmatrix}$

Note that

$$\begin{bmatrix} A_{11}^{l-1,q,r} & A_{12}^{l,q-1,r} & A_{13}^{l,q,r-1} \\ A_{21}^{l-1,q,r} & A_{22}^{l,q-1,r} & A_{23}^{l,q,r-1} \\ A_{31}^{l-1,q,r} & A_{32}^{l,q-1,r} & A_{33}^{l,q,r-1} \end{bmatrix} B = A^{l-1,q,r} B^{1,0,0} + A^{l,q-1,r} B^{0,1,0} +$$

$$+ A^{l,q,r-1} B^{0,0,1}$$

and

$$B_{ijk} B = \sum_{l=0}^{n_1-i} \sum_{q=0}^{n_2-j} \sum_{r=0}^{n_3-k} a_{i+l,j+q,k+r} \left( A^{l-1,q,r} B^{1,0,0} + \right.$$

$$\left. + A^{l,q-1,r} B^{0,1,0} + A^{l,q,r-1} B^{0,0,1} \right) \qquad /2.48/$$

Substitution of /2.48/ into /2.46/ yields /2.47/. □

### Example 2.3

Using /2.47/ evaluate $G/z_1, z_2, z_3/$ for 3-D system with

$$A = \begin{bmatrix} A_{11} & A_{12} & A_{13} \\ A_{21} & A_{22} & A_{23} \\ A_{31} & A_{32} & A_{33} \end{bmatrix} = \begin{bmatrix} 1 & 0 & | & 1 & | & 0 \\ 2 & -1 & | & 0 & | & 1 \\ \hline 0 & 1 & | & 2 & | & -1 \\ \hline 1 & -1 & | & 0 & | & 1 \end{bmatrix}, \quad B = \begin{bmatrix} B_1 \\ B_2 \\ B_3 \end{bmatrix} = \begin{bmatrix} 1 & -1 \\ 0 & 1 \\ \hline 1 & 0 \\ \hline 1 & -1 \end{bmatrix}$$

$$C = \begin{bmatrix} C_1 & C_2 & C_3 \end{bmatrix} = \begin{bmatrix} 1 & 0 & | & 1 & | & 1 \\ -1 & 1 & | & 0 & | & 1 \end{bmatrix}, \quad D = 0.$$

In this case we have

$$B^{1,0,0} = \begin{bmatrix} B_1 \\ 0 \\ 0 \end{bmatrix} = \begin{bmatrix} 1 & -1 \\ 0 & 1 \\ 0 & 0 \\ 0 & 0 \end{bmatrix}, \quad B^{0,1,0} = \begin{bmatrix} 0 \\ B_2 \\ 0 \end{bmatrix} = \begin{bmatrix} 0 & 0 \\ 0 & 0 \\ 1 & 0 \\ 0 & 0 \end{bmatrix}, \quad B^{0,0,1} = \begin{bmatrix} 0 \\ 0 \\ B_3 \end{bmatrix} = \begin{bmatrix} 0 & 0 \\ 0 & 0 \\ 0 & 0 \\ 1 & -1 \end{bmatrix}$$

and

$$A^{1,0,0} = \begin{bmatrix} 1 & 0 & 1 & 0 \\ 2 & -1 & 0 & 1 \\ 0 & 0 & 0 & 0 \\ 0 & 0 & 0 & 0 \end{bmatrix}, \quad A^{0,1,0} = \begin{bmatrix} 0 & 0 & 0 & 0 \\ 0 & 0 & 0 & 0 \\ 0 & 1 & 2 & -1 \\ 0 & 0 & 0 & 0 \end{bmatrix}, \quad A^{0,0,1} = \begin{bmatrix} 0 & 0 & 0 & 0 \\ 0 & 0 & 0 & 0 \\ 0 & 0 & 0 & 0 \\ 1 & -1 & 0 & 1 \end{bmatrix}$$

Using /2.47a/ we evaluate

$$A^{1,1,0} = A^{1,0,0}A^{0,1,0} + A^{0,1,0}A^{1,0,0} + A^{0,0,1}A^{1,1,-1} = \begin{bmatrix} 0 & 1 & 2 & -1 \\ 0 & 0 & 0 & 0 \\ 2 & -1 & 0 & 1 \\ 0 & 0 & 0 & 0 \end{bmatrix}$$

$$A^{0,1,1} = A^{1,0,0}A^{-1,1,1} + A^{0,1,0}A^{0,0,1} + A^{0,0,1}A^{0,1,0} = \begin{bmatrix} 0 & 0 & 0 & 0 \\ 0 & 0 & 0 & 0 \\ -1 & 1 & 0 & -1 \\ 0 & 0 & 0 & 0 \end{bmatrix}$$

$$A^{1,0,1} = A^{1,0,0}A^{0,0,1} + A^{0,1,0}A^{1,-1,1} + A^{0,0,1}A^{1,0,0} = \begin{bmatrix} 0 & 0 & 0 & 0 \\ 1 & -1 & 0 & 1 \\ 0 & 0 & 0 & 0 \\ -1 & 1 & 1 & -1 \end{bmatrix}$$

$$A^{1,1,1} = A^{1,0,0}A^{0,1,1} + A^{0,1,0}A^{1,0,1} + A^{0,0,1}A^{1,1,0} = \begin{bmatrix} -1 & 1 & 0 & -1 \\ 0 & 0 & 0 & 0 \\ 2 & -2 & -1 & 2 \\ 0 & 1 & 2 & -1 \end{bmatrix}$$

$$A^{2,0,0} = A^{1,0,0}A^{1,0,0} + A^{0,1,0}A^{2,-1,0} + A^{0,0,1}A^{2,0,-1} = \begin{bmatrix} 1 & 0 & 1 & 0 \\ 0 & 1 & 2 & -1 \\ 0 & 0 & 0 & 0 \\ 0 & 0 & 0 & 0 \end{bmatrix}$$

$$A^{2,0,1} = A^{1,0,0}A^{1,0,1} + A^{0,1,0}A^{2,-1,1} + A^{0,0,1}A^{2,0,0} = \begin{bmatrix} 0 & 0 & 0 & 0 \\ -2 & 2 & 1 & -2 \\ 0 & 0 & 0 & 0 \\ 1 & -1 & -1 & 1 \end{bmatrix}$$

$$A^{2,1,0} = A^{1,0,0}A^{1,1,0} + A^{0,1,0}A^{2,0,0} + A^{0,0,1}A^{2,1,-1} = \begin{bmatrix} 2 & 0 & 2 & 0 \\ 0 & 2 & 4 & -2 \\ 0 & 1 & 2 & -1 \\ 0 & 0 & 0 & 0 \end{bmatrix}$$

$$A^{2,1,1} = A^{1,0,0}A^{1,1,1} + A^{0,1,0}A^{2,0,1} + A^{0,0,1}A^{2,1,0} = \begin{bmatrix} 1 & -1 & -1 & 1 \\ -2 & 3 & 2 & -3 \\ -3 & 3 & 2 & -3 \\ 2 & -2 & -2 & 2 \end{bmatrix}$$

$$p/z_1,z_2,z_3/ = \begin{vmatrix} z_1-1 & 0 & -1 & 0 \\ -2 & z_1+1 & 0 & -1 \\ 0 & -1 & z_2-2 & 1 \\ -1 & 1 & 0 & z_3-1 \end{vmatrix} = z_1^2 z_2 z_3 - z_1^2 z_2 - 2z_1^2 z_3 +$$

$$+ 2z_1^2 + z_1 z_2 - z_2 z_3 - z_1 .$$

From /2.47/ we obtain

$$G/z_1,z_2,z_3/ = \frac{1}{p/z_1,z_2,z_3/} \left[ \sum_{i=0}^{2} \sum_{j=0}^{1} \sum_{k=0}^{1} \sum_{l=0}^{2-i} \sum_{q=0}^{1-j} \sum_{r=0}^{1-k} a_{i+l,j+q,k+r} C \right.$$

$$\left. \left( A^{l,q-1,r} B^{0,1,0} + A^{l-1,q,r} B^{1,0,0} + A^{l,q,r-1} B^{0,0,1} \right) z_1^i z_2^j z_3^k \right] =$$

$$= \frac{1}{p/z_1,z_2,z_3/} \begin{bmatrix} N_{11} & N_{12} \\ N_{21} & N_{22} \end{bmatrix}$$

where

$$N_{11} = z_1^2 z_3 + z_1^2 z_2 - 4z_1^2 + z_1 z_2 z_3 + z_2 z_3 - z_1 z_3 - 2z_2 + 1 ,$$

$$N_{12} = -z_1^2 z_2 + 3z_1^2 - z_1 z_2 z_3 + 3z_1 z_3 - z_1 z_2 - z_2 z_3 + 3z_1 + 3z_2 - 3 ,$$

$$N_{21} = z_1^2 z_2 - 2z_1^2 - z_1 z_2 z_3 + z_1 z_2 + 3z_1 z_3 + z_2 z_3 - 3z_1 - 4z_2 - z_3 + 2 ,$$

$$N_{22} = -z_1^2 z_2 + 2z_1^2 + 2z_1 z_2 z_3 - 5z_1 z_2 - 4z_1 z_3 - 2z_2 z_3 + 7z_1 + 6z_2 + 3z_3 - 6 .$$

## 2.3 TRANSFER FUNCTION MATRIX OF THE FORNASINI-MARCHESINI'S MODELS

Let us consider the first Fornasini-Marchesini's model /F-MM I/ defined by the equations /1.18/, /1.19/. Taking the 2-D Z transform of /1.18/, /1.19/ for zero boundary conditions and making use of /A.2.2/ we obtain

$$z_1 z_2 \hat{X}/z_1,z_2/ = \left( \hat{A}_0 + \hat{A}_1 z_1 + \hat{A}_2 z_2 \right) \hat{X}/z_1,z_2/ + \hat{B} U/z_1,z_2/ \qquad /2.49/$$

and

$$Y/z_1,z_2/ = \hat{C} \hat{X}/z_1,z_2/ \qquad /2.50/$$

where $\hat{X}/z_1,z_2/$, $U/z_1,z_2/$ and $Y/z_1,z_2/$ are the 2-D Z transforms of $\hat{x}/i,j/$, $u/i,j/$ and $y/i,j/$, respectively.

From /2.49/ we have

$$\hat{X}/z_1,z_2/ = \left[I_n z_1 z_2 - \hat{A}_0 - \hat{A}_1 z_1 - \hat{A}_2 z_2\right]^{-1} \hat{B} U/z_1,z_2/ \qquad /2.51/$$

Substitution of /2.51/ into /2.50/ yields

$$Y/z_1,z_2/ = G/z_1,z_2/ U/z_1,z_2/ \qquad /2.52/$$

where

$$G/z_1,z_2/ = \hat{C}\left[I_n z_1 z_2 - \hat{A}_0 - \hat{A}_1 z_1 - \hat{A}_2 z_2\right]^{-1} \hat{B} \qquad /2.52/$$

is the transfer function matrix of F-MM I.

Let us consider the second Fornasini-Marchesini's model /F-MM II/ defined by the equations /1.20/, /1.21/. Taking the 2-D Z transform of /1.20/, /1.21/ for zero boundary conditions and making use of /A2.2/ we obtain

$$z_1 z_2 X/z_1,z_2/ = \left[A_1 z_2 + A_2 z_1\right] X/z_1,z_2/ + \left[B_{01} z_1 + B_{10} z_2\right] U/z_1,z_2/ \qquad /2.53/$$

and

$$Y/z_1,z_2/ = CX/z_1,z_2/ + DU/z_1,z_2/ \qquad /2.54/$$

where $X/z_1,z_2/$, $U/z_1,z_2/$ and $Y/z_1,z_2/$ are the 2-D Z transforms of $x/i,j/$, $u/i,j/$ and $y/i,j/$, respectively.

From /2.53/ we have

$$X/z_1,z_2/ = \left[I_n z_1 z_2 - A_1 z_2 - A_2 z_1\right]^{-1} \left[B_{01} z_1 + B_{10} z_2\right] U/z_1,z_2/ \qquad /2.55/$$

Substitution of /2.55/ into /2.54/ yields

$$Y/z_1,z_2/ = G/z_1,z_2/ U/z_1,z_2/$$

where

$$G/z_1,z_2/ = C\left[I_n z_1 z_2 - A_2 z_1 - A_1 z_2\right]^{-1} \left[B_{01} z_1 + B_{10} z_2\right] + D \qquad /2.56/$$

is the transfer matrix of F-MM II.

The transfer function matrix of RM can be obtained from /2.56/ by substitution of

$$A_2 = \begin{bmatrix} 0 & 0 \\ A_{21} & A_{22} \end{bmatrix}, \quad A_1 = \begin{bmatrix} A_{11} & A_{12} \\ 0 & 0 \end{bmatrix}, \quad B_{01} = \begin{bmatrix} 0 \\ B_2 \end{bmatrix}, \quad B_{10} = \begin{bmatrix} B_1 \\ 0 \end{bmatrix}.$$

Taking into account that $A_1 + A_2 = A$ and $B_{10} + B_{01} = B$ we obtain

$$G/z_1,z_2/ = C\left[\begin{bmatrix} I_{n_1} z_1 z_2 & 0 \\ 0 & I_{n_2} z_1 z_2 \end{bmatrix} - \begin{bmatrix} A_{11} z_2 & A_{12} z_2 \\ A_{21} z_1 & A_{22} z_1 \end{bmatrix}\right]^{-1} \begin{bmatrix} B_1 z_2 \\ B_2 z_1 \end{bmatrix} + D =$$

$$= C \left[ \begin{bmatrix} I_{n_1} z_1 & 0 \\ 0 & I_{n_2} z_2 \end{bmatrix} - \begin{bmatrix} A_{11} & A_{12} \\ A_{21} & A_{22} \end{bmatrix} \right]^{-1} \begin{bmatrix} B_1 \\ B_2 \end{bmatrix} + D =$$

$$= C \left[ Z - A \right]^{-1} B + D$$

where $Z$ is defined by /2.6/.

## 2.4 MATRIX FRACTION DESCRIPTION

Consider a transfer function matrix of the form

$$G/z_1,z_2/ = \begin{bmatrix} G_{11}/z_1,z_2/ & \cdots & G_{1m}/z_1,z_2/ \\ \vdots & & \vdots \\ G_{11}/z_1,z_2/ & \cdots & G_{1m}/z_1,z_2/ \end{bmatrix} \qquad /2.57/$$

Let $D_k/z_1,z_2/$ be the least common denominator of $G_{1k}/z_1,z_2/$, $G_{2k}/z_1,z_2/$, ..., $G_{1k}/z_1,z_2/$ for $k = 1, 2, \ldots, m$. We can write /2.57/ in the form

$$G/z_1,z_2/ = \begin{bmatrix} \dfrac{N_{11}/z_1,z_2/}{D_1/z_1,z_2/} & \cdots & \dfrac{N_{1m}/z_1,z_2/}{D_m/z_1,z_2/} \\ \vdots & & \vdots \\ \dfrac{N_{11}/z_1,z_2/}{D_1/z_1,z_2/} & \cdots & \dfrac{N_{1m}/z_1,z_2/}{D_m/z_1,z_2/} \end{bmatrix} = N_R/z_1,z_2/ \; D_R/z_1,z_2/^{-1} \qquad /2.58/$$

where

$$N_R/z_1,z_2/ = \begin{bmatrix} N_{11}/z_1,z_2/ & \cdots & N_{1m}/z_1,z_2/ \\ \vdots & & \vdots \\ N_{11}/z_1,z_2/ & \cdots & N_{1m}/z_1,z_2/ \end{bmatrix} \qquad /2.59a/$$

$$D_R/z_1,z_2/ = \begin{bmatrix} D_1/z_1,z_2/ & 0 & \cdots & 0 \\ 0 & D_2/z_1,z_2/ & \cdots & 0 \\ \vdots & & & \vdots \\ 0 & 0 & \cdots & D_m/z_1,z_2/ \end{bmatrix} \qquad /2.59b/$$

$N_R/z_1,z_2/$ is called the right numerator, and $D_R/z_1,z_2/$ is the right denominator of $G/z_1,z_2/$.

Let $\bar{D}_k/z_1,z_2/$ be the least common denominator of $G_{k1}/z_1,z_2/$, $G_{k2}/z_1,z_2/$, ..., $G_{km}/z_1,z_2/$ for $k = 1, 2, \ldots, l$. The transfer function matrix

/2.57/ can also be written in the form

$$G/z_1,z_2/ = \begin{bmatrix} \dfrac{\bar{N}_{11}/z_1,z_2/}{D_1/z_1,z_2/} & \cdots & \dfrac{\bar{N}_{1m}/z_1,z_2/}{D_1/z_1,z_2/} \\ \cdots & \cdots & \cdots \\ \dfrac{\bar{N}_{11}/z_1,z_2/}{D_1/z_1,z_2/} & \cdots & \dfrac{\bar{N}_{1m}/z_1,z_2/}{D_1/z_1,z_2/} \end{bmatrix} = D_L^{-1}/z_1,z_2/\ N_L/z_1,z_2/ \quad /2.60/$$

where

$$D_L/z_1,z_2/ = \begin{bmatrix} \bar{D}_1/z_1,z_2/ & 0 & \cdots & 0 \\ 0 & \bar{D}_2/z_1,z_2/ & \cdots & 0 \\ \cdots & \cdots & \cdots & \cdots \\ 0 & 0 & \cdots & \bar{D}_1/z_1,z_2/ \end{bmatrix} \quad /2.61a/$$

$$N_L/z_1,z_2/ = \begin{bmatrix} \bar{N}_{11}/z_1,z_2/ & \cdots & \bar{N}_{1m}/z_1,z_2/ \\ \cdots & \cdots & \cdots \\ \bar{N}_{11}/z_1,z_2/ & \cdots & \bar{N}_{1m}/z_1,z_2/ \end{bmatrix} \quad /2.61b/$$

$N_L/z_1,z_2/$ and $D_L/z_1,z_2/$ are called the left numerator and denominator of $G/z_1,z_2/$, respectively.

Let $W_R/z_1,z_2/$ be a right divisor of $N_R/z_1,z_2/$ and $D_R/z_1,z_2/$, i.e.

$$N_R/z_1,z_2/ = \hat{N}_R/z_1,z_2/\ W_R/z_1,z_2/\ ,\quad D_R/z_1,z_2/ = \hat{D}_R/z_1,z_2/\ W_R/z_1,z_2/.$$
$$D_R/z_1,z_2/ = \hat{D}_R/z_1,z_2/\ W_R/z_1,z_2/ \quad /2.62/$$

Substitution of /2.62/ into /2.58/ yields

$$G/z_1,z_2/ = N_R/z_1,z_2/\ D_R^{-1}/z_1,z_2/ =$$
$$= \hat{N}_R/z_1,z_2/\ W_R/z_1,z_2/\ \left[\hat{D}_R/z_1,z_2/\ W_R/z_1,z_2/\right]^{-1} =$$
$$= \hat{N}_R/z_1,z_2/\ \hat{D}_R^{-1}/z_1,z_2/ \quad /2.63/$$

Similarly, if $W_L/z_1,z_2/$ is a left divisor of $N_L/z_1,z_2/$ and $D_L/z_1,z_2/$, i.e.

$$N_L/z_1,z_2/ = W_L/z_1,z_2/\ \hat{N}_L/z_1,z_2/\ ,\quad D_L/z_1,z_2/ = W_L/z_1,z_2/\ \hat{D}_L/z_1,z_2/$$

then we obtain

$$G/z_1,z_2/ = D_L^{-1}/z_1,z_2/\ N_L/z_1,z_2/ =$$
$$= \left[W_L/z_1,z_2/\ \hat{D}_L/z_1,z_2/\right]^{-1} W_L/z_1,z_2/\ \hat{N}_L/z_1,z_2/ =$$

$$= \hat{D}_L^{-1}/z_1,z_2/ \; \hat{N}_L/z_1,z_2/ \qquad /2.64/$$

Therefore, we have many different right /left/ matrix fraction descriptions for a given $G/z_1,z_2/$.

For example,

$$G/z_1,z_2/ = \begin{bmatrix} \dfrac{z_1}{z_1z_2+1} & \dfrac{z_1z_2}{z_1^2z_2+z_1+3} \\ \dfrac{z_2}{z_1z_2+1} & \dfrac{z_1^2}{z_1^2z_2+z_1+3} \\ \dfrac{1}{z_1z_2+1} & \dfrac{z_2}{z_1^2z_2+z_1+3} \end{bmatrix} = \begin{bmatrix} z_1 & z_1z_2 \\ z_2 & z_1^2 \\ 1 & z_2 \end{bmatrix} \begin{bmatrix} z_1z_2+1 & 0 \\ 0 & z_1^2z_2+z_1+3 \end{bmatrix}^{-1} =$$

$$= \begin{bmatrix} z_1^2 & z_1z_2^2 \\ z_1z_2 & z_1z_2^2+z_2 \\ z_1 & z_2^2+1 \end{bmatrix} \begin{bmatrix} z_1^2z_2+z_1 & z_1z_2+1 \\ 0 & z_1^2z_2^2+z_1z_2+3z_2 \end{bmatrix}^{-1}$$

## 2.5 PROPER TRANSFER FUNCTION MATRICES

Any $l \times m$ transfer function matrix $G/z_1,z_2/$ may be written as

$$G/z_1,z_2/ = \dfrac{N/z_1,z_2/}{d/z_1,z_2/} \qquad /2.65/$$

where $N/z_1,z_2/$ is an $l \times m$ polynomial matrix in $z_1,z_2$ and

$$d/z_1,z_2/ = \sum_{i=0}^{n_1} \sum_{j=0}^{n_2} d_{ij} z_1^i z_2^j \qquad /2.66/$$

Following [1], the polynomial /2.66/ with $d_{n_1 n_2} \neq 0$ will be called accaptable.

Definition 2.1

The transfer function matrix /2.65/ is called proper /strictly proper/

if $d/z_1,z_2/$ is acceptable and

$$\deg_{z_1} n_{ij}/z_1,z_2/ \leqslant n_1 \quad / \deg_{z_1} n_{ij}/z_1,z_2/ < n_1 / \qquad /2.67a/$$

and

$$\deg_{z_2} n_{ij}/z_1,z_2/ \leqslant n_2 \quad / \deg_{z_2} n_{ij}/z_1,z_2/ < n_2 / \qquad /2.67b/$$

for $/1,1/ \leqslant /i,j/ \leqslant /l,m/$,

where $n_{ij}/z_1,z_2/$ are the elements of $N/z_1,z_2/$.

The transfer function matrix $G/z_1,z_2/$ may be also expressed in the form of the power series expansion:

$$G/z_1,z_2/ = \sum_{i=k}^{\infty} \sum_{j=l}^{\infty} g_{ij} z_1^{-i} z_2^{-j} \qquad /2.65'/$$

where $k,l$ are some integers.

### Definition 2.2

A system with the transfer function matrix /2.65'/ is said to be causal /strictly causal/ if $g_{kl}=0$ for $k<0$ or $l<0$ / $k\leqslant 0$ or $l\leqslant 0$ /.

Note that the system is causal if the output at $/k,l/$ is dependent only on past inputs, i.e. on $u/i,j/$ for $/i,j/ < /k,l/$.
From /2.65/ and Definition 2.2 it follows that $G/z_1,z_2/$ is proper /strictly proper/ if and only if the associated system is causal /strictly causal/.

Let $R[z_1,z_2]$ be the set of polynomials in $z_1,z_2$ with real coefficients. Note that the elements of $R[z_1,z_2]$ can be considered as polynomials in $z_2$ (or $z_1$) with coefficients in $R[z_1]$ ($R[z_2]$, respectively), that is

$$R[z_1,z_2] = R[z_1][z_2] = R[z_2][z_1].$$

Let $R^{l\times m}[z_1,z_2]$ denote the set of $l\times m$ polynomial matrices with elements from $R[z_1,z_2]$. Therefore, $N/z_1,z_2/ \in R^{l\times m}[z_1,z_2]$ may be viewed as an element of $R^{l\times m}[z_1][z_2]$ or $R^{l\times m}[z_2][z_1]$.
$R(z_1,z_2)$ is a field of rational functions with real coefficients and $R^{l\times m}(z_1,z_2)$ is a set of matrices with entries in $R(z_1,z_2)$.

## Definition 2.3 [1]

A non-singular matrix $D/z_1,z_2/ \in R^{p \times p}[z_1,z_2]$ is said to be column reduced over $R[z_1][z_2]$ $\left(R[z_2][z_1]\right)$ iff its leading column coefficient matrix of $z_2$ $(z_1)$, represented by $D^1_{hc}/z_1/ \in R^{p \times p}[z_1]$ $\left(D^2_{hc}/z_2/ \in R^{p \times p}[z_2]\right)$ is column reduced over $R[z_1]$ $\left(R[z_2]\right)$, i.e. when its leading coefficient matrix $D^1_{hc}$ $\left(D^2_{hc}\right)$ is non-singular.

## Example 2.4

For
$$D/z_1,z_2/ = \begin{bmatrix} z_1^2 z_2 & 2z_1 \\ -z_2 & z_1 z_2^2 \end{bmatrix} = \begin{bmatrix} z_1^2 & 0 \\ -1 & z_1 \end{bmatrix} \begin{bmatrix} z_2 & 0 \\ 0 & z_2^2 \end{bmatrix} + \begin{bmatrix} 0 & 2z_1 \\ 0 & 0 \end{bmatrix} \qquad /2.68/$$

the matrix
$$D^1_{hc}/z_1/ = \begin{bmatrix} z_1^2 & 0 \\ -1 & z_1 \end{bmatrix} = \begin{bmatrix} 1 & 0 \\ 0 & 1 \end{bmatrix} \begin{bmatrix} z_1^2 & 0 \\ 0 & z_1 \end{bmatrix} + \begin{bmatrix} 0 & 0 \\ -1 & 0 \end{bmatrix} \qquad /2.69/$$

is the leading coefficient matrix in $z_2$ of $D/z_1,z_2/$. The matrix /2.69/ is column reduced over $R[z_1]$ since its leading column coefficient matrix

$$D^1_{hc} = \begin{bmatrix} 1 & 0 \\ 0 & 1 \end{bmatrix}$$

is non-singular. Thus, by Definition 2.3 the matrix /2.68/ is column reduced over $R[z_1][z_2]$.

Let us denote by $R(z_2)[z_1]$ $\left(R(z_1)[z_2]\right)$ the ring of polynomials in $z_1$ $(z_2)$ with coefficients in the field of rational functions $R(z_2)$ $\left(R(z_1)\right)$.

## Definition 2.4

A non-singular 2-D polynomial matrix $D/z_1,z_2/ \in R^{p \times p}[z_1,z_2]$ is said to be column reduced over $R(z_1)[z_2]$ $\left(R(z_2)[z_1]\right)$ iff its leading column coefficient matrix of $z_2$ $(z_1)$, $D^1_{hc}/z_1/ \in R^{p \times p}[z_1]$ $\left(D^2_{hc}/z_2/ \in R^{p \times p}[z_2]\right)$ is non-singular.

Note that in Definition 2.4 we require only that $\det D^1_{hc}/z_1/ \neq 0$ $\left(\det D^2_{hc}/z_2/ \neq 0\right)$, whereas in Definition 2.3 we further required that $D^1_{hc}/z_1/$ $\left(D^2_{hc}/z_2/\right)$ be column reduced over $R[z_1]$ $\left(R[z_2]\right)$, i.e. $\det D^1_{hc} \neq 0$ $\left(\det D^2_{hc} \neq 0\right)$.

Example 2.5

For

$$D/z_1,z_2/ = \begin{bmatrix} z_1^2 z_2^2 + z_1 z_2^2 + z_1 & -z_1 z_2^3 + 1 \\ z_1^2 z_2^2 & z_2^3 - z_1 z_2^3 + z_2^2 \end{bmatrix} = $$

$$= \begin{bmatrix} z_1^2 + z_1 & -z_1 \\ z_1^2 & -z_1 + 1 \end{bmatrix} \begin{bmatrix} z_2^2 & 0 \\ 0 & z_2^3 \end{bmatrix} + \begin{bmatrix} z_1 & 1 \\ 0 & z_2^2 \end{bmatrix}$$

/2.70/

the matrix

$$D_{hc}^1/z_1/ = \begin{bmatrix} z_1^2 + z_1 & -z_1 \\ z_1^2 & -z_1 + 1 \end{bmatrix} = \begin{bmatrix} 1 & -1 \\ 1 & -1 \end{bmatrix} \begin{bmatrix} z_1^2 & 0 \\ 0 & z_1 \end{bmatrix} + \begin{bmatrix} z_1 & 0 \\ 0 & 1 \end{bmatrix}$$

is non-singular $\left(\det D_{hc}^1/z_1/ = z_1 \neq 0\right)$. So the matrix /2.70/ is column reduced over $R(z_1)[z_2]$. Note that

$$\det D_{hc}^1 = \begin{vmatrix} 1 & -1 \\ 1 & -1 \end{vmatrix} = 0.$$

Thus, by Definition 2.3 the matrix /2.70/ is not column reduced over $R[z_1][z_2]$.

Let us denote by $k_i$ the i-th column degree of $D/z_1,z_2/$ in $z_1$, and by $l_i$ the i-th column degree of $D/z_1,z_2/$ in $z_2$. Note that we can always write

$$D/z_1,z_2/ = D_{hc}^1/z_1/ \, S_2/z_2/ + L_1/z_1,z_2/ \qquad /2.71a/$$

and

$$D/z_1,z_2/ = D_{hc}^2/z_2/ \, S_1/z_1/ + L_2/z_1,z_2/ \qquad /2.71b/$$

where

$$D_{hc}^1/z_1/ = \begin{bmatrix} d_{11}^1/z_1/ & \cdots & d_{1p}^1/z_1/ \\ \cdots & \cdots & \cdots \\ d_{p1}^1/z_1/ & \cdots & d_{pp}^1/z_1/ \end{bmatrix}, \quad D_{hc}^2/z_2/ = \begin{bmatrix} d_{11}^2/z_2/ & \cdots & d_{1p}^2/z_2/ \\ \cdots & \cdots & \cdots \\ d_{p1}^2/z_2/ & \cdots & d_{pp}^2/z_2/ \end{bmatrix}$$

$$S_2/z_2/ = \begin{bmatrix} z_2^{l_1} & 0 & \cdots & 0 \\ 0 & z_2^{l_2} & \cdots & 0 \\ \cdots & \cdots & \cdots & \cdots \\ 0 & 0 & \cdots & z_2^{l_p} \end{bmatrix}, \quad S_1/z_1/ = \begin{bmatrix} z_1^{k_1} & 0 & \cdots & 0 \\ 0 & z_1^{k_2} & \cdots & 0 \\ \cdots & \cdots & \cdots & \cdots \\ 0 & 0 & \cdots & z_1^{k_p} \end{bmatrix}$$

$$\deg_{z_2}^{c_i} L_1/z_1,z_2/ < l_i, \qquad \deg_{z_1}^{c_i} L_2/z_1,z_2/ < k_i \qquad /i=1,2,\ldots,p/$$

$\deg_{z_2}^{c_i} L_1/z_1,z_2/$ denotes the degree of the i-th column of $L_1/z_1,z_2/$ with respect to $z_2$.

Defining

$$S/z_1,z_2/ = S_1/z_1/\, S_2/z_2/ = \begin{bmatrix} z_1^{k_1}z_2^{l_1} & 0 & \cdots & 0 \\ 0 & z_1^{k_2}z_2^{l_2} & \cdots & 0 \\ & & & \\ 0 & 0 & \cdots & z_1^{k_p}z_2^{l_p} \end{bmatrix}$$

we can also write $D/z_1,z_2/$ in the following form

$$D/z_1,z_2/ = D_{hc}S/z_1,z_2/ + L/z_1,z_2/ \qquad /2.72/$$

where

$$\deg_{z_1}^{c_i} L/z_1,z_2/ \leqslant k_i \quad \text{and} \quad \deg_{z_2}^{c_i} L/z_1,z_2/ \leqslant l_i \quad \text{for} \quad i=1,2,\ldots,p.$$

Definition 2.5

A non-singular 2-D polynomial matrix $D/z_1,z_2/ \in R^{p \times p}[z_1,z_2]$ is said to be column reduced over $R[z_1,z_2]$ iff its leading column coefficient matrix $D_{hc}$ is non-singular.

For example, the matrix /2.68/ is column reduced over $R[z_1,z_2]$ since

$$D/z_1,z_2/ = \begin{bmatrix} z_1^2 z_2 & 2z_1 \\ -z_2 & z_1 z_2^2 \end{bmatrix} = \begin{bmatrix} 1 & 0 \\ 0 & 1 \end{bmatrix} \begin{bmatrix} z_1^2 z_2 & 0 \\ 0 & z_1 z_2^2 \end{bmatrix} + \begin{bmatrix} 0 & 2z_1 \\ -z_2 & 0 \end{bmatrix}$$

and

$$\det D_{hc} = \begin{vmatrix} 1 & 0 \\ 0 & 1 \end{vmatrix} \neq 0 \,.$$

From /2.71/ it follows that

$$\det D/z_1,z_2/ = \left(\det D_{hc}^1/z_1/\right)z_2^{n_2} + \text{terms of lower degree in } z_2 \qquad /2.73a/$$

and

$$\det D/z_1,z_2/ = \left(\det D_{hc}^2/z_2/\right)z_1^{n_1} + \text{terms of lower degree in } z_1 \qquad /2.73b/$$

where

$$n_1 = \sum_{i=1}^{p} k_i, \quad n_2 = \sum_{i=1}^{p} l_i \qquad /2.74/$$

## Theorem 2.4

Let $D/z_1,z_2/ \in R^{p \times p}[z_1,z_2]$.
If

$$\det D/z_1,z_2/ = \sum_{i=0}^{m_1} \sum_{j=0}^{m_2} a_{ij} z_1^i z_2^j \qquad /2.75/$$

is acceptable, i.e. $a_{m_1 m_2} \neq 0$, then the following statements are equivalent:

/i/ $m_1 = n_1$ and $m_2 = n_2$

/ii/ $D/z_1,z_2/$ is column reduced over $R(z_1)[z_2]$ and $R(z_2)[z_1]$, i.e.
$$\det D_{hc}^1/z_1/ \neq 0 \quad \text{and} \quad \det D_{hc}^2/z_2/ \neq 0 \qquad /2.76/$$

/iii/ $D/z_1,z_2/$ is column reduced over $R[z_1][z_2]$ and $R[z_2][z_1]$, i.e.
$$\det D_{hc}^1 \neq 0 \quad \text{and} \quad \det D_{hc}^2 \neq 0 \qquad /2.77/$$

/iv/ $D/z_1,z_2/$ is column reduced over $R[z_1,z_2]$, i.e.
$$\det D_{hc} \neq 0 . \qquad /2.78/$$

**Proof:** From the assumption $a_{m_1 m_2} \neq 0$ and /2.73/ it follows that /2.76/ holds if and only if $m_1 = n_1$ and $m_2 = n_2$. So the conditions /i/ and /ii/ are equivalent. The conditions /2.76/ and $a_{m_1 m_2} \neq 0$ imply that

$$\deg \det D_{hc}^1/z_1/ = n_1 \quad \text{and} \quad \deg \det D_{hc}^2/z_2/ = n_2$$

which can be satisfied if and only if /2.77/ holds. So the conditions /ii/ and /iii/ are equivalent. To show the equivalence of /ii/ and /iv/ let us consider the equations /2.71/. Substituting

$$D_{hc}^1/z_1/ = D_{hc}^1 S_1/z_1/ + L_1/z_1/ \quad \text{and} \quad D_{hc}^2/z_2/ = D_{hc}^2 S_2/z_2/ + L_2/z_2/$$

into /2.71/ we obtain

$$D/z_1,z_2/ = D_{hc}^1 S_1/z_1/ S_2/z_2/ + L_1/z_1/ S_2/z_2/ + L_1/z_1,z_2/ =$$
$$= D_{hc}^2 S_2/z_2/ S_1/z_1/ + L_2/z_2/ S_1/z_1/ + L_2/z_1,z_2/ =$$
$$= D_{hc} S/z_1,z_2/ + L/z_1,z_2/$$

where

$$D_{hc} = D_{hc}^1 = D_{hc}^2 \qquad /2.79/$$

$$L/z_1,z_2/ = L_1/z_1/ S_2/z_2/ + L_1/z_1,z_2/ = L_2/z_2/ S_1/z_1/ + L_2/z_2/. \qquad /2.79/$$

From /2.79/ it follows that the conditions /2.77/ and /2.78/ are equivalent. $\square$

Since there are no zero divisors in the real field, we have the following

### Lemma 2.1

A product of two polynomials with real coefficients is acceptable if and only if both factors are acceptable.

### Theorem 2.5

Let
$$H/z_1,z_2/ = N/z_1,z_2/ D^{-1}/z_1,z_2/ \qquad /2.80/$$

where $N/z_1,z_2/ \in R^{m \times p}[z_1,z_2]$, $D/z_1,z_2/ \in R^{p \times p}[z_1,z_2]$

If $D/z_1,z_2/$ is column reduced over $R[z_1,z_2]$ and $\det D/z_1,z_2/$ is acceptable. Then $H/z_1,z_2/$ is proper (strictly proper) if and only if

$$\deg_{z_1}^{c_i} N/z_1,z_2/ \leqslant \deg_{z_1}^{c_i} D/z_1,z_2/ \quad \left( \deg_{z_1}^{c_i} N/z_1,z_2/ < \deg_{z_1}^{c_i} D/z_1,z_2/ \right)$$
and
$$\deg_{z_2}^{c_i} N/z_1,z_2/ \leqslant \deg_{z_2}^{c_i} D/z_1,z_2/ \quad \left( \deg_{z_2}^{c_i} N/z_1,z_2/ < \deg_{z_2}^{c_i} D/z_1,z_2/ \right) \qquad /2.81/$$

for $i = 1,2,\dots,p$.

**Proof:** From /2.80/ it follows that

$$h_{ij}/z_1,z_2/ = \frac{p_{ij}/z_1,z_2/}{\det D/z_1,z_2/} \qquad \text{for } i,j = 1,2,\dots,p \qquad /2.82/$$

where $p_{ij}/z_1,z_2/$ is the $(i,j)$th element of $N/z_1,z_2/ \operatorname{adj} D/z_1,z_2/$.
If $D/z_1,z_2/$ is column reduced over $R[z_1,z_2]$ and $\det D/z_1,z_2/$ is acceptable, then /2.81/ implies

$$\deg_{z_1} p_{ij}/z_1,z_2/ \leqslant \deg_{z_1} \det D/z_1,z_2/$$
and
$$\deg_{z_2} p_{ij}/z_1,z_2/ \leqslant \deg_{z_2} \det D/z_1,z_2/$$

for $i,j = 1,2,\dots,p$.

It follows from Lemma 2.1 that possible cancellations in /2.82/ by a common factor of $p_{ij}/z_1,z_2/$ and $\det D/z_1,z_2/$ does not affect the properness of $h_{ij}/z_1,z_2/$. So, by Definition 2.2, $h_{ij}/z_1,z_2/$ is a proper 2-D rational function and thus $H/z_1,z_2/$ is proper.

To prove necessity note that if $h_{ij}/z_1,z_2/$ is a proper 2-D rational function, then from /2.82/ the conditions /2.81/ follow. □

Let us consider the equation

$$H/z_1,z_2/\, D/z_1,z_2/ = N/z_1,z_2/ \qquad /2.83/$$

where $D/z_1,z_2/ \in R^{r \times p}[z_1,z_2]$ $(r \geqslant p)$ and $N/z_1,z_2/ \in R^{m \times p}[z_1,z_2]$ are given and $H/z_1,z_2/ \in R^{m \times r}[z_1,z_2]$ is unknown. It is assumed that $D/z_1,z_2/$ has at least one $p \times p$ minor whose determinant is acceptable.

The following question is arising. Under which conditions the equation /2.83/ has a proper (strictly proper) solution $H/z_1,z_2/$?

Select the rows of $D/z_1,z_2/$ which contain an acceptable minor $D_a/z_1,z_2/$. Let us assume that

$$D_a/z_1,z_2/ = \begin{bmatrix} I_p & 0 \end{bmatrix} D/z_1,z_2/ \qquad /2.84/$$

Next form the matrix

$$F/z_1,z_2/ = \begin{bmatrix} D_a/z_1,z_2/ \\ N/z_1,z_2/ \end{bmatrix}$$

and use elementary column operations to transform it to the forms

$$F_1/z_1,z_2/ = \begin{bmatrix} D_1/z_1,z_2/ \\ N_1/z_1,z_2/ \end{bmatrix} \quad \text{and} \quad F_2/z_1,z_2/ = \begin{bmatrix} D_2/z_1,z_2/ \\ N_2/z_1,z_2/ \end{bmatrix}$$

such that $D_1/z_1,z_2/$ is column reduced over $R(z_2)[z_1]$ and $D_2/z_1,z_2/$ is column reduced over $R(z_1)[z_2]$, as shown in the Appendix.

### Theorem 2.6

The equation /2.83/ has the proper (strictly proper) solution

$$H/z_1,z_2/ = N/z_1,z_2/\, D_a^{-1}/z_1,z_2/ \begin{bmatrix} I_p & 0 \end{bmatrix} \qquad /2.85/$$

if and only if

$$\deg_{z_1}^{c_i} N_1/z_1,z_2/ \leqslant \deg_{z_1}^{c_i} D_1/z_1,z_2/ \quad \left( \deg_{z_1}^{c_i} N_1/z_1,z_2/ < \deg_{z_1}^{c_i} D_1/z_1,z_2/ \right) \qquad /2.86a/$$

$$\deg_{z_2}^{c_i} N_2/z_1,z_2/ \leqslant \deg_{z_2}^{c_i} D_2/z_1,z_2/ \quad \left(\deg_{z_2}^{c_i} N_2/z_1,z_2/ < \deg_{z_2}^{c_i} D_2/z_1,z_2/\right) \quad /2.86b/$$

for $i = 1, 2, \ldots, p$.

**Proof:** Note that there exists a matrix $U_1/z_1,z_2/ \in R^{p \times p}[z_1,z_2]$ unimodular over $R(z_2)[z_1]$ $\left(\det U_1/z_1,z_2/ \in R[z_2]\setminus\{0\}\right)$ such that $D_1/z_1,z_2/ = D_a/z_1,z_2/ U_1/z_1,z_2/$ is column reduced over $R(z_2)[z_1]$. From the formula describing $F_1/z_1,z_2/$ we have $N_1/z_1,z_2/ = N/z_1,z_2/ U_1/z_1,z_2/$. From Lemma 2.1 it follows that $\det D_1/z_1,z_2/ = \det D_a/z_1,z_2/ \det U_1/z_1,z_2/$ is acceptable. If the condition /2.86a/ is satisfied, then /2.85/ is a proper (strictly proper) solution in $z_1$ to /2.83/ because

$$H/z_1,z_2/ = N/z_1,z_2/ D_a^{-1}/z_1,z_2/ \begin{bmatrix} I_p & \vdots & 0 \end{bmatrix} =$$
$$= N_1/z_1,z_2/ D_1^{-1}/z_1,z_2/ \begin{bmatrix} I_p & \vdots & 0 \end{bmatrix} \qquad /2.87a/$$

and

$$H/z_1,z_2/ D/z_1,z_2/ = N/z_1,z_2/ D_a^{-1}/z_1,z_2/ \begin{bmatrix} I_p & \vdots & 0 \end{bmatrix} D/z_1,z_2/ = N/z_1,z_2/.$$

Similarly, there exists a matrix $U_2/z_1,z_2/ \in R^{p \times p}[z_1,z_2]$ unimodular over $R(z_1)[z_2]$ $\left(\det U_2/z_1,z_2/ \in R[z_1]\setminus\{0\}\right)$ such that $D_2/z_1,z_2/ = D_a/z_1,z_2/ U_2/z_1,z_2/$ is column reduced over $R(z_1)[z_2]$. From Lemma 2.1 it follows that $\det D_2/z_1,z_2/ = \det D_a/z_1,z_2/ \det U_2/z_1,z_2/$ is acceptable. If the condition /2.86b/ is satisfied, then /2.85/ is a proper strictly proper solution in $z_2$ to /2.82/ because

$$H/z_1,z_2/ = N/z_1,z_2/ D_a^{-1}/z_1,z_2/ \begin{bmatrix} I_p & \vdots & 0 \end{bmatrix} =$$
$$= N_2/z_1,z_2/ D_2^{-1}/z_1,z_2/ \begin{bmatrix} I_p & \vdots & 0 \end{bmatrix} \qquad /2.87b/$$

Therefore /2.85/ is a proper (strictly proper) solution to the equation /2.83/.

To show sufficiency note that the properness of $H/z_1,z_2/$ (given by /2.87/) implies /2.86/ ☐

## 2.6 REALIZATION PROBLEM

### 2.6.1 Realization of the Roesser's model.

If the matrices

$$A = \begin{bmatrix} A_{11} & A_{12} \\ A_{21} & A_{22} \end{bmatrix}, \quad B = \begin{bmatrix} B_1 \\ B_2 \end{bmatrix}, \quad C = \begin{bmatrix} C_1 & C_2 \end{bmatrix}, \quad D \qquad /2.88/$$

of Roesser's model /RM/ are given, then using the formula /2.8/

$$G/z_1,z_2/ = C[Z-A]^{-1} B + D \qquad /2.89/$$

where Z is defined by /2.6/, we can evaluate the unique transfer function matrix $G/z_1,z_2/$ of RM.
Consider a proper transfer function matrix of the form

$$G/z_1,z_2/ = \begin{bmatrix} G_{11}/z_1,z_2/ & \cdots & G_{1m}/z_1,z_2/ \\ \cdots & \cdots & \cdots \\ G_{11}/z_1,z_2/ & \cdots & G_{1m}/z_1,z_2/ \end{bmatrix} \qquad /2.90/$$

The realization problem for 2-D systems can be formulated as follows:
Given the matrices /2.90/, find matrices /2.88/ such that /2.89/ holds.

### Definition 2.6
The matrices /2.88/ are said to be a realization $\bigl($denoted by $\{A,B,C,D\}\bigr)$ of the proper transfer function matrix /2.90/ iff they satisfy /2.89/.

### Definition 2.7
A realization $\{A,B,C,D\}$ is called the minimal one iff the matrix A has minimal dimension amongst all dimensions of all possible realizations.

### Definition 2.8
A realization $\{A,B,C,D\}$ with $A_{11}$, $A_{22}$ of dimensions $n_1 \times n_1$ and $n_2 \times n_2$, respectively, is called a realization of dimension $/n_1,n_2/$.

### Theorem 2.7
The transfer function matrix of RM can be written in the form

$$G/z_1,z_2/ = C/z_2/\bigl[I_{n_1} z_1 - A/z_2/\bigr]^{-1} B/z_2/ + D/z_2/ \qquad /2.91/$$

where

$$A/z_2/ = A_{11} + A_{12}\left[I_{n_2}z_2 - A_{22}\right]^{-1}A_{21}$$

$$B/z_2/ = B_1 + A_{12}\left[I_{n_2}z_2 - A_{22}\right]^{-1}B_2$$

$$C/z_2/ = C_1 + C_2\left[I_{n_2}z_2 - A_{22}\right]^{-1}A_{21} \quad /2.92/$$

$$D/z_2/ = C_2\left[I_{n_2}z_2 - A_{22}\right]^{-1}B_2 + D$$

**Proof:** From the ralation

$$\begin{bmatrix} I_{n_1}z_1 - A_{11} & -A_{12} \\ -A_{21} & I_{n_2}z_2 - A_{22} \end{bmatrix} =$$

$$\begin{bmatrix} I_{n_1}z_1 - A_{11} - A_{12}\left[I_{n_2}z_2-A_{22}\right]^{-1}A_{21} & , -A_{12}\left[I_{n_2}z_2-A_{22}\right]^{-1} \\ 0 & I_{n_2} \end{bmatrix}\begin{bmatrix} I_{n_1} & 0 \\ -A_{21} & , I_{n_2}z_2-A_{22} \end{bmatrix}$$

it follows that

$$\begin{bmatrix} I_{n_1}z_1 - A_{11} & A_{12} \\ -A_{21} & I_{n_2}z_2 - A_{22} \end{bmatrix}^{-1} =$$

$$= \begin{bmatrix} I_{n_1} & 0 \\ -A_{21} & , I_{n_2}z_2-A_{22} \end{bmatrix}^{-1}\begin{bmatrix} I_{n_1}z_1-A_{11}-A_{12}\left[I_{n_2}z_2-A_{22}\right]^{-1}A_{21} & , -A_{12}\left[I_{n_2}z_2-A_{22}\right]^{-1} \\ 0 & I_{n_2} \end{bmatrix}^{-1} =$$

$$= \begin{bmatrix} I_{n_1} & \\ \left[I_{n_2}z_2-A_{22}\right]^{-1}A_{21} & , \left[I_{n_2}z_2-A_{22}\right]^{-1} \end{bmatrix}\begin{bmatrix} \left[I_{n_1}z_1-A/z_2/\right]^{-1} & , \left[I_{n_1}z_1-A/z_2/\right]^{-1}A_{12}\left[I_{n_2}z_2-A_{22}\right]^{-1} \\ 0 & I_{n_2} \end{bmatrix}$$

$$/2.93/$$

Substitution of /2.93/ into /2.89/ yields

$$G/z_1,z_2/ = \begin{bmatrix} C_1 & C_2 \end{bmatrix}\begin{bmatrix} I_{n_1} & 0 \\ \left[I_{n_2}z_2-A_{22}\right]^{-1}A_{21} & , \left[I_{n_2}z_2-A_{22}\right]^{-1} \end{bmatrix}\begin{bmatrix} \left[I_{n_1}z_1-A/z_2/\right]^{-1} \\ 0 \end{bmatrix}$$

$$\begin{bmatrix} [I_{n_1}z_1 - A/z_2/]^{-1} A_{12}[I_{n_2}z_2 - A_{22}]^{-1} \\ I_{n_2} \end{bmatrix} \begin{bmatrix} B_1 \\ B_2 \end{bmatrix} + D =$$

$$= \left[ C_1 + C_2[I_{n_2}z_2 - A_{22}]^{-1} A_{21} \right] [I_{n_1}z_1 - A/z_2/]^{-1} \left[ B_1 + A_{12}[I_{n_2}z_2 - A_{22}]^{-1} B_2 \right] +$$

$$+ C_2[I_{n_2}z_2 - A_{22}]^{-1} B_2 + D =$$

$$= C/z_2/ [I_{n_1}z_1 - A/z_2/]^{-1} B/z_2/ + D/z_2/.$$

$\square$

From /2.92/ it follows that

$$\begin{bmatrix} A/z_2/ & B/z_2/ \\ C/z_2/ & D/z_2/ \end{bmatrix} = \begin{bmatrix} A_{12} \\ C_2 \end{bmatrix} [I_{n_2}z_2 - A_{22}]^{-1} \begin{bmatrix} A_{21} & B_2 \end{bmatrix} + \begin{bmatrix} A_{11} & B_1 \\ C_1 & D \end{bmatrix} \qquad /2.94'/$$

Note that the matrix /2.90/ may be written in the form

$$G/z_1, z_2/ = \frac{N/z_1, z_2/}{d/z_1, z_2/} =$$

$$= \frac{N_{r_1}/z_2/}{d_{r_1}/z_2/} + \frac{\bar{N}_{r_1-1}/z_2/ z_1^{r_1-1} + \dots + \bar{N}_1/z_2/ z_1 + \bar{N}_0/z_2/}{z_1^{r_1} + \bar{d}_{r_1-1}/z_2/ z_1^{r_1-1} + \dots + \bar{d}_1/z_2/ z_1 + \bar{d}_0/z_2/}$$

/2.95/

where

$$N/z_1, z_2/ = N_{r_1}/z_2/ z_1^{r_1} + \dots + N_1/z_2/ z_1^1 + N_0/z_2/$$

$$d/z_1, z_2/ = d_{r_1}/z_2/ z_1^{r_1} + \dots + d_1/z_2/ z_1 + d_0/z_2/$$

$$N_j/z_2/ \in R^{1 \times m}[z_2], \quad d_j/z_2/ \in R[z_2] \qquad j = 0, 1, \dots, r_1$$

$$\bar{d}_i/z_2/ = \frac{d_i/z_2/}{d_{r_1}/z_2/} \in R(z_2)$$

$$\bar{N}_i/z_2/ = \frac{N_i/z_2/ - N_{r_1}/z_2/ \bar{d}_i/z_2/}{d_{r_1}/z_2/} \in R^{1 \times m}(z_2)$$

$i = 0, 1, 2, \dots, r_1 - 1$.

Considering $G/z_1,z_2/$ as an element of $R^{1\times m}(z_2)[z_1]$ and using one of the well-known algorithms for realization of 1-D systems [5,9, 12, 4] we can find a realization $\{A/z_2/, B/z_2/, C/z_2/, D/z_2/\}$ such that /2.91/ holds. For example, it is easy to check that

$$A/z_2/ = \begin{bmatrix} 0 & 0 & \cdots & 0 & -\bar{d}_0/z_2/\ I_1 \\ I_1 & 0 & \cdots & 0 & -\bar{d}_1/z_2/\ I_1 \\ 0 & I_1 & \cdots & 0 & -\bar{d}_2/z_2/\ I_1 \\ \cdots & \cdots & \cdots & \cdots & \cdots \\ 0 & 0 & \cdots & 0 & -\bar{d}_{r_1-2}/z_2/\ I_1 \\ 0 & 0 & \cdots & I_1 & -\bar{d}_{r_1-1}/z_2/\ I_1 \end{bmatrix} \qquad B/z_2/ = \begin{bmatrix} \bar{N}_0/z_2/ \\ \bar{N}_1/z_2/ \\ \bar{N}_2/z_2/ \\ \cdots \\ \bar{N}_{r_1-2}/z_2/ \\ \bar{N}_{r_1-1}/z_2/ \end{bmatrix}$$

$$C/z_2/ = \begin{bmatrix} 0 & 0 & \cdots & 0 & I_1 \end{bmatrix} \qquad D/z_2/ = \frac{N_{r_1}/z_2/}{d_{r_1}/z_2/} \qquad /2.96/$$

is a realization of /2.95/.

A realization $\{A/z_2/, B/z_2/, C/z_2/, D/z_2/\}$ of $G/z_1,z_2/$ is called a first level realization.
From /2.94/ it follows that

$$\begin{bmatrix} A_{11} & B_1 \\ C_1 & D \end{bmatrix} = \lim_{z_2 \to \infty} \begin{bmatrix} A/z_2/ & B/z_2/ \\ C/z_2/ & D/z_2/ \end{bmatrix} \qquad /2.97/$$

since $\lim_{z_2 \to \infty} [I_{n_2} z_2 - A_{22}]^{-1} = 0$.

Thus, the matrices $A_{11}$, $B_1$, $C_1$ and $D$ can be found using /2.97/

To find the matrices $A_{12}$, $A_{21}$, $A_{22}$, $B_2$ and $C_2$ we can use one of the well-known algorithms [5, 9, 12, 4] for realization of the 1-D transfer function matrix

$$G_{sp}/z_2/ = \begin{bmatrix} A_{12} \\ C_2 \end{bmatrix} [I_{n_2} z_2 - A_{22}]^{-1} \begin{bmatrix} A_{21} & B_2 \end{bmatrix} \qquad /2.98/$$

The matrices $A_{11}$, $A_{12}$, $A_{21}$, $A_{22}$, $B_1$, $B_2$, $C_1$, $C_2$ and $D$ are called the second level realization of $G/z_1,z_2/$.

From above considerations it follows that every proper $G/z_1,z_2/$ has a realization /2.88/. Note that the roles of $z_1$ and $z_2$ can be reversed.
From /2.91/ and /2.89/ it follows that

$$D/z_2/ = \lim_{z_1 \to \infty} G/z_1, z_2/ \qquad /2.99/$$

and

$$D = \lim_{\substack{z_1 \to \infty \\ z_2 \to \infty}} G/z_1, z_2/ \qquad /2.100/$$

since $\lim_{z_1 \to \infty} \left[ I_{n_1} z_1 - A/z_2/ \right]^{-1} = 0$ and $\lim_{\substack{z_1 \to \infty \\ z_2 \to \infty}} \left[ Z - A \right]^{-1} = 0$.

A realization /2.88/ of /2.90/ can be found by the use of the following

Algorithm 2.2

Step 1. Write $G/z_1, z_2/$ in the form /2.95/ and evaluate $D/z_2/ = \dfrac{N_{r_1}/z_2/}{d_{r_1}/z_2/}$.
To find $D/z_2/$ formula /2.99/ can also be used.

Step 2. Find $A/z_2/$, $B/z_2/$ and $C/z_2/$ for

$$G_{sp}/z_1, z_2/ = \frac{\bar{N}_{r_1-1}/z_2/ z_1^{r_1-1} + \ldots + \bar{N}_1/z_2/ z_1 + \bar{N}_0/z_2/}{z_1^{r_1} + \bar{d}_{r_1-1}/z_2/ z_1^{r_1-1} + \ldots + \bar{d}_1/z_2/ z_1 + \bar{d}_0/z_2/} \qquad /2.101/$$

using one of the algorithms for realization of 1-D systems. For example, the realization given by /2.96/ can be used.

Step 3. Find the matrices $A_{11}$, $B_1$, $C_1$ and $D$ using /2.97/.

Step 4. Evaluate $A_{12}$, $A_{21}$, $A_{22}$, $B_2$ and $C_2$ as a realization of the 1-D transfer function matrix $G_{sp}/z_2/$

Example 2.6

Find a realization $A, B, C, D$ for the transfer function

$$G/z_1, z_2/ = \frac{z_1 z_2 + 2}{z_1^2 (z_2^2 + 1) + z_1 z_2 + 1} \qquad /2.102/$$

Step 1.

$$G/z_1, z_2/ = \frac{\dfrac{z_2}{z_2^2 + 1} z_1 + \dfrac{2}{z_2^2 + 1}}{z_1^2 + \dfrac{z_2}{z_2^2 + 1} z_1 + \dfrac{1}{z_2^2 + 1}} = \frac{\bar{N}_1/z_2/ z_1 + \bar{N}_0/z_2/}{z_1^2 + d_1/z_2/ z_1 + d_0/z_2/} \qquad /2.103/$$

where

$$\bar{N}_1/z_2/ = \frac{z_2}{z_2^2 + 1}, \quad \bar{N}_0/z_2/ = \frac{2}{z_2^2 + 1}, \quad d_1/z_2/ = \frac{z_2}{z_2^2 + 1},$$

$$d_0/z_2/ = \frac{1}{z_2^2 + 1}. \quad \text{and} \quad D_2/z_2/ = 0.$$

Step 2. It is easy to check that the matrices

$$A/z_2/ = \begin{bmatrix} 0 & 1 \\ -d_0/z_2/ & -d_1/z_2/ \end{bmatrix}, \quad B/z_2/ = \begin{bmatrix} 0 \\ 1 \end{bmatrix}, \quad C/z_2/ = \begin{bmatrix} \bar{N}_0/z_2/ & \bar{N}_1/z_2/ \end{bmatrix}$$

are a realization of /2.103/.

Step 3. Using /2.97/ we obtain

$$\begin{bmatrix} A_{11} & B_1 \\ C_1 & D \end{bmatrix} = \lim_{z_2 \to \infty} \begin{bmatrix} A/z_2/ & B/z_2/ \\ C/z_2/ & D/z_2/ \end{bmatrix} = \begin{bmatrix} 0 & 1 & | & 0 \\ 0 & 0 & | & 1 \\ \hline 0 & 0 & | & 0 \end{bmatrix}$$

Step 4. It is easy to check that the matrices

$$A_{12} = \begin{bmatrix} 0 & 0 & 0 & 0 \\ -1 & 0 & 0 & -1 \end{bmatrix}, \quad A_{21} = \begin{bmatrix} 0 & 0 \\ 1 & 0 \\ 0 & 0 \\ 0 & 1 \end{bmatrix}, \quad A_{22} = \begin{bmatrix} 0 & 1 & 0 & 0 \\ -1 & 0 & 0 & 0 \\ 0 & 0 & 0 & 1 \\ 0 & 0 & -1 & 0 \end{bmatrix}, \quad B_2 = \begin{bmatrix} 0 \\ 0 \\ 0 \\ 0 \end{bmatrix},$$

$$C_2 = \begin{bmatrix} 2 & 0 & 0 & 1 \end{bmatrix}$$

are a realization of the matrix

$$\begin{bmatrix} A/z_2/ & B/z_2/ \\ C/z_2/ & D/z_2/ \end{bmatrix} - \begin{bmatrix} A_{11} & B_1 \\ C_1 & D \end{bmatrix} = \frac{1}{z_2^2 + 1} \begin{bmatrix} 0 & 0 & 0 \\ -1 & -z_2 & 0 \\ 2 & z_2 & 0 \end{bmatrix}$$

Hence

$$A = \begin{bmatrix} A_{11} & A_{12} \\ A_{21} & A_{22} \end{bmatrix} = \begin{bmatrix} 0 & 1 & | & 0 & 0 & 0 & 0 \\ 0 & 0 & | & -1 & 0 & 0 & -1 \\ \hline 0 & 0 & | & 0 & 1 & 0 & 0 \\ 1 & 0 & | & -1 & 0 & 0 & 0 \\ 0 & 0 & | & 0 & 0 & 0 & 1 \\ 0 & 1 & | & 0 & 0 & -1 & 0 \end{bmatrix}, \quad B = \begin{bmatrix} B_1 \\ B_2 \end{bmatrix} = \begin{bmatrix} 0 \\ 1 \\ \hline 0 \\ 0 \\ 0 \\ 0 \end{bmatrix},$$

/2.104/

$$C = \begin{bmatrix} C_1 & C_2 \end{bmatrix} = \begin{bmatrix} 0 & 0 & | & 2 & 0 & 0 & 1 \end{bmatrix}, \quad D = \begin{bmatrix} 0 \end{bmatrix}.$$

**Theorem 2.8** [11]

A proper transfer function $G/z_1,z_2/ = \dfrac{N/z_1,z_2/}{d/z_1,z_2/}$, $N/z_1,z_2/ \in R[z_1,z_2]$, $d/z_1,z_2/ \in R[z_1,z_2]$ has a realization $\{A,B,C,D\}$ of dimension $/n_1, 2n_2/$, where $n_1 = \deg_{z_1} d/z_1,z_2/$ and $n_2 = \deg_{z_2} d/z_1,z_2/$.

**Proof:** From /2.96/ and /2.97/ it follows that for a single-output system $l=1$, $r_1=n_1$, $A/z_2/$ and $A_{11}$ have dimensions $n_1 \times n_1$. It is well known from 1-D systems theory [5, 9] that the minimal possible dimension of $A_{22}$ is given by McMillan degree of /2.98/ and the McMillan degree of /2.98/ is equal to degree of the least common denominator of all minors of /2.98/. Note that $d_{n_1}^2/z_2/$ is the common denominator for all minors of the matrix

$$G_{sp}/z_2/ = \begin{bmatrix} A/z_2/ & B/z_2/ \\ C/z_2/ & D/z_2/ \end{bmatrix} - \begin{bmatrix} A_{11} & B_1 \\ C_1 & D \end{bmatrix} = \begin{bmatrix} 0 & 0 & \cdots & 0 & -\bar{d}_0/z_2/ & \bar{N}_0/z_2/ \\ 1 & 0 & \cdots & 0 & -\bar{d}_1/z_2/ & \bar{N}_1/z_2/ \\ 0 & 1 & \cdots & 0 & -\bar{d}_2/z_2/ & \bar{N}_2/z_2/ \\ \vdots & & & & & \vdots \\ 0 & 0 & \cdots & 0 & -\bar{d}_{n_1-2}/z_2/ & \bar{N}_{n_1-2}/z_2/ \\ 0 & 0 & \cdots & 1 & -\bar{d}_{n_1-1}/z_2/ & \bar{N}_{n_1-1}/z_2/ \\ 0 & 0 & \cdots & 0 & 1 & \dfrac{N_{n_1}/z_2/}{d_{n_1}/z_2/} \end{bmatrix}$$

So, we have McMillan degree of $G_{sp}/z_2/$ not greater than $2\deg_{z_2} d_{n_1}^2/z_2/ = 2n_2$. □

For example, the denominator of /2.102/ equals

$$d/z_1,z_2/ = z_1^2(z_2^2 + 1) + z_1 z_2 + 1$$

and

$$\deg_{z_1} d/z_1,z_2/ = n_1 = \deg_{z_2} d/z_1,z_2/ = n_2 = 2.$$

The realization /2.104/ of /2.102/ is of dimension $/n_1, 2n_2/ = /2,4/$.

## 2.6.2 Realization of the Attasi's model.

Consider the Attasi's model /AM/ in the form suggested in [6]

$$x/i+1,j+1/ = A_1 x/i,j+1/ + A_2 x/i+1,j/ + A_0 x/i,j/ + Bu/i,j/$$
$$y/i,j/ = Cx/i,j/ \qquad i,j \geqslant 0 \qquad /2.105/$$

where

$$A_0 = -A_1 A_2 \quad \text{or} \quad A_0 = -A_2 A_1$$

$x/i,j/ \in R^n$ is the state vector,

$u/i,j/ \in R^m$ is the input vector,

$y/i,j/ \in R^l$ is the output vector,

$A_0$, $A_1$, $A_2$, B, C are constant real matrices of appropriate dimensions.

The transfer function $G/z_1,z_2/$ of /2.105/ is given by

$$G/z_1,z_2/ = C\left[z_1 z_2 I_n - z_2 A_1 - z_1 A_2 - A_0\right]^{-1} B \qquad /2.106/$$

Note that if $A_0 = -A_1 A_2$ then /2.106/ may be written as

$$G/z_1,z_2/ = C\left[z_2 I_n - A_2\right]^{-1} \left[z_1 I_n - A_1\right]^{-1} B \qquad /2.107/$$

and if $A_0 = -A_2 A_1$ then /2.106/ may be written in the form

$$G/z_1,z_2/ = C\left[z_1 I_n - A_1\right]^{-1} \left[z_2 I_n - A_2\right]^{-1} B \qquad /2.108/$$

### Definition 2.9

A realization $\{A_1, A_2, B, C\}$ is said to be a minimal one iff the dimension of $A_1$ $(A_2)$ is minimal amongst all dimensions of all possible realizations.

### Theorem 2.9a [6]

AM with $A_0 = - A_1 A_2$ is minimal if and only if

$$\text{rank } S_n = \text{rank } H_n = n \qquad /2.109a/$$

where

$$S_n = \left[B \quad A_1 B \quad \ldots \quad A_1^{n-1} B\right],$$

$$H_n = \begin{bmatrix} C \\ CA_2 \\ \vdots \\ CA_2^{n-1} \end{bmatrix}$$

**Proof:** From /2.107/ it follows that

$$G/z_1, z_2/ = G_1/z_1/ \, G_2/z_2/ \qquad /2.110/$$

where

$$G_1/z_1/ = I_n \bigl[z_1 I_n - A_1\bigr]^{-1} B, \qquad G_2/z_2/ = C \bigl[z_2 I_n - A_2\bigr]^{-1} I_n$$

Therefore, AM can be viewed as a composite system consisting of the cascade connection of two 1-D systems whose transfer functions are $G_1/z_1/$ and $G_2/z_2/$. It is well known [6, 5, 10, 13] that the 1-D system with $G_1/z_1/$ is minimal if and only if the pair $A_1, B$ is controllable, since the pair $A_1, I_n$ is observable. Similarly, the 1-D system with $G_2/z_2/$ is minimal if and only if the pair $A_2, C$ is observable, since the pair $A_2, I_n$ is controllable. Therefore the composite system consisting of the cascade connection of two 1-D systems is minimal if and only if the condition /2.109a/ holds. □

Similarly we can prove the dual

**Theorem 2.9b** [6]

AM with $A_0 = -A_2 A_1$ is minimal if and only if

$$\operatorname{rank} \bar{S}_n = \operatorname{rank} \bar{H}_n = n \qquad /2.109b/$$

where

$$S_n = \begin{bmatrix} B & A_2 B & \cdots & A_2^{n-1} B \end{bmatrix}, \qquad H_n = \begin{bmatrix} C \\ CA_1 \\ \vdots \\ CA_1^{n-1} \end{bmatrix}$$

Taking into account that

$$\bigl[z_k I_n - A_k\bigr]^{-1} = \sum_{i=1}^{\infty} A_k^{i-1} z_k^{-i} \qquad \text{for } k = 1, 2$$

we can write /2.107/ in the form

$$G/z_1, z_2/ = \sum_{i=1}^{\infty} \sum_{j=1}^{\infty} M_{ij} z_1^{-i} z_2^{-j} \qquad /2.111/$$

where

$$M_{ij} = CA_2^{j-1} A_1^{i-1} B \qquad /2.112/$$

are Markov parameters of the model.

### Definition 2.10

The matrix quadruple $\{A_1, A_2, B, C\}$ is said to be a realization of AM with the given Markov parameters $\{M_{ij}: i,j = 1,2,...,N\}$ iff /2.111/ holds.

Let us define the Hankel matrix

$$H_{kl} = \begin{bmatrix} M_{11} & M_{21} & \cdots & M_{l1} \\ M_{12} & M_{22} & \cdots & M_{l2} \\ \vdots & & & \\ M_{1k} & M_{2k} & \cdots & M_{lk} \end{bmatrix} \quad \text{for} \quad k,l \geq 1 \qquad /2.113/$$

From /2.112/ and /2.113/ it follows that

$$H_{kl} = H_k S_l \qquad /2.114/$$

$$H_k = \begin{bmatrix} C \\ CA_2 \\ \vdots \\ CA_2^{k-1} \end{bmatrix}, \quad S_l = \begin{bmatrix} B & A_1 B & \cdots & A_1^{l-1} B \end{bmatrix}$$

### Theorem 2.10

If

$$\text{rank } H_{N-1,N} = \text{rank } H_{N,N-1} = \text{rank } H_{NN} \qquad /2.115/$$

then it is possible to obtain a realization $\{A_1, A_2, B, C\}$ from the given finite set of Markov parameters $\{M_{ij}: i,j = 1,2,...,N\}$.

**Proof:** Assume that /2.15/ holds and rank $H_{NN} = n$. Then $H_{NN} \in R^{lN \times mN}$ can be factored as

$$H_{NN} = \bar{Q}_N \bar{P}_N \qquad /2.116/$$

where

$$\bar{Q}_N = \begin{bmatrix} Q_1 \\ Q_2 \\ \vdots \\ Q_N \end{bmatrix}, \quad \bar{P}_N = \begin{bmatrix} P_1 & P_2 & \cdots & P_N \end{bmatrix}, \quad Q_i \in R^{l \times n}, \quad P_i \in R^{n \times m} \qquad /2.117/$$

$$i = 1, 2, ..., N$$

From comparison of /2.114/ and /2.116/ we have

$$P_i = A_1^{i-1} B, \quad Q_i = C A_2^{i-1} \quad \text{for} \quad i = 1, 2, \ldots, N \qquad /2.118/$$

Note that

$$\begin{bmatrix} P_2 & P_3 & \cdots & P_N \end{bmatrix} = A_1 \begin{bmatrix} B & A_1 B & \cdots & A_1^{N-2} B \end{bmatrix} = A_1 \bar{P}_{N-1} \qquad /2.119/$$

and

$$\begin{bmatrix} Q_2 \\ Q_3 \\ \vdots \\ Q_N \end{bmatrix} = \begin{bmatrix} C \\ C A_2 \\ \vdots \\ C A_2^{N-2} \end{bmatrix} A_2 = \bar{Q}_{N-1} A_2 \qquad /2.120/$$

From /2.115/ it follows that

$$\text{rank } \bar{P}_{N-1} = \text{rank } \bar{Q}_{N-1} = n$$

Therefore, from /2.119/ and /2.120/ we obtain

$$A_1 = \begin{bmatrix} P_2 & P_3 & \cdots & P_N \end{bmatrix} \bar{P}_{N-1}^+$$

$$A_2 = \bar{Q}_{N-1}^+ \begin{bmatrix} Q_2 \\ Q_3 \\ \vdots \\ Q_N \end{bmatrix}, \quad B = P_1, \quad C = Q_1 \qquad /2.121/$$

where $\bar{P}_{N-1}^+$ $\left(\bar{Q}_{N-1}^+\right)$ is a right (left) inverse of $\bar{P}_{N-1}$ $\left(\bar{Q}_{N-1}\right)$, i.e.

$$\bar{P}_{N-1} \bar{P}_{N-1}^+ = I_n, \quad \bar{Q}_{N-1}^+ \bar{Q}_{N-1} = I_n \qquad \square$$

### Theorem 2.11

The dimension of the minimal realization is equal to the rank of $H_{NN}$.

**Proof:** From /2.116/ and Sylvester's inequality it follows that the rank of $\bar{P}_N$ and $\bar{Q}_N$ cannot be less than the rank of $H_{NN}$. On the other hand, the dimension of the realization is equal to the ranks of $\bar{P}_N$ and $\bar{Q}_N$ which are equal to the rank of $H_{NN}$. Therefore, the realization has minimal dimension. $\square$

### Theorem 2.12

If $\{A_1, A_2, B, C\}$ is a realization of dimension n of AM and $T \in R^{n \times n}$ is a nonsingular matrix, then $\{TA_1T^{-1}, TA_2T^{-1}, TB, CT^{-1}\}$ is also a realization of AM.

**Proof:** It can be easily verified that

$$[TA_1T^{-1}]^k = TA_1^k T^{-1} \quad \text{and} \quad [TA_2T^{-1}]^k = TA_2^k T^{-1} \qquad /2.122/$$

for $k = 1, 2, \ldots$ .

Substituting /2.122/ into /2.112/ we obtain

$$M_{ij} = CT^{-1}[TA_2T^{-1}]^{j-1}[TA_1T^{-1}]^{i-1}TB = CT^{-1}TA_2^{j-1}T^{-1}TA_1^{i-1}T^{-1}TB =$$

$$= CA_2^{j-1}A_1^{i-1}B \qquad \square$$

To find a realization $\{A_1, A_2, B, C\}$ of AM from the given $G/z_1, z_2/$ ($M_{ij}$ : $i, j = 1, 2, \ldots, N$) the following algorithm can be used.

## Algorithm 2.3

**Step 1.** Write $G/z_1, z_2/$ in the form /2.111/ and find $M_{ij}$.
To evaluate $M_{ij}$ the formula

$$M_{ij} = \lim_{z_1, z_2 \to \infty} z_1^i z_2^j \left[G/z_1, z_2/ - \sum_{/1,1/\leqslant /k,l/ < /i,j/} M_{kl} z_1^{-k} z_2^{-l}\right] \qquad /2.123/$$

or a long division algorithm may be used.

**Step 2.** Build the Hankel matrix /2.113/ for $k = l = N$ and determine its rank $n$.

**Step 3.** Construct factorization /2.116/.
This can be done by constructing the Hermite form for $H_{NN}$

$$H_{NN} = L^{-1}\begin{bmatrix} D & 0 \\ 0 & 0 \end{bmatrix} R^{-1} \qquad /2.124/$$

where $D$ is $n \times n$ lower triangular matrix with full rank and $L$, $R$ are $lN \times lN$ and $mN \times mN$ non-singular matrices of elementary row and culumn operations, respectively.
We can take

$$\bar{Q}_N = \bar{L}D \text{ and } \bar{P}_N = \bar{R}, \quad \text{or} \quad \bar{Q}_N = \bar{L} \text{ and } \bar{P}_N = D\bar{R} \qquad /2.125/$$

where $\bar{L}$ ($\bar{R}$) is $lN \times n$ ($n \times mN$) matrix built from the first $n$ columns (rows) of $L^{-1}$ ($R^{-1}$).
Alternatively, the independent columns of $H_{NN}$ can be used to find $Q_N$.

Step 4. Evaluate $A_1$, $A_2$, $B$ and $C$ using /2.121/.

Example 2.7

Find a realization $\{A_1, A_2, B, C\}$ of AM with the transfer function

$$G/z_1, z_2/ = \frac{-z_1 + z_2 + 1}{(z_1^2 - z_1 - 3)(z_2^2 - 2z_2 - 1)} \qquad /2.126/$$

Step 1. Using a long division algorithm, or formula /2.123/, we obtain

$$G/z_1, z_2/ = z_1^{-2} z_2^{-1} - z_1^{-1} z_2^{-2} + 2z_1^{-2} z_2^{-2} - 2z_1^{-1} z_2^{-3} + z_1^{-3} z_2^{-1} +$$
$$- z_1^{-3} z_2^{-2} + 5z_1^{-2} z_2^{-3} - z_1^{-3} z_2^{-3} + \ldots$$

Step 2. The Hankel matrix /2.113/ for $k = l = 3$ has the form

$$H_{33} = \begin{bmatrix} M_{11} & M_{21} & M_{31} \\ M_{12} & M_{22} & M_{32} \\ M_{13} & M_{23} & M_{33} \end{bmatrix} = \begin{bmatrix} 0 & 1 & 1 \\ -1 & 2 & -1 \\ -2 & 5 & -1 \end{bmatrix} \qquad /2.127/$$

and rank $H_{33} = n = 2$.

Step 3. It is easy to check that

$$H_{33} = \begin{bmatrix} 0 & 1 & 1 \\ -1 & 2 & -1 \\ -2 & 5 & -1 \end{bmatrix} = \begin{bmatrix} 1 & 0 \\ 0 & 1 \\ 1 & 2 \end{bmatrix} \begin{bmatrix} 0 & 1 & 1 \\ -1 & 2 & -1 \end{bmatrix} = \bar{Q}_3 \bar{P}_3$$

where

$$\bar{Q}_3 = \begin{bmatrix} Q_1 \\ Q_2 \\ Q_3 \end{bmatrix} = \begin{bmatrix} 1 & 0 \\ 0 & 1 \\ 1 & 2 \end{bmatrix}, \quad \bar{P}_3 = \begin{bmatrix} P_1 & P_2 & P_3 \end{bmatrix} = \begin{bmatrix} 0 & 1 & 1 \\ -1 & 2 & -1 \end{bmatrix}.$$

Step 4. Using /2.121/ we obtain

$$A_1 = \begin{bmatrix} P_2 & P_3 \end{bmatrix} \begin{bmatrix} P_1 & P_2 \end{bmatrix}^{-1} = \begin{bmatrix} 1 & 1 \\ 2 & -1 \end{bmatrix} \begin{bmatrix} 0 & 1 \\ -1 & 2 \end{bmatrix}^{-1} = \begin{bmatrix} 3 & -1 \\ 3 & -2 \end{bmatrix},$$

$$A_2 = \begin{bmatrix} Q_1 \\ Q_2 \end{bmatrix}^{-1} \begin{bmatrix} Q_2 \\ Q_3 \end{bmatrix} = \begin{bmatrix} 1 & 0 \\ 0 & 1 \end{bmatrix}^{-1} \begin{bmatrix} 0 & 1 \\ 1 & 2 \end{bmatrix} = \begin{bmatrix} 0 & 1 \\ 1 & 2 \end{bmatrix},$$

$$B = P_1 = \begin{bmatrix} 0 \\ -1 \end{bmatrix}, \quad C = Q_1 = \begin{bmatrix} 1 & 0 \end{bmatrix}.$$

The realization problem for F-MM I and F-MM II is considered in [5,4].

### 2.6.3 Realization of the Roesser's model in canonical form.

#### Definition 2.11
Single-input single-output RM has the canonical form if the submatrices $A_{11}$, $A_{12}$, $A_{21}$, $A_{22}$, $B_1$, $B_2$, $C_1$ and $C_2$ have the following structures

$$A_{11} = \begin{bmatrix} 0 & 1 & 0 & \cdots & 0 \\ 0 & 0 & 1 & \cdots & 0 \\ \cdot & \cdot & \cdot & \cdots & \cdot \\ 0 & 0 & 0 & \cdots & 1 \\ -\hat{a}_{01} & -\hat{a}_{02} & -\hat{a}_{03} & \cdots & -\hat{a}_{0n_1} \end{bmatrix}, \quad A_{12} = \begin{bmatrix} 0 & 0 & \cdots & 0 \\ 0 & 0 & \cdots & 0 \\ \cdot & \cdot & \cdots & \cdot \\ 0 & 0 & \cdots & 0 \\ 1 & 0 & \cdots & 0 \end{bmatrix},$$

$$A_{21} = \begin{bmatrix} \hat{a}_{11} & \cdots & \hat{a}_{1n_1} \\ \hat{a}_{21} & \cdots & \hat{a}_{2n_1} \\ \cdot & \cdots & \cdot \\ \hat{a}_{n_2-1,1} & \cdots & \hat{a}_{n_2-1,n_1} \\ \hat{a}_{n_2 1} & \cdots & \hat{a}_{n_2 n_1} \end{bmatrix}, \quad A_{22} = \begin{bmatrix} -\hat{a}_{10} & 1 & 0 & \cdots & 0 \\ -\hat{a}_{20} & 0 & 1 & \cdots & 0 \\ \cdot & \cdot & \cdot & \cdots & \cdot \\ -\hat{a}_{n_2-1,0} & 0 & 0 & \cdots & 1 \\ -\hat{a}_{n_2 0} & 0 & 0 & \cdots & 0 \end{bmatrix}$$

$$B_1 = \begin{bmatrix} 0 \\ 0 \\ \vdots \\ 0 \\ 1 \end{bmatrix}, \quad B_2 = \begin{bmatrix} \hat{b}_1 \\ \hat{b}_2 \\ \vdots \\ \hat{b}_{n_2} \end{bmatrix} \quad \begin{matrix} C_1 = \begin{bmatrix} \hat{c}_1 & \hat{c}_2 & \cdots & \hat{c}_{n_1} \end{bmatrix} \\ \\ C_2 = \begin{bmatrix} \hat{c}_{21} & 0 & \cdots & 0 & 1 & 0 & \cdots & 0 \end{bmatrix} \\ \phantom{C_2 =} \text{position } m+1 \end{matrix} \quad /2.128/$$

Consider a proper transfer function of the form

$$G/z_1,z_2/ = \frac{\sum_{i=0}^{n}\sum_{j=0}^{m} \tilde{b}_{ij} z_1^i z_2^j}{\sum_{i=0}^{n}\sum_{j=0}^{m} \tilde{a}_{ij} z_1^i z_2^j}, \quad (\tilde{a}_{nm} \neq 0) \qquad /2.129/$$

We shall present two methods for finding the canonical form realization of /2.129/.

## Method 1.

The transfer function /2.129/ can be rewritten in the form

$$G/z_1,z_2/ = \frac{\sum_{i=0}^{n}\sum_{j=0}^{m} b_{n-i,m-j} z_1^{-i} z_2^{-j}}{\sum_{i=0}^{n}\sum_{j=0}^{m} a_{n-i,m-j} z_1^{-i} z_2^{-j}} =$$

$$= \frac{\sum_{i=0}^{n} b_i z_1^{-i}}{\sum_{i=0}^{n} a_i z_1^{-i}} \qquad /2.130/$$

where

$$b_k = b_{k0} + b_{k1} z_2^{-1} + \ldots + b_{km} z_2^{-m} \qquad (k = 0,1,\ldots,n) \qquad /2.130a/$$

$$a_k = a_{k0} + a_{k1} z_2^{-1} + \ldots + a_{km} z_2^{-m} \qquad /2.130b/$$

We assume that $a_0, a_1, \ldots, a_n$ and $b_0, b_1, \ldots, b_n$ are coprime.

Taking into account that $G/z_1,z_2/ = \frac{Y}{U}$, where $Y = Y/z_1,z_2/$ and $U = U/z_1,z_2/$ are 2-D Z transforms of $y/i,j/$ and $u/i,j/$, and defining

$$E = \frac{U}{\sum_{i=0}^{n} a_i z_1^{-i}} \qquad /2.131/$$

we can write /2.130/ in the form

$$Y = \sum_{i=0}^{n} b_i z_1^{-i} E \qquad /2.132/$$

Without loss of generality we can assume $a_{00} = 1$. Let us denote

$$\bar{a}_0 = a_0 - 1 = \sum_{i=1}^{m} a_{0i} z_2^{-i} \qquad /2.133/$$

From /2.131/ and /2.133/ it follows that

$$E = U - \left[\bar{a}_0 + a_1 z_1^{-1} + \ldots + a_n z_1^{-n}\right] E \qquad /2.134/$$

From /2.134/ and /2.132/ the block diagram shown in Fig. 2.1 follows. Note that in addition to the n horizontal delay elements of Fig. 2.1 we need m vertical delay elements to implement the feedback gains $a_i$ /i = 0,1, ..., n/ and m other vertical delay elements to implement the read-out gains $b_i$ /i = 0,1, ..., n/. Thus, the complete block diagram shown in Fig. 2.2 requires n + 2m delay /dynamic/ elements.

Using Fig. 2.2 we can write the following equations

$$x_1^h/i+1,j/ = x_2^h/i,j/$$

$$x_2^h/i+1,j/ = x_3^h/i,j/$$

$$\cdots\cdots\cdots\cdots$$

$$x_{n-1}^h/i+1,j/ = x_n^h/i,j/$$

$$x_n^h/i+1,j/ = -a_{n0}x_1^h/i,j/ - a_{n-1,0}x_2^h/i,j/ - \cdots - a_{10}x_n^h/i,j/ + x_1^{v_1}/i,j/ + u/i,j/$$

$$x_1^{v_1}/i,j+1/ = -a_{n1}x_1^h/i,j/ - a_{n-1,1}x_2^h/i,j/ - \cdots - a_{11}x_n^h/i,j/ - a_{01}x_n^h/i+1,j/ + x_2^{v_1}/i,j/ =$$

$$= -\bar{a}_{n1}x_1^h/i,j/ - \bar{a}_{n-1,1}x_2^h/i,j/ - \cdots - \bar{a}_{11}x_n^h/i,j/ - a_{01}x_1^{v_1}/i,j/ + x_2^{v_1}/i,j/ - a_{01}u/i,j/$$

$$x_2^{v_1}/i,j+1/ = -a_{n2}x_1^h/i,j/ - a_{n-1,2}x_2^h/i,j/ - \cdots - a_{12}x_n^h/i,j/ - a_{02}x_n^h/i+1,j/ + x_3^{v_1}/i,j/ =$$

$$= -\bar{a}_{n2}x_1^h/i,j/ - \bar{a}_{n-1,2}x_2^h/i,j/ - \cdots - \bar{a}_{12}x_n^h/i,j/ - a_{02}x_1^{v_1}/i,j/ + x_3^{v_1}/i,j/ - a_{02}u/i,j/$$

$$\cdots\cdots\cdots\cdots\cdots\cdots\cdots\cdots$$

$$x_m^{v_1}/i,j+1/ = -a_{nm}x_1^h/i,j/ - a_{n-1,m}x_2^h/i,j/ - \cdots - a_{1m}x_n^h/i,j/ - a_{0m}x_n^h/i+1,j/ =$$

$$= -\bar{a}_{nm}x_1^h/i,j/ - \bar{a}_{n-1,m}x_2^h/i,j/ - \cdots - \bar{a}_{1m}x_n^h/i,j/ +$$

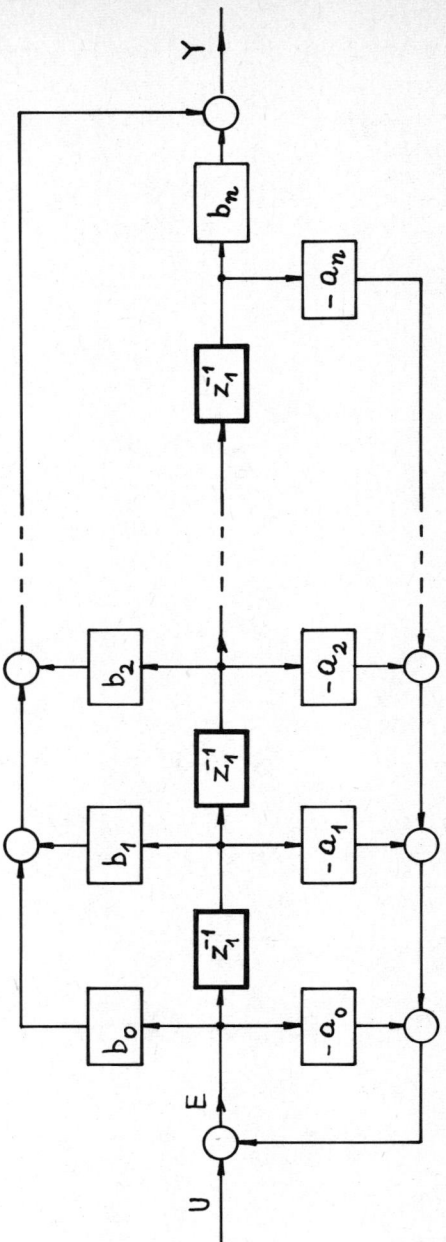

Fig. 2.1.

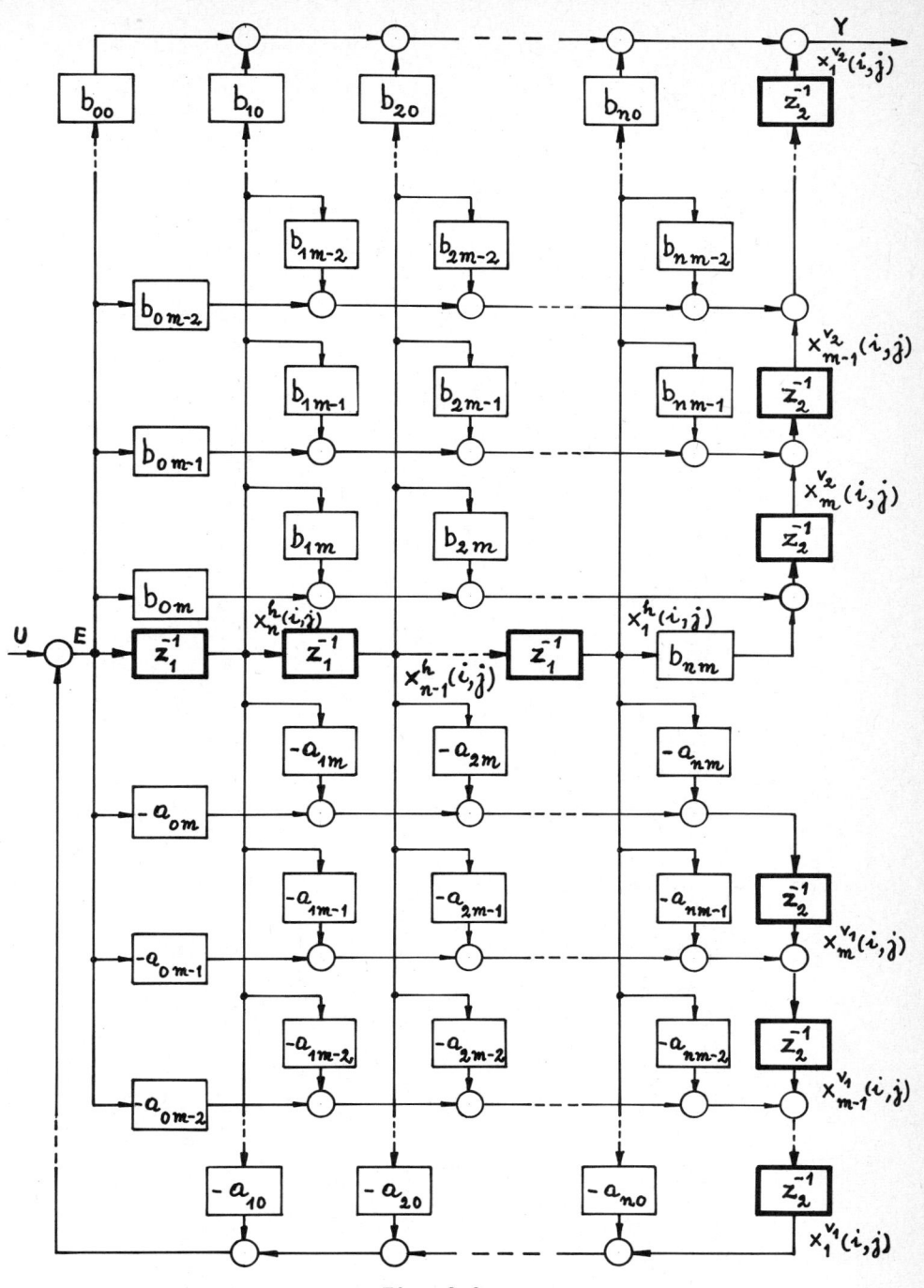

Fig. 2.2.

$$-a_{om}x_1^h/i,j/ - a_{om}u/i,j/$$

$$x_1^{v_2}/i,j+1/ = b_{n1}x_1^h/i,j/ + b_{n-1,1}x_2^h/i,j/ + \ldots + b_{11}x_n^h/i,j/ +$$

$$+ b_{01}x_n^h/i+1,j/ + x_2^{v_2}/i,j/ =$$

$$= \bar{b}_{n1}x_1^h/i,j/ + \bar{b}_{n-1,1}x_2^h/i,j/ + \ldots + \bar{b}_{11}x_n^h/i,j/ +$$

$$+ b_{01}x_1^{v_1}/i,j/ + x_2^{v_2}/i,j/ + b_{01}u/i,j/$$

$$x_2^{v_2}/i,j+1/ = b_{n2}x_1^h/i,j/ + b_{n-1,2}x_2^h/i,j/ + \ldots + b_{12}x_n^h/i,j/ +$$

$$+ b_{02}x_n^h/i+1,j/ + x_3^{v_2}/i,j/ =$$

$$= \bar{b}_{n2}x_1^h/i,j/ + \bar{b}_{n-1,2}x_2^h/i,j/ + \ldots + \bar{b}_{12}x_n^h/i,j/ +$$

$$+ b_{02}x_1^{v_1}/i,j/ + x_3^{v_2}/i,j/ + b_{02}u/i,j/$$

$$\ldots \ldots \ldots \ldots \ldots \ldots \ldots \ldots \ldots \ldots \ldots \ldots \ldots \quad /2.135/$$

$$x_m^{v_2}/i,j+1/ = b_{nm}x_1^h/i,j/ + b_{n-1,m}x_2^h/i,j/ + \ldots + b_{1m}x_n^h/i,j/ +$$

$$+ b_{0m}x_n^h/i+1,j/ =$$

$$= \bar{b}_{nm}x_1^h/i,j/ + \bar{b}_{n-1,m}x_2^h/i,j/ + \ldots + \bar{b}_{1m}x_n^h/i,j/ +$$

$$+ b_{0m}x_1^{v_1}/i,j/ + b_{0m}u/i,j/$$

and

$$y/i,j/ = b_{n0}x_1^h/i,j/ + b_{n-1,0}x_2^h/i,j/ + \ldots + b_{10}x_n^h/i,j/ + b_{00}x_n^h/i+1,j/ +$$

$$+ b_{00}x_n^h/i+1,j/ + x_1^{v_1}/i,j/ =$$

$$= \bar{b}_{n0}x_1^h/i,j/ + \bar{b}_{n-1,0}x_2^h/i,j/ + \ldots + \bar{b}_{10}x_n^h/i,j/ +$$

$$+ b_{00}x_1^{v_1}/i,j/ + x_1^{v_2}/i,j/ + b_{00}u/i,j/$$

where

$$\bar{a}_{ij} = a_{ij} - a_{i0}a_{0j} \quad \text{for} \quad 1 \leqslant i \leqslant n, \; 1 \leqslant j \leqslant m$$

$$\bar{b}_{ij} = b_{ij} - a_{i0}b_{0j} \quad \text{for} \quad 1 \leqslant i \leqslant n, \; 0 \leqslant j \leqslant m$$

For
$$x^h/i,j/ = \begin{bmatrix} x_1^h/i,j/ & x_2^h/i,j/ & \cdots & x_n^h/i,j/ \end{bmatrix}^T$$

$$x^v/i,j/ = \begin{bmatrix} x_1^{v_1}/i,j/ & \cdots & x_m^{v_1}/i,j/ & x_1^{v_2}/i,j/ & \cdots & x_m^{v_2}/i,j/ \end{bmatrix}^T$$

from /2.135/ we obtain

$$A_{11} = \begin{bmatrix} 0 & 1 & 0 & \cdots & 0 \\ 0 & 0 & 1 & \cdots & 0 \\ \cdot & \cdot & \cdot & & \cdot \\ 0 & 0 & 0 & \cdots & 1 \\ -a_{n0} & -a_{n-1,0} & -a_{n-2,0} & \cdots & -a_{10} \end{bmatrix} \quad A_{12} = \begin{bmatrix} 0 & 0 & 0 & \cdots & 0 & 0 & 0 & 0 & \cdots & 0 \\ 0 & 0 & 0 & \cdots & 0 & 0 & 0 & 0 & \cdots & 0 \\ \cdot & \cdot & \cdot & & \cdot & \cdot & \cdot & \cdot & & \cdot \\ 0 & 0 & 0 & \cdots & 0 & 0 & 0 & 0 & \cdots & 0 \\ 1 & 0 & 0 & \cdots & 0 & 0 & 0 & 0 & \cdots & 0 \end{bmatrix}$$

$$A_{21} = \begin{bmatrix} -\bar{a}_{n1} & -\bar{a}_{n-1,1} & -\bar{a}_{n-2,1} & \cdots & -\bar{a}_{11} \\ -\bar{a}_{n2} & -\bar{a}_{n-1,2} & -\bar{a}_{n-2,2} & \cdots & -\bar{a}_{12} \\ \cdot & \cdot & \cdot & & \cdot \\ -\bar{a}_{nm} & -\bar{a}_{n-1,m} & -\bar{a}_{n-2,m} & \cdots & -\bar{a}_{1m} \\ \bar{b}_{n1} & \bar{b}_{n-1,m} & \bar{b}_{n-2,m} & \cdots & \bar{b}_{11} \\ \bar{b}_{n2} & \bar{b}_{n-1,2} & \bar{b}_{n-2,2} & \cdots & \bar{b}_{12} \\ \cdot & \cdot & \cdot & & \cdot \\ \bar{b}_{nm} & \bar{b}_{n-1,m} & \bar{b}_{n-2,m} & \cdots & \bar{b}_{1m} \end{bmatrix} \quad A_{22} = \begin{bmatrix} -a_{01} & 1 & 0 & \cdots & 0 & 0 & 0 & 0 & \cdots & 0 \\ -a_{02} & 0 & 1 & \cdots & 0 & 0 & 0 & 0 & \cdots & 0 \\ \cdot & \cdot & \cdot & & \cdot & \cdot & \cdot & \cdot & & \cdot \\ -a_{0m} & 0 & 0 & \cdots & 0 & 1 & 0 & 0 & \cdots & 0 \\ b_{01} & 0 & 0 & \cdots & 0 & 0 & 1 & 0 & \cdots & 0 \\ b_{02} & 0 & 0 & \cdots & 0 & 0 & 0 & 1 & \cdots & 0 \\ \cdot & \cdot & \cdot & & \cdot & \cdot & \cdot & \cdot & & \cdot \\ b_{0m} & 0 & 0 & \cdots & 0 & 0 & 0 & 0 & \cdots & 0 \end{bmatrix}$$

$$C_1 = \begin{bmatrix} \bar{b}_{n0} & \bar{b}_{n-1,0} & \bar{b}_{n-2,0} & \cdots & \bar{b}_{10} \end{bmatrix} \quad C_2 = \begin{bmatrix} b_{00} & 0 & 0 & \cdots & 0 & 0 & 1 & 0 & \cdots & 0 \end{bmatrix}$$
$$\text{position } m+1$$

$$B_1 = \begin{bmatrix} 0 \\ 0 \\ \vdots \\ 0 \\ 1 \end{bmatrix} \quad B_2 = \begin{bmatrix} -a_{01} \\ -a_{02} \\ \vdots \\ -a_{0m} \\ b_{01} \\ b_{02} \\ \vdots \\ b_{0m} \end{bmatrix} \quad D = \begin{bmatrix} b_{00} \end{bmatrix}$$

/2.136/

The submatrices /2.136/ have the desired canonical forms. /2.128/.

Example 2.8

Find a realization in the canonical form for

$$G/z_1,z_2/ = \frac{2z_1^2 z_2 + 3z_1^2 + z_1 z_2 - z_1}{z_1^2 z_2 + z_1^2 + 2z_1 + z_2} \qquad /2.137/$$

We rewrite /2.137/ in the form

$$G/z_1,z_2/ = \frac{2 + 3z_2^{-1} + (1 - z_2^{-1})z_1^{-1}}{1 + z_2^{-1} + 2z_2^{-1} z_1^{-1} + z_1^{-2}} \qquad /2.137'/$$

Hence $a_{00} = 1$, $a_{01} = 1$, $a_{10} = 0$, $a_{11} = 2$, $a_{20} = 1$, $a_{21} = 0$, $b_{00} = 2$, $b_{01} = 3$, $b_{10} = 1$, $b_{11} = -1$.

Using /2.136/ we obtain

$$A = \begin{bmatrix} A_{11} & A_{12} \\ A_{21} & A_{22} \end{bmatrix} = \left[\begin{array}{cc|cc} 0 & 1 & 0 & 0 \\ -\bar{a}_{20} & -\bar{a}_{10} & 1 & 0 \\ \hline -\bar{a}_{21} & -\bar{a}_{11} & -a_{01} & 0 \\ \bar{b}_{21} & \bar{b}_{11} & b_{01} & 0 \end{array}\right] = \left[\begin{array}{cc|cc} 0 & 1 & 0 & 0 \\ -1 & 0 & 1 & 0 \\ \hline 1 & -2 & -1 & 0 \\ -3 & -1 & 2 & 0 \end{array}\right]$$

$$C = \begin{bmatrix} C_1 & C_2 \end{bmatrix} = \begin{bmatrix} \bar{b}_{20} & \bar{b}_{10} & b_{00} & 1 \end{bmatrix} = \begin{bmatrix} -2 & 1 & 2 & 1 \end{bmatrix}$$

$$B = \begin{bmatrix} B_1 \\ B_2 \end{bmatrix} = \begin{bmatrix} 0 \\ 1 \\ \hline -a_{01} \\ b_{01} \end{bmatrix} = \begin{bmatrix} 0 \\ 1 \\ \hline -1 \\ 3 \end{bmatrix}, \quad D = \begin{bmatrix} 2 \end{bmatrix}.$$

**Method 2.**

The transfer function /2.129/ can be rewritten in the form

$$G/z_1,z_2/ = \frac{p_0/z_2/ + p_1/z_2/ z_1 + \ldots + p_n/z_2/ z_1^n}{d_0/z_2/ + d_1/z_2/ z_1 + \ldots + d_n/z_2/ z_1^n} \qquad /2.138/$$

where

$$p_i/z_2/ = \sum_{j=0}^{m} \bar{b}_{ij} z_2^j, \quad d_i/z_2/ = \sum_{j=0}^{m} \bar{a}_{ij} z_2^j \qquad /\bar{a}_{nm} \neq 0/ \qquad /2.139/$$

From causality assumption it follows that

$$\begin{aligned} \deg d_n/z_2/ &\geq \deg d_i/z_2/ \quad \text{for} \quad i = 0,1,\ldots,n-1 \\ \deg d_n/z_2/ &\geq \deg p_i/z_2/ \quad \text{for} \quad i = 0,1,\ldots,n \end{aligned} \qquad /2.140/$$

## Case I

Let $d/z_2/$ be a greatest common divisor of $p_0/z_2/$, $p_1/z_2/$, ..., $p_n/z_2/$ and $\deg d/z_2/ = l \geqslant 0$. Thus

$$p/z_1,z_2/ = p_0/z_2/ + p_1/z_2/ z_1 + \ldots + p_n/z_2/ z_1^n = d/z_2/ \bar{p}/z_1,z_2/$$

where

$$\bar{p}/z_1,z_2/ = \bar{p}_0/z_2/ + \bar{p}_1/z_2/ z_1 + \ldots + \bar{p}_n/z_2/ z_1^n$$

is called the primitive part of $p/z_1,z_2/$ and $d/z_2/$ is called the content of $p/z_1,z_2/$.

Without loss of generality we can assume $\tilde{a}_{nm} = 1$.

We factor $d_n/z_2/$ as follows

$$d_n/z_2/ = d_{n1}/z_2/ \, d_{n2}/z_2/ \qquad /2.141/$$

where $d_{n1}/z_2/$ and $d_{n2}/z_2/$ are monic polynomials satisfying

$$\deg d_{n2}/z_2/ \geqslant \deg d/z_2/, \quad \deg d_{n1}/z_2/ \geqslant \max_i \deg \bar{p}_i/z_2/ \qquad /2.141'/$$

If $d/z_2/$ is a polynomial with an odd degree and the factorization /2.141/ with real polynomial $d_{n2}/z_2/$ does not exist, then we proceed as follows. Let $\bar{d}/z_2/$ be a common factor of $p_0/z_2/$, $p_1/z_2/$, ..., $p_n/z_2/$ such that $\deg \bar{d}/z_2/ = l-1$. In this case there exists a factorization

$$\bar{d}/z_2/ = d_{n1}/z_2/ \, d_{n2}/z_2/$$

with real $d_{n1}/z_2/$, $d_{n2}/z_2/$ such that $\deg d_{n2}/z_2/ = \deg \bar{d}/z_2/$ and we can use $\bar{d}/z_2/$ instead of $d/z_2/$.

It is easy to check that the matrices

$$A/z_2/ = \begin{bmatrix} 0 & 1 & 0 & \cdots & 0 \\ 0 & 0 & 1 & \cdots & 0 \\ \cdot & \cdot & \cdot & & \cdot \\ 0 & 0 & 0 & \cdots & 1 \\ -\dfrac{d_0/z_2/}{d_n/z_2/} & -\dfrac{d_1/z_2/}{d_n/z_2/} & -\dfrac{d_2/z_2/}{d_n/z_2/} & \cdots & -\dfrac{d_{n-1}/z_2/}{d_n/z_2/} \end{bmatrix}, \quad B/z_2/ = \begin{bmatrix} 0 \\ 0 \\ \cdots \\ 0 \\ \dfrac{d/z_2/}{d_{n2}/z_2/} \end{bmatrix}$$

$$C/z_2/ = \begin{bmatrix} \dfrac{\bar{p}_0/z_2/}{d_{n1}/z_2/} & \cdots & \dfrac{\bar{p}_{n-1}/z_2/}{d_{n1}/z_2/} \end{bmatrix} + \dfrac{\bar{p}_n/z_2/}{d_{n1}/z_2/} \begin{bmatrix} -\dfrac{d_0/z_2/}{d_n/z_2/} & \cdots & -\dfrac{d_{n-1}/z_2/}{d_n/z_2/} \end{bmatrix}$$

$$D/z_2/ = \frac{p_n/z_2/}{d_n/z_2/} \qquad /2.142/$$

are a first level realization of /2.138/.

Note that the matrices

$$\bar{A}/z_2/ = \left[ -\frac{d_0/z_2/}{d_n/z_2/} \quad \cdots \quad -\frac{d_{n-1}/z_2/}{d_n/z_2/} \right], \quad \bar{B}/z_2/ = \frac{d/z_2/}{d_{n2}/z_2/}$$

$$\bar{C}/z_2/ = \left[ \frac{\bar{p}_0/z_2/}{d_{n1}/z_2/} \quad \cdots \quad \frac{\bar{p}_{n-1}/z_2/}{d_{n1}/z_2/} \right], \quad \bar{D}/z_2/ = \frac{\bar{p}_n/z_2/}{d_{n1}/z_2/} \qquad /2.143/$$

are proper transfer function matrices and using one of the well-known algorithms [8, 12, 15, 17] we can find the realizations

$$\{A_{\bar{A}}, B_{\bar{A}}, C_{\bar{A}}, D_{\bar{A}}\} \quad \text{for} \quad \bar{A}/z_2/$$

$$\{A_{\bar{B}}, B_{\bar{B}}, C_{\bar{B}}, D_{\bar{B}}\} \quad \text{for} \quad \bar{B}/z_2/$$

$$\{A_{\bar{C}}, B_{\bar{C}}, C_{\bar{C}}, D_{\bar{C}}\} \quad \text{for} \quad \bar{C}/z_2/ \qquad /2.144/$$

and

$$\{A_{\bar{D}}, B_{\bar{D}}, C_{\bar{D}}, D_{\bar{D}}\} \quad \text{for} \quad \bar{D}/z_2/$$

in a way such that

$$A_{\bar{A}} = A_{\bar{B}} = A_D, \quad A_{\bar{C}} = A_{\bar{D}} \qquad /2.145a/$$

and

$$C_{\bar{A}} = C_{\bar{B}} = C_D, \quad C_{\bar{C}} = C_{\bar{D}} \qquad /2.145b/$$

In Section 2.6.1 it was shown that

$$\begin{bmatrix} A/z_2/ & B/z_2/ \\ C/z_2/ & D/z_2/ \end{bmatrix} = \begin{bmatrix} A_{12} \\ C_2 \end{bmatrix} \left[ I_{n_2} z_2 - A_{22} \right]^{-1} \begin{bmatrix} A_{21} & B_2 \end{bmatrix} + \begin{bmatrix} A_{11} & B_1 \\ C_1 & D \end{bmatrix} \qquad /2.146/$$

From /2.146/ it follows that

$$\begin{bmatrix} A_{11} & B_1 \\ C_1 & D \end{bmatrix} = \lim_{z_2 \to \infty} \begin{bmatrix} A/z_2/ & B/z_2/ \\ C/z_2/ & D/z_2/ \end{bmatrix} \qquad /2.147/$$

and

$$A_{11} = \begin{bmatrix} 0 & 1 & 0 & \cdots & 0 \\ 0 & 0 & 1 & \cdots & 0 \\ \vdots & & & & \vdots \\ 0 & 0 & 0 & \cdots & 1 \\ -a_{01} & -a_{02} & -a_{03} & \cdots & -a_{0n} \end{bmatrix}, \quad B_1 = \begin{bmatrix} 0 \\ 0 \\ \vdots \\ 0 \\ D_{\bar{B}} \end{bmatrix} \qquad /2.148/$$

$$C = \begin{bmatrix} C_1 & C_2 & \cdots & C_n \end{bmatrix}, \quad D = D_D$$

where

$$\begin{bmatrix} -a_{01} & -a_{02} & \cdots & -a_{0n} \end{bmatrix} = D_{\bar{A}} = \lim_{z_2 \to \infty} \bar{A}/z_2/, \quad D_{\bar{B}} = \lim_{z_2 \to \infty} \bar{B}/z_2/$$

$$\begin{bmatrix} c_1 & c_2 & \cdots & c_n \end{bmatrix} = \lim_{z_2 \to \infty} C/z_2/ = D_{\bar{C}} + D_{\bar{D}} D_{\bar{A}}, \quad D_D = \lim_{z_2 \to \infty} D/z_2/$$

/2.149/

It can be easily shown by substituting into /2.146/ that

$$A_{12} = \begin{bmatrix} 0 & | & 0 \\ \hline C_{\bar{A}} & | & \end{bmatrix} \in R^{n \times (2m-1)}, \quad A_{21} = \begin{bmatrix} B_{\bar{A}} \\ B_{\bar{D}} D_{\bar{A}} + B_{\bar{C}} \end{bmatrix} \in R^{(2m-1) \times n}$$

$$A_{22} = \begin{bmatrix} A_{\bar{A}} & | & 0 \\ \hline B_{\bar{D}} C_{\bar{A}} & | & A_{\bar{C}} \end{bmatrix} \in R^{(2m-1) \times (2m-1)}, \quad B_2 = \begin{bmatrix} B_{\bar{B}} \\ B_{\bar{D}} D_{\bar{B}} \end{bmatrix} \in R^{(2m-1) \times 1}$$

$$C_2 = \begin{bmatrix} D_{\bar{D}} C_{\bar{A}} & | & C_{\bar{C}} \end{bmatrix} \in R^{1 \times (2m-1)} \qquad /2.150/$$

Therefore we have shown that if $d/z_2/$ is the content of $p/z_1,z_2/$ and deg $d/z_2/ = 1$ then there exists a realization of the form /2.147/ and /2.150/ with dimension $n+2m-1$. This realization is possibly in complex numbers /depending upon the factorization /2.141// but there exists always a real realization with dimension $n+2m-1+1$ /$1 \geqslant 1$/.

Note that by interchanging $z_1$ and $z_2$ the same result can be obtained for /2.138/. Thus, using this method it is possible to find a realization with dimension $\min(n+2m-1, m+2n-1)$ or $\min(n+2m-1+1, m+2n-1+1)$.

Case II

Assume that $d/z_2/$ is the content of $d/z_1,z_2/$ and deg $\bar{d}/z_2/ = 1$.
Let $d_i/z_2/ = \bar{d}/z_2/\bar{d}_i/z_2/$ for $i = 0,1,\ldots,n$.
It is easy to verify that the matrices

$$A/z_2/ = \begin{bmatrix} 0 & 1 & 0 & \cdots & 0 \\ 0 & 0 & 1 & \cdots & 0 \\ \vdots & & & & \vdots \\ 0 & 0 & 0 & \cdots & 1 \\ -\dfrac{\bar{d}_0/z_2/}{\bar{d}_n/z_2/} & -\dfrac{\bar{d}_1/z_2/}{\bar{d}_n/z_2/} & -\dfrac{\bar{d}_2/z_2/}{\bar{d}_n/z_2/} & \cdots & -\dfrac{\bar{d}_{n-1}/z_2/}{\bar{d}_n/z_2/} \end{bmatrix}, \quad B/z_2/ = \begin{bmatrix} 0 \\ 0 \\ \vdots \\ 0 \\ 1 \end{bmatrix}$$

$$C/z_2/ = \begin{bmatrix} \dfrac{p_0/z_2/}{d_n/z_2/} & \cdots & \dfrac{p_{n-1}/z_2/}{d_n/z_2/} \end{bmatrix} + \dfrac{p_n/z_2/}{d_n/z_2/} \begin{bmatrix} -\dfrac{\bar{d}_0/z_2/}{\bar{d}_n/z_2/} & \cdots & -\dfrac{\bar{d}_{n-1}/z_2/}{\bar{d}_n/z_2/} \end{bmatrix}$$

$$D/z_2/ = \dfrac{p_n/z_2/}{d_n/z_2/} \qquad\qquad /2.151/$$

are a first level realization of /2.138/.

Note that the matrices

$$\hat{A}/z_2/ = \begin{bmatrix} -\dfrac{\bar{d}_0/z_2/}{\bar{d}_n/z_2/} & \cdots & -\dfrac{\bar{d}_{n-1}/z_2/}{\bar{d}_n/z_2/} \end{bmatrix}, \quad \hat{B}/z_2/ = 1$$

$$\hat{C}/z_2/ = \begin{bmatrix} \dfrac{p_0/z_2/}{d_n/z_2/} & \cdots & \dfrac{p_{n-1}/z_2/}{d_n/z_2/} \end{bmatrix}, \quad \hat{D}/z_2/ = \dfrac{p_n/z_2/}{d_n/z_2/} \qquad /2.152/$$

are proper transfer function matrices.

Further we proceed in completely the same way as in Case I and we obtain an analogous result. Note that the realization we obtain in this way is always real. Summarizing we have:

If $\bar{d}/z_2/$ is the content of $d/z_1,z_2/$ and $\deg \bar{d}/z_2/ = \bar{1}$, then there exists a real realization with dimension $\min(n+2m-1, m+2n-1)$.

### Example 2.9

Using the Method 2 find a realization for

$$G/z_1,z_2/ = \dfrac{1 + z_2 + 2z_1 z_2 + 2z_1 z_2^2}{1 + z_1 z_2 + z_1^2 z_2^2} \qquad /2.153/$$

Writing /2.153/ in the form /2.138/ we obtain

$$G/z_1,z_2/ = \dfrac{p_0/z_2/ + p_1/z_2/ z_1}{d_0/z_2/ + d_1/z_2/ z_1 + d_2/z_2/ z_1^2} \qquad /2.154/$$

where

$$p_0/z_2/ = 1 + z_2, \quad p_1/z_2/ = 2z_2(1+z_2)$$

$$d_0/z_2/ = 1, \quad d_1/z_2/ = z_2, \quad d_2/z_2/ = z_2^2 \qquad /2.155/$$

From /2.155/ it follows that the content of $p/z_1,z_2/$ is $d/z_2/ = 1 + z_2$ and $\bar{p}_0/z_2/ = 1, \bar{p}_1/z_2/ = 2z_2$. We factor $d_2/z_2/$ as

$$d_2/z_2/ = d_{21}/z_2/ d_{22}/z_2/ = z_2 z_2$$

Thus $d_{21}/z_2/ = d_{22}/z_2/ = z_2$.
Hence from /2.143/ we obtain

$$\bar{A}/z_2/ = \left[ -\frac{d_0/z_2/}{d_2/z_2/} \quad -\frac{d_1/z_2/}{d_2/z_2/} \right] = \left[ -\frac{1}{z_2^2} \quad -\frac{1}{z_2} \right]$$

$$\bar{B}/z_2/ = \frac{d/z_2/}{d_{22}/z_2/} = \frac{1}{z_2} + 1$$

$$\bar{C}/z_2/ = \left[ \frac{\bar{p}_0/z_2/}{d_{21}/z_2/} \quad \frac{\bar{p}_1/z_2/}{d_{21}/z_2/} \right] = \left[ \frac{1}{z_2} \quad 2 \right] \qquad /2.156/$$

$$\bar{D}/z_2/ = \frac{\bar{p}_2/z_2/}{d_{22}/z_2/} = 0$$

Realizations of /2.156/ satisfying /2.145/ are as follows

$$A_{\bar{A}} = \begin{bmatrix} 0 & 1 \\ 0 & 0 \end{bmatrix}, \quad B_{\bar{A}} = \begin{bmatrix} 0 & -1 \\ -1 & 0 \end{bmatrix}, \quad C_{\bar{A}} = \begin{bmatrix} 1 & 0 \end{bmatrix}, \quad D_{\bar{A}} = 0$$

$$A_{\bar{B}} = \begin{bmatrix} 0 & 1 \\ 0 & 0 \end{bmatrix}, \quad B_{\bar{B}} = \begin{bmatrix} 1 \\ 0 \end{bmatrix}, \quad C_{\bar{B}} = \begin{bmatrix} 1 & 0 \end{bmatrix}, \quad D_{\bar{B}} = \begin{bmatrix} 1 \end{bmatrix} \qquad /2.157/$$

$$A_{\bar{C}} = \begin{bmatrix} 0 \end{bmatrix}, \quad B_{\bar{C}} = \begin{bmatrix} 1 & 0 \end{bmatrix}, \quad C_{\bar{C}} = \begin{bmatrix} 1 \end{bmatrix}, \quad D_{\bar{C}} = \begin{bmatrix} 0 & 2 \end{bmatrix}$$

$$A_{\bar{D}} = \begin{bmatrix} 0 \end{bmatrix}, \quad B_{\bar{D}} = \begin{bmatrix} 0 \end{bmatrix}, \quad C_{\bar{D}} = \begin{bmatrix} 1 \end{bmatrix}, \quad D_{\bar{D}} = \begin{bmatrix} 0 \end{bmatrix}$$

Using /2.148/ and /2.150/ we obtain the desired realization

$$A_{11} = \begin{bmatrix} 0 & 1 \\ 0 & 0 \end{bmatrix}, \quad A_{12} = \begin{bmatrix} 0 & 0 & 0 \\ 1 & 0 & 0 \end{bmatrix}, \quad B_1 = \begin{bmatrix} 0 \\ 1 \end{bmatrix}$$

$$A_{21} = \begin{bmatrix} 0 & -1 \\ -1 & 0 \\ 1 & 0 \end{bmatrix}, \quad A_{22} = \begin{bmatrix} 0 & 1 & 0 \\ 0 & 0 & 0 \\ 0 & 0 & 0 \end{bmatrix}, \quad B_2 = \begin{bmatrix} 1 \\ 0 \\ 0 \end{bmatrix} \qquad /2.158/$$

$$C_1 = \begin{bmatrix} 0 & 2 \end{bmatrix}, \quad C_2 = \begin{bmatrix} 0 & 0 & 1 \end{bmatrix}, \quad D = \begin{bmatrix} 0 \end{bmatrix}$$

PROBLEMS.

1. Find the transfer function matrix $G(z_1,z_2)$ of Attasi's model.

   Hint: Apply the 2-D Z transform to /1.16/ and /1.17/.

   Answer: $G(z_1,z_2) = \bar{C}\left[I_n z_1 z_2 + \bar{A}_1\bar{A}_2 - \bar{A}_1 z_1 - \bar{A}_2 z_2\right]^{-1}\bar{B}$

2. Show that

$$\frac{\partial p(z_1,z_2)}{\partial z_1} = \operatorname{tr} B^{1,0}(z_1,z_2)$$

   and

$$\frac{\partial p(z_1,z_2)}{\partial z_2} = \operatorname{tr} B^{0,1}(z_1,z_2)$$

   where $p(z_1,z_2) = \det[Z - A]$, $B(z_1,z_2) = \operatorname{adj}[Z - A]$
   and $Z$ is defined by /2.6/.

   Hint: Use /2.40/.

3. Show that if $\{A_0, A_1, A_2, B, C\}$ is a realization of dimension $n$ of F-MM I and $T \in R^{n \times n}$ is a non-singular matrix, then $\{TA_0T^{-1}, TA_1T^{-1}, TA_2T^{-1}, TB, CT^{-1}\}$ is also a realization of F-MM I.

   Hint: Show that $CT^{-1}\left[I_n - TA_0T^{-1}z_1z_2 - TA_1T^{-1}z_1 - TA_2T^{-1}z_2\right]^{-1}TB =$
   $= C\left[I_n - A_0 z_1 z_2 - A_1 z_1 - A_2 z_2\right]^{-1}B$.

4. Show that

$$\left[I_n - A_1 z_1 - A_2 z_2\right]^{-1} = \sum_{i,j \geq 0} A^{i,j} z_1^i z_2^j$$

   where

   $A^{0,0} = I_n$, $A^{1,0} = A_1$, $A^{0,1} = A_2$, $A^{i,j} = A_1 A^{i-1,j} + A_2 A^{i,j-1}$;
   $A^{i,j} = 0$ for $i < 0$ or $j < 0$. /2.159/

   Hint: Use the formula

$$\left[I_n - A_1 z_1 - A_2 z_2\right]^{-1} = \sum_{i \geq 0}\left[A_1 z_1 + A_2 z_2\right]^i$$

5. Show that $A^{i,j}$ for $i+j \geq n$ are linear combinations of $A^{i,j}$ for $i+j < n$, where $A^{i,j}$ is defined by /2.159/.

   Hint: Use the identity

$$\sum_{i,j>0} A^{i,j} z_1^i z_2^j \det\left[I_n z_1^{-1} z_2^{-1} - A_1 z_1^{-1} - A_2 z_2^{-1}\right] =$$

$$= z_1^{-1} z_2^{-1} \operatorname{adj}\left[I_n z_1^{-1} z_2^{-1} - A_1 z_2^{-1} - A_2 z_1^{-1}\right].$$

6. Find a realization $\{A_0, A_1, A_2, B, C\}$ for F-MM I with

$$G/z_1, z_2/ = \frac{z_1^{-1} + z_2^{-1} + 2}{z_1^{-2} z_2^{-2} + 2 z_1^{-2} z_2^{-1} + z_1^{-1} z_2^{-1} + 1}$$

Hint: Use the procedure given in [5].

Answer:

$$A_0 = \begin{bmatrix} 0 & 0 & 0 & 1 \\ 0 & 0 & 0 & 0 \\ 0 & 0 & 0 & 0 \\ -1 & 0 & 0 & -1 \end{bmatrix}, \quad A_1 = \begin{bmatrix} 0 & 0 & 0 & 0 \\ 0 & 0 & 0 & 0 \\ 0 & 0 & 0 & 1 \\ 0 & 0 & 0 & 0 \end{bmatrix}, \quad A_2 = \begin{bmatrix} 0 & 0 & 0 & 0 \\ 0 & 0 & 0 & 1 \\ 0 & 0 & 0 & 0 \\ 0 & 0 & 0 & -2 \end{bmatrix},$$

$$B = \begin{bmatrix} 0 \\ 0 \\ 0 \\ 1 \end{bmatrix}, \quad C = \begin{bmatrix} 2 & 1 & 1 & 0 \end{bmatrix}.$$

# REFERENCES.

[1] Chiasson J.N., Brierley S.D., Żak S.H., Lee E.B.: On 2-D proper transfer functions. Int.J.Contr. vol.35, no.1, 1982, pp.159-174.

[2] Eising R.: 2-D Systems; An Algebraic Approach. Mathematisch Centrum, Amsterdam, 1979.

[3] Eising R.: Realization and stabilization of 2-D systems. IEEE Trans. Autom. Contr. vol. AC-23, no.5, Oct. 1978, pp.793-799.

[4] Fornasini E., Marchesini G.: Doubly-Indexed Dynamical Systems; State-Space Models and Structural Properties. Math. Systems Theory vol.12, 1978, pp.59-72.

[5] Fornasini E., Marchesini G.: State-Space Realization Theory of Two-Dimensional Filters. IEEE Trans. Autom. Control vol. AC-21, no.4, Aug. 1976, pp.484-491.

[6] Hinamoto t., Fairman F.W.,: Realization of the Attasi state space model for 2-D filters. Int.J.Systems Sci. vol.15, no.2, 1984, pp.215-228.

[7] Kaczorek T.: Theory of Multivariable Dynamical Systems. Wydawnictwa Naukowo-Techniczne, Warszawa 1983 /in Polish/.

[8] Kailath T.: Linear Systems. Prentice-Hall, Englewood Cliffs 1980.

[9] Koo C.S., Chen C.T.: Fadeeva's algorithm for spatial dynamical equations. Proc. IEEE vol.65, June 1977, pp.975-976.

[10] Mertzios B.G., Paraskevopoulos P.N.: Transfer function matrix of 2-D systems. IEEE Trans. Autom. Contr. vol. AC-26, no.3, June 1981, pp. 722-724.

[11] Paraskevopoulos P.N., Mertzios B.G.: Transfer function factorization of SISO 2-D systems using state feedback. Int.J.Systems Sci. vol.12, no.9, 1981, pp.1135-1147.

[12] Rosenbrock H.H.: State Space and Multivariable Theory. John Wiley, New York 1970.

[13] Sontag E.D.: On first-order equations for multidimensional filters. IEEE Transactions on Acoustics, Speech and Signal Processing, vol. ASSP-26, no.5, Oct. 1978, pp.480-482.

[14] Tzafestas S.G., Pimenides T.G.: Transfer function computation and factorization of 3-dimensional systems in state-space. IEE Proc. vol.130, Pt.D, no.5, Sept. 1983, pp.231-242.

[15] Wolovich W.A.: Linear Multivariable Systems. Springer-Verlag, New York 1974.

[16] Zadeh L.A., Desoer C.A.: Linear Systems Theory. McGraw-Hill, New York 1963.

# 3 CONTROLLABILITY AND OBSERVABILITY

## 3.1 LOCAL CONTROLLABILITY AND OBSERVABILITY OF ROESSER'S MODEL.

### 3.1.1 Local controllability.

Let us consider the Roesser's model described by the equations

$$x' = Ax + Bu \qquad /3.1/$$
$$y = Cx + Du \qquad /3.2/$$

where

$$x' = \begin{bmatrix} x^h/i+1,j/ \\ x^v/i,j+1/ \end{bmatrix}, \quad x = \begin{bmatrix} x^h/i,j/ \\ x^v/i,j/ \end{bmatrix}$$

$$A = \begin{bmatrix} A_{11} & A_{12} \\ A_{21} & A_{22} \end{bmatrix}, \quad B = \begin{bmatrix} B_1 \\ B_2 \end{bmatrix}, \quad C = \begin{bmatrix} C_1 & C_2 \end{bmatrix}$$

$x^h/i,j/ \in R^{n_1}$ is the horizontal state vector,

$x^v/i,j/ \in R^{n_2}$ is the vertical state vector,

$u = u/i,j/ \in R^m$ is the input vector,

$y = y/i,j/ \in R^l$ is the output vector,

$i \in Z$ is the horizontal coordinate, $j \in Z$ is the vertical coordinate, Z is the set of nonnegative integers,
$A_{11}$, $A_{12}$, $A_{21}$, $A_{22}$, $B_1$, $B_2$, $C_1$, $C_2$ and D are real matrices of appropriate dimensions.

The boundary conditions for RM are given by

$$x^h/0,j/ \quad \text{and} \quad x^v/i,0/ \quad \text{for} \quad i,j \geqslant 0 \qquad /3.1a/$$

For $/h,k/ < /r,p/$ we define the rectangle $[/h,k/, /r,p/]$ as follows

$$[/h,k/, /r,p/] = \left\{ /i,j/ \in Z \times Z, \ /h,k/ \leqslant /i,j/ \leqslant /r,p/ \right\} \qquad /3.3/$$

In a particular case we assume $[h,r] = \{h, h+1, \ldots, r\}$.

## Definition 3.1

RM is said to be locally controllable in the rectangle $[/0,0/, /h,k/]$ iff for every boundary conditions $x^h/0,j/$, $j \in [0,k]$, $x^v/i,0/$, $i \in [0,h]$ and every vector $s \in R^n$ $/n = n_1 + n_2/$ there exists a sequence of inputs $u/i,j/$ $/0,0/ \leqslant /i,j/ < /h,k/$ such that $x/h,k/ = s$.

## Theorem 3.1

RM is locally controllable in the rectangle $[/0,0/, /h,k/]$ if and only if

$$\text{rank } C/h,k/ = n \qquad /3.4/$$

where

$$C/h,k/ = [M/0,1/ \quad M/1,0/ \quad \ldots \quad M/i,j/ \quad \ldots \quad M/h,k/] \qquad /3.5a/$$

and

$$M/i,j/ = A^{i-1,j} B^{1,0} + A^{i,j-1} B^{0,1} \qquad /3.5b/$$

**Proof:** Using the formula $/1.28/$ and taking into account that the desired final state is $x/h,k/ = s$ we can write

$$s - \sum_{i=0}^{h} A^{h-i,k} \begin{bmatrix} 0 \\ x^v/i,0/ \end{bmatrix} - \sum_{j=0}^{k} A^{h,k-j} \begin{bmatrix} x^h/0,j/ \\ 0 \end{bmatrix} =$$

$$= \sum_{/0,0/ \leqslant /i,j/ < /h,k/} M/h-i,k-j/ \; u/i,j/ =$$

$$= [M/0,1/ \quad M/1,0/ \quad \ldots \quad M/i,j/ \quad \ldots \quad M/h,k/] \begin{bmatrix} u/h,k-1/ \\ u/h-1,k/ \\ \ldots \\ u/h-i,k-j/ \\ \ldots \\ u/0,0/ \end{bmatrix} \qquad /3.6/$$

From $/3.6/$ it follows that RM is locally controllable in the rectangle $[/0,0/, /h,k/]$ if and only if the condition $/3.4/$ is satisfied. □

## Theorem 3.2

RM is locally controllable in the rectangle $[/0,0/, /h,k/]$ if and only if the matrix

$$W/h,k/ = C/h,k/ \; C^T/h,k/ \qquad /3.7/$$

is positive definite.

**Proof:** Let us define a vector $v = C^T/h,k/\, r$, $r \in R^n$ and

$$\|v\|^2 = v^T v = r^T C/h,k/\, C^T/h,k/\, r = r^T W/h,k/\, r \qquad /3.8/$$

Note that $W/h,k/$ is positive definite if and only if condition /3.4/ holds. Therefore, by Theorem 3.1, RM is locally controllable in the rectangle $[/0,0/,\ /h,k/]$ if and only if $W/h,k/$ is positive definite. □

### Example 3.1

Check the local controllability in the rectangle $[/0,0/,\ /1,1/]$ of RM with

$$A = \begin{bmatrix} A_{11} & A_{12} \\ A_{21} & A_{22} \end{bmatrix} = \begin{bmatrix} 1 & 0 & | & 0 \\ 0 & 1 & | & 1 \\ \hline 1 & 0 & | & 0 \end{bmatrix}, \quad B = \begin{bmatrix} B_1 \\ B_2 \end{bmatrix} = \begin{bmatrix} 1 \\ 0 \\ \hline 1 \end{bmatrix} \qquad /3.9/$$

Using /3.5b/ we calculate

$$M/0,1/ = A^{-1,1}\, B^{1,0} + A^{0,0}\, B^{0,1} = B^{0,1} = \begin{bmatrix} 0 \\ 0 \\ 1 \end{bmatrix}$$

$$M/1,0/ = A^{0,0}\, B^{1,0} + A^{1,-1}\, B^{0,1} = B^{1,0} = \begin{bmatrix} 1 \\ 0 \\ 0 \end{bmatrix}$$

and

$$M/1,1/ = A^{0,1}\, B^{1,0} + A^{1,0}\, B^{0,1} = \begin{bmatrix} 0 & 0 & 0 \\ 0 & 0 & 0 \\ 1 & 0 & 0 \end{bmatrix}\begin{bmatrix} 1 \\ 0 \\ 0 \end{bmatrix} + \begin{bmatrix} 1 & 0 & 0 \\ 0 & 1 & 1 \\ 0 & 0 & 0 \end{bmatrix}\begin{bmatrix} 0 \\ 0 \\ 1 \end{bmatrix} = \begin{bmatrix} 0 \\ 1 \\ 1 \end{bmatrix}$$

Hence

$$C/1,1/ = \begin{bmatrix} M/0,1/ & M/1,0/ & M/1,1/ \end{bmatrix} = \begin{bmatrix} 0 & 1 & 0 \\ 0 & 0 & 1 \\ 1 & 0 & 1 \end{bmatrix}$$

and $\det C/1,1/ = 1$.

Therefore, condition /3.4/ is satisfied and RM with /3.9/ is locally controllable in the rectangle $[/0,0/,\ /1,1/]$. It is easy to check that the matrix

$$W/1,1/ = C/1,1/\, C^T/1,1/ = \begin{bmatrix} 1 & 0 & 0 \\ 0 & 1 & 1 \\ 0 & 1 & 2 \end{bmatrix}$$

is positive definite.

### Remark 3.1

From Theorem 1.3 it follows that RM is locally controllable in the rectangle $[/0,0/, /h,k/]$, $/h,k/ > /n_1,n_2/$, only if it is locally controllable in the rectangle $[/0,0/, /n_1,n_2/]$.

### 3.1.2 Local observability.

### Definition 3.2

RM is said to be locally observable in the rectangle $[/0,0/, /h,k/]$ iff there is no local initial state $x/0,0/ \neq 0$ such that for zero inputs $u/i,j/ = 0$, $/0,0/ \leqslant /i,j/ \leqslant /h,k/$, and zero boundary conditions $x^h/0,j/ = 0$, $j \in [1,k]$, $x^v/i,0/ = 0$, $i \in [1,h]$, the output is also zero $y/i,j/ = 0$, $/0,0/ \leqslant /i,j/ \leqslant /h,k/$.

### Theorem 3.3

RM is locally observable in the rectangle $[/0,0/, /h,k/]$ if and only if

$$\text{rank } O/h,k/ = n \qquad /3.10/$$

where

$$O/h,k/ = \begin{bmatrix} C \\ CA^{1,0} \\ CA^{0,1} \\ \ldots \\ CA^{i,j} \\ \ldots \\ CA^{h,k} \end{bmatrix} \qquad /3.11/$$

**Proof:** From $/1.31/$ for $u/i,j/ = 0$, $/0,0/ \leqslant /i,j/ \leqslant /h,k/$ and $x^h/0,j/ = 0$, $j \in [1,k]$, $x^v/i,0/ = 0$, $i \in [1,h]$ we obtain

$$y/i,j/ = CA^{i,j} x/0,0/ \quad \text{for} \quad i,j \geqslant 0 \qquad /3.12/$$

Taking into account that $y/i,j/ = 0$, $/0,0/ \leqslant /i,j/ \leqslant /h,k/$ we can write

$$O/h,k/ \, x/0,0/ = 0 \qquad /3.13/$$

Note that if $/3.10/$ holds then $/3.13/$ implies $x/0,0/ = 0$, and if rank $O/h,k/ < n$

then there exists $x/0,0/ \neq 0$ such that /3.13/ holds.
$\square$

**Theorem 3.4**

RM is locally observable in the rectangle $[/0,0/, /h,k/]$ if and only if the matrix

$$V/h,k/ = 0^T/h,k/\, 0/h,k/ \qquad /3.14/$$

is positive definite.

The proof is similar to one of Theorem 3.2.

**Example 3.2**

Check the local observability in the rectangle $[/0,0/, /1,1/]$ of RM with

$$A = \begin{bmatrix} A_{11} & A_{12} \\ A_{21} & A_{22} \end{bmatrix} = \left[\begin{array}{cc|c} 1 & 0 & 0 \\ 0 & 1 & 1 \\ \hline 1 & 0 & 0 \end{array}\right]$$

$$C = \begin{bmatrix} C_1 & C_2 \end{bmatrix} = \left[\begin{array}{cc|c} 0 & -1 & 1 \end{array}\right] \qquad /3.15/$$

Using /1.25/ we calculate

$$CA^{1,0} = \begin{bmatrix} 0 & -1 & 1 \end{bmatrix} \begin{bmatrix} 1 & 0 & 0 \\ 0 & 1 & 1 \\ 0 & 0 & 0 \end{bmatrix} = \begin{bmatrix} 0 & -1 & -1 \end{bmatrix}$$

$$CA^{0,1} = \begin{bmatrix} 0 & -1 & 1 \end{bmatrix} \begin{bmatrix} 0 & 0 & 0 \\ 0 & 0 & 0 \\ 1 & 0 & 0 \end{bmatrix} = \begin{bmatrix} 1 & 0 & 0 \end{bmatrix}$$

and

$$CA^{1,1} = C\begin{bmatrix} A^{1,0}A^{0,1} + A^{0,1}A^{1,0} \end{bmatrix} = \begin{bmatrix} 0 & -1 & 1 \end{bmatrix}\begin{bmatrix} 1 & 0 & 0 \\ 0 & 1 & 1 \\ 0 & 0 & 0 \end{bmatrix}\begin{bmatrix} 0 & 0 & 0 \\ 0 & 0 & 0 \\ 1 & 0 & 0 \end{bmatrix} +$$

$$+ \begin{bmatrix} 0 & 0 & 0 \\ 0 & 0 & 0 \\ 1 & 0 & 0 \end{bmatrix}\begin{bmatrix} 1 & 0 & 0 \\ 0 & 1 & 1 \\ 0 & 0 & 0 \end{bmatrix} = \begin{bmatrix} 0 & 0 & 0 \end{bmatrix}$$

Hence

$$0/1,1/ = \begin{bmatrix} C \\ CA^{1,0} \\ CA^{0,1} \\ CA^{1,1} \end{bmatrix} = \begin{bmatrix} 0 & -1 & 1 \\ 0 & -1 & -1 \\ 1 & 0 & 0 \\ 0 & 0 & 0 \end{bmatrix}$$

and rank $0/1,1/ = 3$. Therefore, the condition /3.10/ is satisfied and

RM with /3.15/ is locally observable in the rectangle $[/0,0/, /1,1/]$. It is easy to check that the matrix

$$V/1,1/ = O^T/1,1/ \, O/1,1/ = \begin{bmatrix} 1 & 0 & 0 \\ 0 & 2 & 0 \\ 0 & 0 & 2 \end{bmatrix}$$

is positive definite.

### Remark 3.2

From Theorem 1.3 it follows that RM is locally observable in the rectangle $[/0,0/, /h,k/]$, $/h,k/ > /n_1,n_2/$, only if it is locally observable in the rectangle $[/0,0/, /n_1,n_2/]$.

## 3.2 SEPARATE LOCAL CONTROLLABILITY AND OBSERVABILITY OF ROESSER'S MODEL

### Definition 3.3

RM is said to be separately locally controllable in the rectangle $[/0,0/, /h,k/]$ iff for every boundary conditions $x^h/0,j/$, $j \in [0,k]$; $x^v/i,0/$, $i \in [0,h]$ and arbitrary vectors $s_1 \in R^{n_1}$, $s_2 \in R^{n_2}$ there exists a sequence of inputs $u/i,j/$, $/0,0/ \leqslant /i,j/ < /h_1,k_1/$ such that $x^h/h_1,k_1/ = s_1$ and there exists a sequence of inputs $u/i,j/$, $/0,0/ \leqslant /i,j/ < /h_2,k_2/$ such that $x^v/h_2,k_2/ = s_2$, where $h = \max(h_1,h_2)$, $k = \max(k_1,k_2)$.

### Theorem 3.5

RM with $A_{12} = 0$ is separately locally controllable in the rectangle $[/0,0/, /h,k/]$ if and only if

$$\mathrm{rank} \begin{bmatrix} B_1 & A_{11}B_1 & \cdots & A_{11}^{h_1-1} B_1 \end{bmatrix} = n_1 \qquad /3.16/$$

and

$$\mathrm{rank} \begin{bmatrix} \bar{B}_{h_2} & A_{22}\bar{B}_{h_2} & \cdots & A_{22}^{k_2-1} \bar{B}_{h_2} \end{bmatrix} = n_2 \qquad /3.17/$$

where

$$\bar{B}_{h_2} = \begin{bmatrix} B_2 & A_{21}B_1 & A_{21}A_{11}B_1 & \cdots & A_{21}A_{11}^{h_2-1} B_1 \end{bmatrix} \qquad /3.18/$$

Proof: From /3.1/ for $A_{12} = 0$ we have

$$x^h/i+1,j/ = A_{11}x^h/i,j/ + B_1u/i,j/ \qquad /3.19/$$

$$x^v/i,j+1/ = A_{21}x^h/i,j/ + A_{22}x^v/i,j/ + B_2u/i,j/ \qquad /3.20/$$

The solution to /3.19/ has the following form

$$x^h/i,j/ = A_{11}^i x^h/0,j/ + \sum_{t=0}^{i-1} A_{11}^{i-t-1} B_1 u/t,j/ \qquad /3.21/$$

Substitution of /3.21/ into /3.20/ yields

$$x^v/i,j+1/ = A_{21}A_{11}^i x^h/0,j/ + \sum_{t=0}^{i-1} A_{21}A_{11}^{i-t-1} B_1 u/t,j/ + A_{22}x^v/i,j/ +$$

$$+ B_2 u/i,j/ =$$

$$= A_{22}x^v/i,j/ + \bar{B}_i \bar{u}/i,j/ + A_{21}A_{11}^i x^h/0,j/ \qquad /3.22/$$

where

$$\bar{u}/i,j/ = \begin{bmatrix} u/i,j/ \\ u/i-1,j/ \\ \cdots \\ u/0,j/ \end{bmatrix}$$

The solution to /3.22/ has the form

$$x^v/i,j/ = A_{22}^j x^v/i,0/ + \sum_{s=0}^{j-1} A_{22}^{j-s-1} \bar{B}_i \bar{u}/i,s/ +$$

$$+ \sum_{s=0}^{j-1} A_{22}^{j-s-1} A_{21} A_{11}^i x^h/0,s/ \qquad /3.23/$$

Using /3.21/, /2.23/ and taking into account that $x^h/h_1,k_1/ = s_1$, $x^v/h_2,k_2/ = s_2$ we can write

$$s_1 - A_{11}^{h_1} x^h/0,k_1/ = \sum_{t=0}^{h_1-1} A_{11}^{h_1-t-1} B_1 u/t,k_1/ =$$

$$= \begin{bmatrix} B_1 & A_{11}B_1 & \cdots & A_{11}^{h_1-1} B_1 \end{bmatrix} \begin{bmatrix} u/h_1-1,k_1/ \\ u/h_1-2,k_1/ \\ \cdots \\ u/0,k_1/ \end{bmatrix} \qquad /3.24/$$

and

$$s_2 - A_{22}^{k_2} x^v/h_2,0/ - \sum_{s=0}^{k_2-1} A_{22}^{k_2-s-1} A_{21} A_{11}^{h_2} x^h/0,s/ =$$

$$= \sum_{s=0}^{k_2-1} A_{22}^{k_2-s-1} \bar{B}_{h_2} \bar{u}/h_2,s/ =$$

$$= \begin{bmatrix} \bar{B}_{h_2} & A_{22}\bar{B}_{h_2} & \cdots & A_{22}^{k_2-1}\bar{B}_{h_2} \end{bmatrix} \begin{bmatrix} \bar{u}/h_2,k_2-1/ \\ \bar{u}/h_2,k_2-2/ \\ \cdots \\ \bar{u}/h_2,0/ \end{bmatrix} \qquad /3.25/$$

From /3.24/ and /3.25/ it follows that RM with $A_{12} = 0$ is separately locally controllable in the rectangle $[/0,0/, /h,k/]$ if and only if conditions /3.16/ and /3.17/ hold. □

### Definition 3.4

RM is said to be separately locally observable in the rectangle $[/0,0/, /h,k/]$ iff there is no local initial horizontal state $x^h/0,0/ \neq 0$ such that for zero inputs $u/i,j/ = 0$, $/0,0/ \leqslant /i,j/ \leqslant /h,k/$ and zero boundary conditions $x^h/0,j/ = 0$, $j \in [1,k]$; $x^v/i,0/ = 0$, $i \in [0,h]$ the output is also zero $y/i,j/ = 0$, $/0,0/ \leqslant /i,j/ \leqslant /h,k/$ and if there is no local initial vertical state $x^v/0,0/ \neq 0$ such that for zero inputs $u/i,j/ = 0$, $/0,0/ \leqslant /i,j/ \leqslant /h,k/$ and zero boundary conditions $x^h/0,j/ = 0$, $j \in [0,k]$; $x^v/i,0/ = 0$, $i \in [1,h]$ the output is also zero $y/i,j/ = 0$, $/0,0/ \leqslant /i,j/ \leqslant /h,k/$.

### Theorem 3.6

RM with $A_{12} = 0$ is separately locally observable in the rectangle $[/0,0/, /h,k/]$ if and only if

$$\text{rank} \begin{bmatrix} C_2 \\ C_2 A_{22} \\ \cdots \\ C_2 A_{22}^k \end{bmatrix} = n_2 \qquad /3.26/$$

and

$$\text{rank} \begin{bmatrix} \bar{C} \\ \bar{C} A_{11} \\ \cdots \\ \bar{C} A_{11}^h \end{bmatrix} = n_1 \qquad /3.27/$$

where

$$\bar{C} = \begin{bmatrix} C_1 \\ C_2 A_{21} \\ C_2 A_{22} A_{21} \\ \cdots \\ C_2 A_{22}^{k-1} A_{21} \end{bmatrix} \qquad /3.28/$$

Proof: From /3.1/ for $u/i,j/ = 0$ and $A_{12} = 0$ we have

$$x^h/i+1,j/ = A_{11} x^h/i,j/ \qquad /3.29/$$

$$x^v/i,j+1/ = A_{21} x^h/i,j/ + A_{22} x^v/i,j/ \qquad /3.30/$$

Substituting the solution of /3.29/ in the form

$$x^h/i,j/ = A_{11}^i x^h/0,j/ \qquad /3.31/$$

into /3.30/ we obtain

$$x^v/i,j+1/ = A_{21} A_{11}^i x^h/0,j/ + A_{22} x^v/i,j/ \qquad /3.32/$$

The solution to /3.32/ has the form

$$x^v/i,j/ = A_{22}^j x^v/i,0/ + \sum_{l=0}^{j-1} A_{22}^{j-l-1} A_{21} A_{11}^i x^h/0,l/ \qquad /3.33/$$

When $x^v/0,0/ \neq 0$ and $x^h/0,j/ = 0$, $j \geqslant 0$, $x^v/i,0/ = 0$, $i \geqslant 1$, from /3.2/, /3.31/ and /3.33/ we obtain

$$y/0,j/ = C_2 A_{22}^j x^v/0,0/$$

and for $y/0,j/ = 0$, $j \in [0,k]$

$$\begin{bmatrix} C_2 \\ C_2 A_{22} \\ \cdots \\ C_2 A_{22}^k \end{bmatrix} x^v/0,0/ = 0 \qquad /3.34/$$

If /3.26/ holds then /3.34/ implies $x^v/0,0/ = 0$ and if /3.26/ is not satisfied then there exists $x^v/0,0/ \neq 0$ such that /3.34/ holds.

When $x^h/0,0/ \neq 0$ and $x^h/0,j/ = 0$, $j \geqslant 1$, $x^v/i,0/ = 0$, $i \geqslant 0$, from /3.2/, /3.31/ and /3.33/ we obtain

$$y/i,j/ = C_1 A_{11}^i x^h/0,j/ + C_2 A_{22}^{j-1} A_{21} A_{11}^i x^h/0,0/$$

and for $y/i,j/ = 0$, $/0,0/ \leqslant /i,j/ \leqslant /h,k/$

$$\begin{bmatrix} \bar{C} \\ \bar{C} A_{11} \\ \cdots \\ \bar{C} A_{11}^h \end{bmatrix} x^h/0,0/ = 0 \qquad \qquad /3.35/$$

If /3.27/ holds then /3.35/ implies $x^h/0,0/ = 0$ and if /3.27/ is not satisfied then there exists $x^h/0,0/ \neq 0$ such that /3.35/ holds. □

## 3.3 MODAL CONTROLLABILITY AND OBSERVABILITY OF ROESSER'S MODEL

### 3.3.1 Coprimeness of polynomial matrices.

Consider two polynomial matrices $P/z_1,z_2/$, $Q/z_1,z_2/$ with the same number of rows. Let us denote by $R^{m \times l}[z_1,z_2]$ the set of $m \times l$ polynomial matrices in $z_1$ and $z_2$ with real coefficients.

### Definition 3.5

Two matrices $P/z_1,z_2/ \in R^{m \times p}[z_1,z_2]$, $Q/z_1,z_2/ \in R^{m \times q}[z_1,z_2]$ are called left coprime with respect to /w.r.t./ $C[z_1,z_2]$ /C is the field of complex numbers/ iff for every left common factor $D/z_1,z_2/ \in R^{m \times n}[z_1,z_2]$ such that $P/z_1,z_2/ = D/z_1,z_2/ \bar{P}/z_1,z_2/$ and $Q/z_1,z_2/ = D/z_1,z_2/ \bar{Q}/z_1,z_2/$ we have $\det D/z_1,z_2/ = d \neq 0$, where d is a complex number.

Two matrices $R/z_1,z_2/$, $V/z_1,z_2/$ with the same number of columns are called right coprime iff the transposed matrices $R^T/z_1,z_2/$, $V^T/z_1,z_2/$ are left coprime.

Let us assume that $A/z_1,z_2/$ and $B/z_1,z_2/$ are polynomial matrices of dimensions $n \times n$ and $n \times m$, respectively, and

$$\det A/z_1,z_2/ = \prod_{i=1}^{k} a_i/z_1,z_2/ \qquad /3.36/$$

where $a_i/z_1,z_2/$, $i=1,2,\ldots,k$ are irreducible polynomials over the field F /R or C/.

Note that $\det A/z_1,z_2/ = 0$ defines an algebraic curve V /i.e. an algebraic variety of dimension 1/ and $a_i/z_1,z_2/$, $i=1,2,\ldots,k$ generates the irreducible algebraic curve $V_i$ such that the decomposition

$$V = \bigcup_i V_i \qquad /3.37/$$

is unique.

### Theorem 3.7 [14]

The matrices $A/z_1,z_2/ \in R^{n \times n}[z_1,z_2]$, $B/z_1,z_2/ \in R^{n \times m}[z_1,z_2]$ are left coprime if and only if

$$\text{rank}\begin{bmatrix} A/z_1,z_2/ & B/z_1,z_2/ \end{bmatrix} = n \qquad /3.38/$$

for any generic point $/\bar{z}_1,\bar{z}_2/$ of $V_i$.

**Proof**: Suppose that $A/z_1,z_2/$, $B/z_1,z_2/$ are not left coprime. Then

$$\begin{bmatrix} A/z_1,z_2/ & B/z_1,z_2/ \end{bmatrix} = D/z_1,z_2/\begin{bmatrix} \bar{A}/z_1,z_2/ & \bar{B}/z_1,z_2/ \end{bmatrix} \qquad /3.39/$$

where $\det D/z_1,z_2/$ has at least one irreducible nontrivial factor $d_i/z_1,z_2/$. Note that $d/z_1,z_2/$ is also one of the irreducible factors appearing in the decomposition of $\det A/z_1,z_2/$ since

$$\det A/z_1,z_2/ = \det \bar{A}/z_1,z_2/ \det D/z_1,z_2/$$

Let $V_i$ be an irreducible algebraic curve associated with $d_i/z_1,z_2/$ and let $/\bar{z}_1,\bar{z}_2/$ be a generic point of $V_i$. Then $\det D/\bar{z}_1,\bar{z}_2/ = 0$ and

$$\text{rank}\begin{bmatrix} A/\bar{z}_1,\bar{z}_2/ & B/\bar{z}_1,\bar{z}_2/ \end{bmatrix} < n \qquad /3.40/$$

Conversely, assume that there is an irreducible algebraic curve V and $/\bar{z}_1,\bar{z}_2/$ is a generic point of V such that /3.40/ holds. Using the Hermite form algorithm [14] w.r.t. $F[z_2][z_1]$ we obtain

$$\begin{bmatrix} A/z_1,z_2/ & B/z_1,z_2/ \end{bmatrix} U/z_1,z_2/ = \begin{bmatrix} D/z_1,z_2/ & 0 \end{bmatrix} \qquad /3.41/$$

where $\det U/z_1,z_2/ = u/z_2/$ and $U/\bar{z}_1,\bar{z}_2/$ has full rank.
From /3.41/ it follows that /3.40/ implies $\det D/\bar{z}_1,\bar{z}_2/ = 0$.

Hence the greatest common left divisor of $A/z_1,z_2/$ and $B/z_1,z_2/$ is non-unimodular and the matrices $A/z_1,z_2/$ and $B/z_1,z_2/$ are not left coprime.

In a similar way the following dual theorem can be proved [14]. □

### Theorem 3.8

The matrices $A/z_1,z_2/ \in R^{n \times n}[z_1,z_2]$, $B/z_1,z_2/ \in R^{m \times n}[z_1,z_2]$ are right coprime if and only if

$$\text{rank} \begin{bmatrix} A/z_1,z_2/ \\ B/z_1,z_2/ \end{bmatrix} = n \qquad /3.42/$$

for any generic point $/\bar{z}_1,\bar{z}_2/$ of $V_i$.

Let us denote by $R^{m \times n}(z_1)[z_2]$ the set of $m \times n$ polynomial matrices in $z_2$ with coefficients in the set of real rational functions in $z_1$.

### Theorem 3.9

The matrices $A/z_1,z_2/ \in R^{n \times n}[z_1,z_2]$, $B/z_1,z_2/ \in R^{n \times m}[z_1,z_2]$ are 2-D left coprime w.r.t. $F[z_1,z_2]$ if and only if

$A/z_1,z_2/$, $B/z_1,z_2/$ are /1-D/ left coprime w.r.t. $F(z_2)|z_1|$

and

$A/z_1,z_2/$, $B/z_1,z_2/$ are /1-D/ left coprime w.r.t. $F(z_1)|z_2|$

Proof: Assume that $A/z_1,z_2/$, $B/z_1,z_2/$ are /2-D/ left coprime w.r.t. $F[z_1,z_2]$ and they are not /1-D/ left coprime w.r.t. $F(z_2)[z_1]$. Then there exists a polynomial matrix $U$, with $\det U \in F[z_2]$, such that

$$\begin{bmatrix} A/z_1,z_2/ & B/z_1,z_2/ \end{bmatrix} U = \begin{bmatrix} D/z_1,z_2/ & 0 \end{bmatrix}$$

where $\deg_{z_1} \det D/z_1,z_2/ > 0$. Thus, $A/z_1,z_2/$, $B/z_1,z_2/$ have nontrivial left common factor $D/z_1,z_2/$ and they are not left coprime w.r.t. $F[z_1,z_2]$. To prove sufficiency let assume $A/z_1,z_2/$, $B/z_1,z_2/$ be /1-D/ left coprime w.r.t. $F(z_2)[z_1]$ and /1-D/ left coprime w.r.t. $F(z_1)[z_2]$. Suppose that $A/z_1,z_2/$, $B/z_1,z_2/$ are not /2-D/ left coprime, i.e. $A/z_1,z_2/$, $B/z_1,z_2/$ have a nontrivial common factor, say $\bar{D}/z_1,z_2/$.

Then there are two cases:

1° $\det \bar{D}/z_1,z_2/$ contains $z_1$, thus $A/z_1,z_2/$ and $B/z_1,z_2/$ are not /1-D/ left coprime w.r.t. $F(z_2)[z_1]$ since $D/z_1,z_2/$ is a nontrivial 1-D common factor;

2° $\det \bar{D}/z_1,z_2/$ does not contain $z_1$, hence it must contain $z_2$, so $A/z_1,z_2/$, $B/z_1,z_2/$ are not /1-D/ left coprime w.r.t. $F(z_2)[z_1]$.

Hence both cases reach contradictions and the proof is completed. □

In a similar way the following dual theorem can be proved [14]:

**Theorem 3.10**

The matrices $A/z_1,z_2/ \in R^{n \times n}[z_1,z_2]$, $B/z_1,z_2/ \in R^{m \times n}[z_1,z_2]$ are /2-D/ right coprime w.r.t. $F[z_1,z_2]$ if and only if

$A/z_1,z_2/$, $B/z_1,z_2/$ are /1-D/ right coprime w.r.t. $F(z_2)[z_1]$

and

$A/z_1,z_2/$, $B/z_1,z_2/$ are /1-D/ right coprime w.r.t. $F(z_1)[z_2]$.


### 3.3.2 Modal controllability and observability.

**Definition 3.6** [14]

RM is modally controllable iff the matrices

$$A/z_1,z_2/ = \begin{bmatrix} I_{n_1}z_1 - A_{11} & -A_{12} \\ -A_{21} & I_{n_2}z_2 - A_{22} \end{bmatrix}, \quad B = \begin{bmatrix} B_1 \\ B_2 \end{bmatrix} \qquad /3.43/$$

are left coprime w.r.t. $C[z_1,z_2]$.

**Definition 3.7** [14]

RM is modally observable iff the matrices

$$A/z_1,z_2/ = \begin{bmatrix} I_{n_1}z_1 - A_{11} & -A_{12} \\ -A_{21} & I_{n_2}z_2 - A_{22} \end{bmatrix}, \quad C = \begin{bmatrix} C_1 & C_2 \end{bmatrix} \qquad /3.44/$$

are right coprime w.r.t. $C[z_1,z_2]$.

From Definition 3.6 /3.7/ and Theorem 3.7 /3.8/ the following theorems follow:

**Theorem 3.11**

RM is modally controllable if and only if

$$\text{rank} \begin{bmatrix} I_{n_1}z_1 - A_{11} & -A_{12} & B_1 \\ -A_{21} & I_{n_2}z_2 - A_{22} & B_2 \end{bmatrix} = n \qquad /3.45/$$

for any generic point $/\bar{z}_1,\bar{z}_2/$ of $V_i$, where $V_i$ is an irreducible curve defined by the irreducible polynomial $a_i/z_1,z_2/$ and

$$\det \begin{bmatrix} I_{n_1}z_1 - A_{11} & -A_{12} \\ -A_{21} & I_{n_2}z_2 - A_{22} \end{bmatrix} = \prod_{i=1}^{k} a_i/z_1,z_2/$$

### Theorem 3.12

RM is modally observable if and only if

$$\operatorname{rank} \begin{bmatrix} I_{n_1}z_1 - A_{11} & -A_{12} \\ -A_{21} & I_{n_2}z_2 - A_{22} \\ C_1 & C_2 \end{bmatrix} = n \qquad /3.46/$$

for any generic point $/\bar{z}_1,\bar{z}_2/$ of $V_i$, where $V_i$ is defined in the same way as in Theorem 3.11.

### Example 3.3

Test the modal controllability of RM with

$$A = \begin{bmatrix} A_{11} & A_{12} \\ A_{21} & A_{22} \end{bmatrix} = \begin{bmatrix} 1 & 0 & | & 0 \\ 0 & -1 & | & -1 \\ \hline 1 & 0 & | & 0 \end{bmatrix}, \quad B = \begin{bmatrix} B_1 \\ B_2 \end{bmatrix} = \begin{bmatrix} 1 \\ 0 \\ \hline 0 \end{bmatrix} \qquad /3.47/$$

$$\det \begin{bmatrix} I_{n_1}z_1 - A_{11} & -A_{12} \\ -A_{21} & I_{n_2}z_2 - A_{22} \end{bmatrix} = \begin{vmatrix} z_1 - 1 & 0 & 0 \\ 0 & z_1 + 1 & 1 \\ -1 & 0 & z_2 \end{vmatrix} =$$

$$= (z_1 + 1)(z_1 - 1) z_2 \qquad /3.48/$$

Hence

$V_1$ is defined by $z_1 = -1$, $z_2$ arbitrary

$V_2$ is defined by $z_1 = 1$, $z_2$ arbitrary

$V_3$ is defined by $z_2 = 0$, $z_1$ arbitrary

and for $V_1$

$$\operatorname{rank} \begin{bmatrix} -2 & 0 & 0 & 1 \\ 0 & 0 & 1 & 0 \\ -1 & 0 & z_2 & 0 \end{bmatrix} = 3$$

for $V_2$

$$\text{rank} \begin{bmatrix} 0 & 0 & 0 & 1 \\ 0 & 2 & 1 & 0 \\ -1 & 0 & z_2 & 0 \end{bmatrix} = 3,$$

for $V_3$

$$\text{rank} \begin{bmatrix} z_1 - 1 & 0 & 0 & 1 \\ 0 & z_1 + 1 & 1 & 0 \\ -1 & 0 & 0 & 0 \end{bmatrix} = 3.$$

Therefore, RM with /3.47/ is modally controllable.

In Section 2.1 it was shown that the transfer function matrix of RM can be written in the form

$$G/z_1,z_2/ = \begin{bmatrix} C_1 & C_2 \end{bmatrix} \begin{bmatrix} I_{n_1}z_1 - A_{11} & -A_{12} \\ -A_{21} & I_{n_2}z_2 - A_{22} \end{bmatrix}^{-1} \begin{bmatrix} B_1 \\ B_2 \end{bmatrix} + D =$$

$$= C/z_2/\left[I_{n_1}z_1 - A/z_2/\right]^{-1} B/z_2/ + D/z_2/ = \qquad /3.49/$$

$$= C/z_1/\left[I_{n_2}z_2 - A/z_1/\right]^{-1} B/z_1/ + D/z_1/$$

where

$$A/z_2/ = A_{11} + A_{12}\left[I_{n_2}z_2 - A_{22}\right]^{-1} A_{21}$$

$$B/z_2/ = B_1 + A_{12}\left[I_{n_2}z_2 - A_{22}\right]^{-1} B_2$$

$$C/z_2/ = C_1 + C_2\left[I_{n_2}z_2 - A_{22}\right]^{-1} A_{21} \qquad /3.50a/$$

$$D/z_2/ = C_2\left[I_{n_2}z_2 - A_{22}\right]^{-1} B_2 + D$$

and

$$A/z_1/ = A_{22} + A_{21}\left[I_{n_1}z_1 - A_{11}\right]^{-1} A_{12}$$

$$B/z_1/ = B_2 + A_{21}\left[I_{n_1}z_1 - A_{11}\right]^{-1} B_1$$

$$C/z_1/ = C_2 + C_1\left[I_{n_1}z_1 - A_{11}\right]^{-1} A_{12} \qquad /3.50b/$$

$$D/z_1/ = C_1\left[I_{n_1}z_1 - A_{11}\right]^{-1} B_1 + D$$

## Theorem 3.13

RM is modally controllable if and only if the matrices $A/z_2/$, $B/z_2/$ are controllable over $R(z_2)$ and the matrices $A/z_1/$, $B/z_1/$ are controllable over $R(z_1)$.

**Proof:** Assume that $A/z_2/$, $B/z_2/$ are not controllable over $R(z_2)$ and $[I_{n_1}z_1 - A/z_2/]$, $B/z_2/$ have a non-unimodular left common factor $D/z_1,z_2/$ such that $\deg_{z_1} \det D/z_1,z_2/ > 0$.

From the equality

$$\begin{bmatrix} I_{n_1}z_1 - A_{11} & -A_{12} & B_1 \\ -A_{21} & I_{n_2}z_2 - A_{22} & B_2 \end{bmatrix} =$$

$$= \begin{bmatrix} I_{n_1} & -A_{12} \\ 0 & I_{n_2}z_2 - A_{22} \end{bmatrix} \begin{bmatrix} I_{n_1}z_1 - A/z_2/ & 0 & B/z_2/ \\ -[I_{n_2}z_2 - A_{22}]^{-1}A_{21} & I_{n_2} & [I_{n_2}z_2 - A_{22}]^{-1}B_2 \end{bmatrix}$$

it follows that the matrices

$$\begin{bmatrix} I_{n_1}z_1 - A_{11} & -A_{12} \\ -A_{21} & I_{n_2}z_2 - A_{22} \end{bmatrix}, \begin{bmatrix} B_1 \\ B_2 \end{bmatrix} \qquad /3.51/$$

have also the non-unimodular left common factor

$$\begin{bmatrix} D/z_1,z_2/ & -A_{12} \\ 0 & I_{n_2}z_2 - A_{22} \end{bmatrix}$$

and they are not left coprime w.r.t. $C(z_2)[z_1]$. By Theorem 3.10 RM is not modally controllable. Necessity of the second part can be proved in an analogous way. To prove sufficiency let us assume that $A/z_2/$, $B/z_2/$ are controllable over $R(z_2)$. Therefore there exist matrices $P$ and $Q$ with entries from $R(z_2)[z_1]$ such that

$$[I_{n_1}z_1 - A/z_2/]P + B/z_2/\,Q = I_{n_1}$$

It is easy to verify that

$$\begin{bmatrix} I_{n_1}z_1 - A_{11} & -A_{12} \\ -A_{21} & I_{n_2}z_2 - A_{22} \end{bmatrix} \begin{bmatrix} P \\ [I_{n_2}z_2 - A_{22}]^{-1}[A_{21}P - B_2Q] \end{bmatrix}$$

$$\begin{aligned}
&\left. \begin{array}{l} \phantom{+}\Big[I_{n_2}z_2 - A_{22}\Big]^{-1}\Big[A_{21}P - B_2Q\Big]A_{12}\Big[I_{n_2}z_2 - A_{22}\Big]^{-1} + \Big[I_{n_2}z_2 - A_{22}\Big]^{-1} \\ \phantom{+} \end{array} \right] + \\
&+ \begin{bmatrix} B_1 \\ B_2 \end{bmatrix} \begin{bmatrix} Q & QA_{12}\Big[I_{n_2}z_2 - A_{22}\Big]^{-1} \end{bmatrix} = \begin{bmatrix} I_{n_1} & 0 \\ 0 & I_{n_2} \end{bmatrix}
\end{aligned}$$

(top row of the large bracket:) $PA_{12}\Big[I_{n_2}z_2 - A_{22}\Big]^{-1}$

This implies left coprimeness of /3.51/ w.r.t. $C(z_2)[z_1]$.
In the same way it can be shown that controllability over $R(z_1)$ of $A/z_1/$, $B/z_1/$ implies the modal controllability of RM. $\square$

By duality a similar result can be proved for modal observability:

### Theorem 3.14

RM is modally observable if and only if the matrices $A/z_2/$, $C/z_2/$ are observable over $R(z_2)$ and the matrices $A/z_1/$, $C/z_1/$ are observable over $R(z_1)$.

Note that the irreducible polynomial $a_i/z_1,z_2/$ can be represented in a parametric form over the field F as

$$z_1 = f_1/t/, \quad z_2 = f_2/t/$$

where t is a transcendental parameter and $f_1$, $f_2$ are algebraic or transcendental functions.

For example, for $a_i/z_1,z_2/ = z_1 z_2 - 1$ we have $f_1/t/ = t$, $f_2/t/ = \frac{1}{t}$.

An irreducible factor $f_1/t/$, $f_2/t/$ for which

$$\begin{bmatrix} f_1/t/\, I_{n_1} - A_{11} & -A_{12} & B_1 \\ -A_{21} & f_2/t/\, I_{n_2} - A_{22} & B_2 \end{bmatrix} \qquad /3.52/$$

is not of a full rank is called an input decoupling factor.

### Theorem 3.15 [19]

RM is modally controllable /observable/ if the pair A,B /C,A/ is controllable /observable/ in a 1-D sense.

Proof: Assume that the pair A,B is not modally controllable in a 2-D sense. Then there exists an input decoupling factor $f_1/t/$, $f_2/t/$ which is an irreducible factor of $\det A/z_1,z_2/$. Thus the rows of /3.52/ are linearly dependent for all values of t, and specifically for $f_1/t/ = f_2/t/$. Therefore, the pair A,B is not controllable in a 1-D sense. $\square$

## Theorem 3.16

RM is modally controllable /observable/ if the pairs $A_{11},B_1$ and $A_{22},B_2$ /$C_1,A_{11}$ and $C_2,A_{22}$/ are controllable /observable/ in a 1-D sense.

**Proof**: The controllability of the pairs $A_{11},B_1$ and $A_{22},B_2$ implies that

$$\text{rank}\begin{bmatrix} I_{n_1}z_1 - A_{11} & B_1 \end{bmatrix} = n_1 \quad \text{for} \quad z_1 \in C$$

$$\text{rank}\begin{bmatrix} I_{n_2}z_2 - A_{22} & B_2 \end{bmatrix} = n_2 \quad \text{for} \quad z_2 \in C$$

and

$$\text{rank}\begin{bmatrix} I_{n_1}z_1 - A_{11} & -A_{12} & B_1 \\ -A_{21} & I_{n_2}z_2 - A_{22} & B_2 \end{bmatrix} = n_1 + n_2 \quad \text{for} \quad z_1,z_2 \in C$$

Thus, by Theorem 3.11 RM is modally controllable. $\square$

## Example 3.4

Test the controllability of RM with

$$A = \begin{bmatrix} A_{11} & A_{12} \\ A_{21} & A_{22} \end{bmatrix} = \begin{bmatrix} 1 & 0 & 0 \\ 1 & 2 & 1 \\ \hline -1 & 1 & 2 \end{bmatrix}, \quad B = \begin{bmatrix} B_1 \\ B_2 \end{bmatrix} = \begin{bmatrix} 1 \\ 0 \\ \hline 1 \end{bmatrix}$$

Using Theorem 3.15 we obtain the controllability matrix

$$\begin{bmatrix} B & AB & A^2B \end{bmatrix} = \begin{bmatrix} 1 & 1 & 1 \\ 0 & 2 & 6 \\ 1 & 1 & 3 \end{bmatrix}$$

which is of full rank and therefore the RM is modally controllable. We obtain the same result using Theorem 3.16 since the pairs

$$A_{11} = \begin{bmatrix} 1 & 0 \\ 1 & 2 \end{bmatrix}, \quad B_1 = \begin{bmatrix} 1 \\ 0 \end{bmatrix} \quad \text{and} \quad A_{22} = 2, \quad B_2 = 1$$

are controllable in a 1-D sense.

The 1-D controllability matrix of the pair $A,B$ can be written in the following block form

$$C_n = \begin{bmatrix} B & AB & A^2B & \cdots & A^{n-1}B \end{bmatrix} = \begin{bmatrix} C_{n_1} \\ \hline C_{n_2} \end{bmatrix}\begin{matrix}\}n_1 \\ \}n_2\end{matrix} \qquad /3.53/$$

Similarly, the 1-D observability matrix of the pair C,A can be written in the following block form

$$O_n = \begin{bmatrix} C \\ CA \\ CA^2 \\ \cdots \\ CA^{n-1} \end{bmatrix} = \begin{bmatrix} \overbrace{O_{n_1}}^{n_1} & | & \overbrace{O_{n_2}}^{n_2} \end{bmatrix}$$

**Theorem 3.17** [19]

RM is modally uncontrollable if some rows of one block of the matrix /3.53/ are linearly dependent, while being independent of the rows of the other block.

**Proof:** If RM is modally controllable, then by Theorem 3.11 for each irreducible factor $z_1 = f_1/t/$, $z_2 = f_2/t/$ all rows of /3.52/ are linearly independent. With the substitution $z_1 = z_2$ in /3.52/ linear dependence between the rows of each block is not lost but some linear dependence between some rows of different blocks may appear. Note that if the rows of the 1-D matrix

$$\begin{bmatrix} I_n s - A & B \end{bmatrix}$$

are linearly dependent, then the rows of the matrix

$$\begin{bmatrix} B & AB & \ldots & A^{n-1}B \end{bmatrix}$$

are also linearly dependent. Therefore, RM is modally uncontrollable if some rows of one block of /3.53/ are linearly dependent, while being independent of the rows of the other block. □

Similar result can be obtained for modal observability.

**Example 3.5**

Test the modal controllability of RM with

$$A = \begin{bmatrix} A_{11} & A_{12} \\ A_{21} & A_{22} \end{bmatrix} = \begin{bmatrix} -3 & 0 & | & 1 \\ -1 & -2 & | & 1 \\ \hline 1 & 0 & | & -1 \end{bmatrix}, \quad B = \begin{bmatrix} B_1 \\ B_2 \end{bmatrix} = \begin{bmatrix} 1 \\ 1 \\ \hline 2 \end{bmatrix}$$

In this case the 1-D controllability matrix has the form

$$C_3 = \begin{bmatrix} B & AB & A^2B \end{bmatrix} = \begin{bmatrix} 1 & -1 & 2 \\ 1 & -1 & 2 \\ 2 & -1 & 0 \end{bmatrix}$$

The rows of the matrix
$$C_{n_1} = \begin{bmatrix} 1 & -1 & 2 \\ 1 & -1 & 2 \end{bmatrix}$$
are linearly dependent, while being independent of a row of the matrix
$$C_{n_2} = \begin{bmatrix} 2 & -1 & 0 \end{bmatrix}.$$
Therefore, by Theorem 3.17 the RM is modally uncontrollable.

### 3.3.3 Reachability of the first level realization.

Let $R_c(z)$ be the set of proper rational functions in z with real coefficients and $R_c^{n \times m}(z)$ be the set of n × m matrices with entries in $R_c(z)$.

**Definition 3.8**

The pair $A/z/ \in R_c^{n \times n}(z)$, $B/z/ \in R_c^{n \times m}(z)$ is said to be reachable over $R_c(z)$ iff
$$\text{rank}\begin{bmatrix} B/z/ & A/z/B/z/ & \cdots & [A/z/]^{n-1}B/z/ \end{bmatrix} = n \qquad /3.54/$$

Note that reachability of the pair $A/z/, B/z/$ is equivalent to the existance of a matrix $L/z/ \in R_c^{nm \times n}/z/$ such that
$$\begin{bmatrix} B/z/ & A/z/B/z/ & \cdots & [A/z/]^{n-1}B/z/ \end{bmatrix} L/z/ = I_n \qquad /3.55/$$
since the columns of
$$\begin{bmatrix} B/z/ & A/z/B/z/ & \cdots & [A/z/]^{n-1}B/z/ \end{bmatrix}$$
generate the standard basis vectors for $R_c^n(z)$ if and only if the pair $A/z/, B/z/$ is reachable.

**Theorem 3.18**

The pair $A/z/, B/z/$ is reachable over $R_c(z)$ if and only if the pair $D_A, D_B$ is reachable over R, where
$$D_A = \lim_{z \to \infty} A/z/ \quad \text{and} \quad D_B = \lim_{z \to \infty} B/z/ \qquad /3.56/$$

Proof: Assume that the pair $A/z/, B/z/$ is reachable and /3.55/ holds for some $L/z/$. Then
$$\lim_{z \to \infty} \begin{bmatrix} B/z/ & A/z/B/z/ & \cdots & [A/z/]^{n-1}B/z/ \end{bmatrix} L/z/ =$$

$$= \begin{bmatrix} D_B & D_A D_B & \cdots & D_A^{n-1} D_B \end{bmatrix} D_L = I_n$$

where

$$D_L = \lim_{z \to \infty} L/z/$$

Therefore the pair $D_A, D_B$ is reachable over R. Conversely let assume that

$$\begin{bmatrix} D_B & D_A D_B & \cdots & D_A^{n-1} D_B \end{bmatrix} D_M = I_n$$

for some $D_M \in R^{nm \times n}$.

Then

$$\begin{bmatrix} B/z/ & A/z/B/z/ & \cdots & [A/z/]^{n-1} B/z/ \end{bmatrix} D_M = I_n + M/z/ \qquad /3.57/$$

where $M/z/$ is a strictly proper rational matrix in $R_c^{n \times n}(z)$.

Let $M/z/$ have a realization $\{A_M, B_M, C_M\}$, i.e.

$$M/z/ = C_M [I_n z - A_M]^{-1} B_M$$

Note that $I_n + M/z/$ is invertible over $R_c(z)$ and

$$[I_n + M/z/]^{-1} = I_n - C_M [I_n z - A_M + B_M C_M]^{-1} B_M$$

If we choose

$$L/z/ = D_M [I_n + M/z/]^{-1}$$

then from /3.57/ we obtain /3.55/.

Therefore the pair $A/z/, B/z/$ is reachable over $R_c(z)$. □

## 3.4 SEPARABILITY OF TRANSFER FUNCTION MATRICES

Consider a 2-D linear system with transfer function matrix

$$G/z_1, z_2/ = \frac{N/z_1, z_2/}{d/z_1, z_2/} \qquad /3.58/$$

where $N/z_1, z_2/ \in R^{1 \times m}[z_1, z_2]$ and $d/z_1, z_2/ \in R[z_1, z_2]$ are coprime.

### Definition 3.9

The transfer function matrix /3.58/ is said to be separable iff $d/z_1, z_2/$

can be written as a product of a polynomial $d_1/z_1/$ in $z_1$ and a polynomial $d_2/z_2/$ in $z_2$, i.e.

$$d/z_1,z_2/ = d_1/z_1/\, d_2/z_2/ \qquad /3.59/$$

Note that

$$N/z_1,z_2/ = \sum_{i=0}^{n_2} \bar{N}_i/z_1/\, z_2^i = \begin{bmatrix} \bar{N}_0/z_1/ & \bar{N}_1/z_1/ & \cdots & \bar{N}_{n_2}/z_1/ \end{bmatrix} \begin{bmatrix} I_m \\ I_m z_2 \\ \vdots \\ I_m z_2^{n_2} \end{bmatrix} =$$

$$= N_1/z_1/\, N_2/z_2/ \qquad /3.60/$$

where

$$N_1/z_1/ = \begin{bmatrix} \bar{N}_0/z_1/ & \bar{N}_1/z_1/ & \cdots & \bar{N}_{n_2}/z_1/ \end{bmatrix}, \quad N_2/z_2/ = \begin{bmatrix} I_m \\ I_m z_2 \\ \vdots \\ I_m z_2^{n_2} \end{bmatrix}$$

Substituting /3.60/ and /3.59/ into /3.58/ we obtain

$$G/z_1,z_2/ = G_1/z_1/\, G_2/z_2/ \qquad /3.61/$$

where

$$G_i/z_i/ = \frac{N_i/z_i/}{d_i/z_i/} \qquad i = 1,2 \qquad /3.62/$$

Therefore a separable transfer function matrix $G/z_1,z_2/$ can be always written as a product of $G_1/z_1/$ and $G_2/z_2/$.

### Theorem 3.19

A transfer function matrix $G/z_1,z_2/$ is separable if and only if there exists a first level realization $\{A, B/z_2/, C/z_2/, D/z_2/\}$ where A is a constant matrix /i.e. it does not depend on $z_2/$.

**Proof:** Assume that A is a constant matrix. Then

$$G/z_1,z_2/ = C/z_1/\left[I_{n_2}z_2 - A\right]^{-1} B/z_1/ + D/z_1/ =$$

$$= \frac{C/z_1/\, \mathrm{adj}\left[I_{n_2}z_2 - A\right] B/z_1/}{\det\left[I_{n_2}z_2 - A\right]} + D/z_1/ =$$

$$= G_1/z_1/\ G_2/z_2/$$

where

$$G_1/z_1/ = \left[ C/z_1/\bar{A}_0 B/z_1/ + d_0 D/z_1/ \quad C/z_1/\bar{A}_1 B/z_1/ + d_1 D/z_1/ \ \cdots \ D/z_1/ \right]$$

$$G_2/z_2/ = \frac{1}{d_2/z_2/} \begin{bmatrix} I_m \\ I_m z_2 \\ \cdots \\ I_m z_2^{n_2} \end{bmatrix}$$

$$d_2/z_2/ = \det\left[I_{n_2} z_2 - A\right] = \sum_{i=0}^{n_2} d_i z_2^i \qquad /d_{n_2} = 1/$$

So, if A is a constant matrix, then $G/z_1, z_2/$ is separable.

Using the procedure presented in Section 2.6 it is easy to show that if $G/z_1, z_2/$ is separable then A of the first level realization is a constant matrix. □

### Definition 3.10 [3]

Two realizations $R_1 = \{A_1/z_1/, B_1/z_1/, C_1/z_1/, D_1/z_1/\}$ and $R_2 = \{A_2/z_1/, B_2/z_1/, C_2/z_1/, D_2/z_1/\}$ are called feedback **equivalent** iff there exists a state space isomorphism $T/z_1/$ and a feedback matrix $F/z_1/$ such that

$1^o \ A_1/z_1/ = [T/z_1/]^{-1} [A_2/z_1/ - B_2/z_1/ F/z_1/] T/z_1/$

$2^o \ B_1/z_1/ = [T/z_1/]^{-1} B_2/z_1/$

$3^o \ C_1/z_1/ = C_2/z_1/ T/z_1/$

$4^o \ D_1/z_1/ = D_2/z_1/$

### Definition 3.11

Two transfer function matrices $G_1/z_1, z_2/$ and $G_2/z_1, z_2/$ are called feedback equivalent iff $G_1/z_1, z_2/$ and $G_2/z_1, z_2/$ have feedback equivalent first level realizations $R_1$ and $R_2$.

### Theorem 3.20

Every reachable realization $R_1 = \{A_1/z/, B_1/z/, C/z/, D_1/z/\}$ over $R_c/z/$ is feedback equivalent to a realization $R_2 = \{A_2, B_2/z/, C/z/, D/z/\}$ where $A_2$ is a constant matrix /i.e. it does not depend on z/.

**Proof:** By Theorem 3.18 the reachability of $A_1/z/, B_1/z/$ implies the reachability of the pair $D_{A_1}, D_{B_1}$ where

$$D_{A_1} = \lim_{z \to \infty} A_1/z/ \quad \text{and} \quad D_{B_1} = \lim_{z \to \infty} B_1/z/ \qquad /3.63/$$

From Heymann's Lemma [6] it follows that it is possible to find a matrix $F \in R^{m \times n}$ and a vector $q \in R^m$ such that the pair $D_{A_1} - D_{B_1} F, D_{B_1} q$ is reachable over $R$. By Theorem 3.18 the pair $A_1/z/ - B_1/z/F, B_1/z/q$ is reachable over $R_c$ and can be transformed into control canonical form. Thus there exists a state space isomorphism $T/z/$ such that

$$[T/z/]^{-1} [A_1/z/ - B_1/z/F] T/z/ = \begin{bmatrix} 0 & 1 & 0 & \cdots & 0 \\ 0 & 0 & 1 & \cdots & 0 \\ \cdot & \cdot & \cdot & & \cdot \\ 0 & 0 & 0 & \cdots & 1 \\ -a_0/z/ & -a_1/z/ & -a_2/z/ & \cdots & -a_{n-1}/z/ \end{bmatrix}$$

and

$$[T/z/]^{-1} B_1/z/ q = \begin{bmatrix} 0 \\ 0 \\ \vdots \\ 0 \\ 1 \end{bmatrix} \qquad /3.64/$$

where $a_i \in R_c(z)$ for $i = 0, 1, \ldots, n-1$.

Therefore, there exists a row vector

$$f/z/ = [f_0/z/ \quad f_1/z/ \quad \cdots \quad f_{n-1}/z/]$$

such that the matrix

$$A_2 = [T/z/]^{-1} [A_1/z/ - B_1/z/F] T/z/ - [T/z/]^{-1} B_1/z/ q \, f/z/ \qquad /3.65/$$

has entries in $R$.

The matrix $F/z/$ has the form

$$F/z/ = F + q \, f/z/ [T/z/]^{-1} \qquad /3.66/$$

□

## Theorem 3.21

Every transfer function matrix $G/z_1, z_2/$ is feedback equivalent to a separable transfer function matrix $G_1/z_1/ G_2/z_2/$.

**Proof:** Let $\{A_1/z_2/, B_1/z_2/, C_1/z_2/, D_1/z_2/\}$ be a reachable realization of $G/z_1,z_2/$. Using the procedure described in the proof of Theorem 3.20 we can find $T/z_2/$ and $F/z_2/$ such that the feedback equivalent realization $\{A_2, B_2/z_2/, C_2/z_2/, D_2/z_2/\}$ has the matrix $A_2$ independent of $z_2$. By Theorem 3.19 the feedback equivalent transfer function matrix is separable. □

From the above considerations we have the following algorithm which enables us to find $T/z_2/$ and $F/z_2/$ such that the feedback equivalent transfer function matrix is separable $/G/z_1,z_2/ = G_1/z_1/ G_2/z_2//$ and

$$G_1/z_1/ = \frac{N_1/z_1/}{d_1/z_1/}$$

has a prespecified denominator $d_1/z_1/$.

**Algorithm 3.1**

<u>Step 1.</u> For a given $G/z_1,z_2/$ find a reachable realization
$$\{A_1/z_2/, B_1/z_2/, C_1/z_2/, D_1/z_2/\}$$

<u>Step 2.</u> Evaluate $D_{A_1}$ and $D_{B_1}$ using the formulae

$$D_{A_1} = \lim_{z_2 \to \infty} A_1/z_2/, \quad D_{B_1} = \lim_{z_2 \to \infty} B_1/z_2/ \qquad /3.67/$$

<u>Step 3.</u> Using Heymann's Lemma find $F \in R^{m \times n}$ and $q \in R^m$ such that the pair $D_{A_1} - D_{B_1} F, D_{B_1} q$ is reachable over $R$.

<u>Step 4.</u> Find an invertible matrix $T/z_2/$ which thansforms the pair $A_1/z_2/ - B_1/z_2/ F, B_1/z_2/ q$ into control canonical form.

<u>Step 5.</u> Calculate

$$f/z_2/ = \begin{bmatrix} d_0 - a_0/z_2/ & d_1 - a_1/z_2/ & \cdots & d_{n_1-1} a_{n_1-1}/z_2/ \end{bmatrix} \qquad /3.68/$$

where $d_i /i = 0,1,\ldots,n_1-1/$ are real coefficients of the polynomial

$$d_1/z_1/ = \sum_{i=0}^{n_1} d_i z_1^i \quad /d_{n_1} = 1/$$

<u>Step 6.</u> Using the formula

$$F/z_2/ = F + q f/z_2/ [T/z_2/]^{-1} \qquad /3.69/$$

evaluate $F/z_2/$ and then calculate the matrices

$$A_2 = \left[T/z_2/\right]^{-1} \left[A_1/z_2/ - B_1/z_2/F/z_2/\right] T/z_2/$$

$$B_2/z_2/ = \left[T/z_2/\right]^{-1} B_1/z_2/$$

$$C_2/z_2/ = C_1/z_2/ T/z_2/ \qquad /3.70/$$

$$D_2/z_2/ = D_1/z_2/$$

Step 7. Evaluate $G_1/z_1/$ and $G_2/z_2/$ using the formula

$$G/z_1,z_2/ = C_2/z_2/\left[I_{n_1}z_1 - A_2\right]^{-1} B_2/z_2/ + D_2/z_2/ =$$
$$= G_1/z_1/\, G_2/z_2/ \qquad /.371/$$

Example 3.6

Given

$$G/z_1,z_2/ = \frac{1}{z_1^2(z_2^2-1) - z_1(z_2^2+1)} \left[\, z_1(z_2^2-1) \;\; -(z_2+1) \;\; 1 \,\right] \qquad /3.72/$$

find $F/z_2/$ and $T/z_2/$ such that the feedback equivalent transfer function matrix is separable and $d_1/z_1/ = z_1^2 + 3z_1 + 2$.

Step 1. It is easy to verify that the matrices

$$A_1/z_2/ = \begin{bmatrix} \dfrac{z_2}{z_2+1} & \dfrac{1}{z_2+1} \\ \dfrac{z_2}{z_2-1} & \dfrac{1}{z_2-1} \end{bmatrix}, \quad B_1/z_2/ = \begin{bmatrix} 1 & 0 \\ 0 & \dfrac{1}{z_2-1} \end{bmatrix},$$

$$C_1/z_2/ = \begin{bmatrix} 1 & 0 \end{bmatrix}, \quad D_1/z_2/ = 0 \qquad /3.73/$$

are a reachable realization of /3.72/.

Step 2. Using /3.67/ we obtain the matrices

$$D_{A_1} = \begin{bmatrix} 1 & 0 \\ 1 & 0 \end{bmatrix}, \quad D_{B_1} = \begin{bmatrix} 1 & 0 \\ 0 & 0 \end{bmatrix}$$

which are reachable.

Step 3. For

$$F = \begin{bmatrix} 1 & 0 \\ 0 & 1 \end{bmatrix} \quad \text{and} \quad q = \begin{bmatrix} 1 \\ 0 \end{bmatrix}$$

the pair

$$D_{A_1} - D_{B_1} F = \begin{bmatrix} 0 & 0 \\ 1 & 0 \end{bmatrix}, \quad D_{B_1} q = \begin{bmatrix} 1 \\ 0 \end{bmatrix}$$

is reachable.

Step 4. To find $T/z_2/$ we calculate

$$\bar{A}_1/z_2/ = A_1/z_2/ - B_1/z_2/F = \begin{bmatrix} -\dfrac{1}{z_2+1} & \dfrac{1}{z_2+1} \\ \dfrac{z_2}{z_2-1} & 0 \end{bmatrix}$$

$$\bar{B}_1/z_2/ = B_1/z_2/q = \begin{bmatrix} 1 \\ 0 \end{bmatrix}$$

and

$$\left[ \bar{B}_1/z_2/ \ \bar{A}_1/z_2/\bar{B}_1/z_2/ \right]^{-1} = \begin{bmatrix} 1 & -\dfrac{1}{z_2+1} \\ 0 & \dfrac{z_2}{z_2-1} \end{bmatrix}^{-1} =$$

$$= \begin{bmatrix} 1 & \dfrac{z_2-1}{z_2(z_2+1)} \\ 0 & \dfrac{z_2-1}{z_2} \end{bmatrix} = \begin{bmatrix} w/z_2/ \\ v/z_2/ \end{bmatrix}$$

Hence

$$[T/z_2/]^{-1} = \begin{bmatrix} v/z_2/ \\ v/z_2/\bar{A}_1/z_2/ \end{bmatrix} = \begin{bmatrix} 0 & \dfrac{z_2-1}{z_2} \\ 1 & 0 \end{bmatrix}, \quad T/z_2/ = \begin{bmatrix} 0 & 1 \\ \dfrac{z_2}{z_2-1} & 0 \end{bmatrix}$$

and

$$[T/z_2/]^{-1}\bar{A}_1/z_2/T/z_2/ = \begin{bmatrix} 0 & 1 \\ \dfrac{z_2}{z_2^2-1} & \dfrac{-1}{z_2+1} \end{bmatrix}, \quad [T/z_2/]^{-1}\bar{B}_1/z_2/ = \begin{bmatrix} 0 \\ 1 \end{bmatrix}$$

Step 5. Taking into account that $d_0 = 2$, $d_1 = 3$ and $a_0/z_2/ = -\dfrac{z_2}{z_2^2 - 1}$,

$a_1/z_2/ = \dfrac{1}{z_2 + 1}$ from /3.68/ we obtain

$$f/z_2/ = \begin{bmatrix} d_0 - a_0/z_2/ & d_1 - a_1/z_2/ \end{bmatrix} = \begin{bmatrix} 2 + \dfrac{z_2}{z_2^2 - 1} & 3 - \dfrac{1}{z_2 + 1} \end{bmatrix}$$

Step 6. Using /3.69/ and /3.70/ we obtain

$$F/z_2/ = F + q\, f/z_2/\bigl[T/z_2/\bigr]^{-1} = \begin{bmatrix} 4 - \dfrac{1}{z_2 + 1} & \dfrac{2z_2^2 + z_2 - 2}{z_2(z_2 + 1)} \\ 0 & 1 \end{bmatrix}$$

and

$$A_2 = \bigl[T/z_2/\bigr]^{-1}\bigl[A_1/z_2/ - B_1/z_2/\, F/z_2/\bigr]T/z_2/ = \begin{bmatrix} 0 & 1 \\ -2 & -3 \end{bmatrix}$$

$$B_2/z_2/ = \bigl[T/z_2/\bigr]^{-1} B_1/z_2/ = \begin{bmatrix} 0 & \dfrac{1}{z_2} \\ 1 & 0 \end{bmatrix}$$

$$C_2/z_2/ = C_1/z_2/\, T/z_2/ = \begin{bmatrix} 0 & 1 \end{bmatrix}, \qquad D_2/z_2/ = D_1/z_2/ = 0$$

Step 7.

$$G/z_1, z_2/ = C_2/z_2/\bigl[I_{n_1} z_1 - A_2\bigr]^{-1} B_2/z_2/ = G_1/z_1/\, G_2/z_2/$$

where

$$G_1/z_1/ = \dfrac{1}{z_1^2 + 3z_1 + 2}\begin{bmatrix} 1 & z_1 \end{bmatrix}, \qquad G_2/z_2/ = \dfrac{1}{z_2}\begin{bmatrix} 0 & -2 \\ z_2 & 0 \end{bmatrix}$$

## 3.5 MINIMUM ENERGY CONTROL OF ROESSER'S MODEL

The minimum energy control problem for RM was formulated and solved by Klamka in [13].

Consider RM described by the equations /3.1/ and /3.2/ and the performance index

$$I/u/ = \sum_{/0,0/ \leq /i,j/ < /h,k/} u^T/i,j/ \, Q \, u/i,j/ \qquad /3.74/$$

where Q is the m×m symmetric and positive definite weighting matrix.

The minimum energy control problem for RM can be stated as follows: Given the matrices A,B of RM, the weighting matrix Q and the boundary conditions $x^h/0,j/$, $j \in [1,k]$, $x^v/i,0/$, $i \in [1,h]$; find a sequence of inputs $u/i,j/$, $/0,0/ \leq /i,j/ < /h,k/$ which transfers RM from the initial local state $x_o = x/0,0/$ to the desired final local state $x_f = x/h,k/$ and minimizes the performance index /3.74/.

To solve the problem we define the matrix

$$W_Q/h,k/ = \sum_{/0,0/ \leq /i,j/ \leq /h,k/} M/h-i,k-j/ \, Q^{-1} \, M^T/h-i,k-j/ \qquad /3.75/$$

where the matrix $M/i,j/$ is defined by /3.5b/.

It is easy to prove that the matrix /3.75/ is non-singular if and only if RM is locally controllable in the rectangle $[/0,0/, /h,k/]$ /see Problem 2./.

Thus, we can define the sequence of inputs

$$\hat{u} = \hat{u}/i,j/ = Q^{-1} M^T/h-i,k-j/ \, W_Q^{-1}/h,k/ \, q/h,k/ \qquad /3.76/$$

where

$$q/h,k/ = x_f - \sum_{i=0}^{h} A^{h-i,k} \begin{bmatrix} 0 \\ x^v/i,0/ \end{bmatrix} - \sum_{j=0}^{k} A^{h,k-j} \begin{bmatrix} x^h/0,j/ \\ 0 \end{bmatrix} \qquad /3.77/$$

### Theorem 3.22

Let us assume that

i/ RM is locally controllable in the rectangle $[/0,0/, /h,k/]$,

ii/ $\bar{u} = \bar{u}/i,j/$ is any sequence of inputs defined for $/0,0/ \leq /i,j/ < /h,k/$ which transfers RM from $x_o$ to $x_f$.

Then the sequence of inputs /3.76/ accomplishes the same task and

$$I/\hat{u}/ \leq I/\bar{u}/ \qquad /3.78/$$

Moreover, the minimum value of /3.75/ is given by

$$I/\hat{u}/ = q^T/h,k/ \, W_Q^{-1}/h,k/ \, q/h,k/ \qquad /3.79/$$

Proof: First we shall show that the sequence of inputs /3.76/ provides $x/h,k/ = x_f$. Substituting /3.76/ into /1.28/ and using /3.77/ and /3.75/

we obtain

$$x/h,k/ = \sum_{i=0}^{h} A^{h-i,k} \begin{bmatrix} 0 \\ x^v/i,0/ \end{bmatrix} + \sum_{j=0}^{k} A^{h,k-j} \begin{bmatrix} x^h/0,j/ \\ 0 \end{bmatrix} +$$

$$+ \sum_{/0,0/ \leqslant /i,j/ < /h,k/} M/h-i,k-j/ \ \hat{u}/i,j/ =$$

$$= x_f - q/h,k/ + \sum_{/0,0/ \leqslant /i,j/ < /h,k/} M/h-i,k-j/ \ Q^{-1} M^T/h-i,k-j/$$

$$W_Q^{-1}/h,k/ \ q/h,k/ = x_f$$

Since $\bar{u}/i,j/$ and $\hat{u}/i,j/$ transfer RM from $x_o$ to $x_f$ then

$$\sum_{\substack{/h,k/ > \\ >/i,j/ \geqslant \\ \geqslant /0,0/}} M/h-i,k-j/ \ \bar{u}/i,j/ = \sum_{\substack{/h,k/ > \\ >/i,j/ \geqslant \\ \geqslant /0,0/}} M/h-i,k-j/ \ \hat{u}/i,j/$$

and

$$\sum_{/0,0/ \leqslant /i,j/ < /h,k/} M/h-i,k-j/ \left[\bar{u}/i,j/ - \hat{u}/i,j/\right] = 0 \qquad /3.80/$$

From /3.80/ and /3.76/ it follows that

$$\sum_{/0,0/ \leqslant /i,j/ < /h,k/} \left[\bar{u}/i,j/ - \hat{u}/i,j/\right]^T M^T/h-i,k-j/ \ W_Q^{-1}/h,k/ \ q/h,k/ =$$

$$= \sum_{/0,0/ \leqslant /i,j/ < /h,k/} \left[\bar{u}/i,j/ - \hat{u}/i,j/\right]^T Q \ \hat{u}/i,j/ = 0 \qquad /3.81/$$

Using /3.81/ it is easy to show that

$$\sum_{/0,0/ \leqslant /i,j/ < /h,k/} \bar{u}^T/i,j/ Q \ \bar{u}/i,j/ = \sum_{/0,0/ \leqslant /i,j/ < /h,k/} \hat{u}^T/i,j/ Q \ \hat{u}/i,j/ +$$

$$+ \sum_{/0,0/ \leqslant /i,j/ < /h,k/} \left[\bar{u}/i,j/ - \hat{u}/i,j/\right]^T Q \left[\bar{u}/i,j/ - \hat{u}/i,j/\right] \qquad /3.82/$$

The inequality /3.78/ holds since the last term in /3.82/ is always nonnegative.

To obtain the minimum value of /3.74/ we substitute /3.76/ into /3.74/

$$I/\hat{u}/ = \sum_{/0,0/ \leqslant /i,j/ < /h,k/} \hat{u}^T/i,j/ Q \ \hat{u}/i,j/ =$$

$$= \sum_{/0,0/ \leq /i,j/ \leq /h,k/} \left[ Q^{-1} M/h-i,k-j/ W_Q^{-1}/h,k/ q/h,k/ \right]^T Q \left[ Q^{-1} M/h-i,k-j/ \right.$$

$$\left. W_Q^{-1}/h,k/ q/h,k/ \right]$$

Taking into account /3.75/ and that $Q^{-1}$, $W_Q^{-1}/h,k/$ are symmetric we obtain

$$I/\hat{u}/ = q^T/h,k/ W_Q^{-1}/h,k/ \left[ \sum_{/0,0/ \leq /i,j/ \leq /h,k/} M/h-i,k-j/ Q^{-1} M^T/h-i,k-j/ \right]$$

$$W_Q^{-1}/h,k/ q/h,k/ =$$

$$= q^T/h,k/ W_Q^{-1}/h,k/ q/h,k/ \qquad \square$$

### Example 3.7

Solve the minimum energy control problem for RM with /3.9/, $Q = [1]$, zero boundary conditions and

$$x_o = \begin{bmatrix} x^h/0,0/ \\ \hline x^v/0,0/ \end{bmatrix} = \begin{bmatrix} 0 \\ \hline 0 \end{bmatrix}, \quad x_f = \begin{bmatrix} x^h/1,1/ \\ \hline x^v/1,1/ \end{bmatrix} = \begin{bmatrix} 1 \\ 2 \\ \hline -1 \end{bmatrix} \qquad /3.83/$$

In Example 3.1 it was shown that RM with /3.9/ is locally controllable in the rectangle $[/0,0/, /1,1/]$.

Using /3.75/ and the results of Example 3.1 we calculate

$$W_Q/1,1/ = \sum_{/0,0/ \leq /i,j/ < /1,1/} M/1-i,1-j/ Q^{-1} M^T/1-i,1-j/ =$$

$$= M/1,1/ M^T/1,1/ + M/1,0/ M^T/1,0/ + M/0,1/ M^T/0,1/ =$$

$$= \begin{bmatrix} 1 & 0 & 0 \\ 0 & 1 & 1 \\ 0 & 1 & 2 \end{bmatrix}$$

and

$$W_Q^{-1}/1,1/ = \begin{bmatrix} 1 & 0 & 0 \\ 0 & 2 & -1 \\ 0 & -1 & 1 \end{bmatrix}$$

In this case $q/1,1/ = x_f$ and from /3.76/ we have

$$\hat{u}/0,0/ = Q^{-1} M^T/1,1/ W_Q^{-1}/1,1/ x_f = 2.$$

$$\hat{u}/1,0/ = Q^{-1} M/0,1/^T W_Q^{-1}/1,1/ x_f = -3$$

$$\hat{u}/0,1/ = Q^{-1} M/1,0/^T W_Q^{-1}/1,1/ x_f = 1$$

The minimum value of the performance index is calculated from /3.79/

$$I/\hat{u}/ = q/1,1/^T W_Q^{-1}/1,1/ q/1,1/ =$$

$$= \begin{bmatrix} 1 & 2 & -1 \end{bmatrix} \begin{bmatrix} 1 & 0 & 0 \\ 0 & 2 & -1 \\ 0 & -1 & 1 \end{bmatrix} \begin{bmatrix} 1 \\ 2 \\ -1 \end{bmatrix} = 14$$

The presented method was extended for 3-D linear systems in [9].

## 3.6 LOCAL CONTROLLABILITY AND OBSERVABILITY OF FORNASINI-MARCHESINI'S MODELS

### 3.6.1 Local controllability.

Consider the first Fornasini-Marchesini's model /F-MM I/ described by the equations

$$x/i+1,j+1/ = A_0 x/i,j/ + A_1 x/i+1,j/ + A_2 x/i,j+1/ + Bu/i,j/ \qquad /3.84/$$

$$y/i,j/ = Cx/i,j/ \qquad /3.85/$$

where

i,j are integer-valued vertical and horizontal coordinates, respectively,

$x/i,j/ \in R^n$ is the local state vector,

$u/i,j/ \in R^m$ is the input vector,

$y/i,j/ \in R^l$ is the output vector,

$A_0, A_1, A_2, B, C$ are real matrices of appropriate dimensions.

Definition 3.12

F-MM I is said to be locally controllable in the rectangle $[/0,0/,/h,k/]$ iff for every boundary conditions $x/i,0/$, $i \in [0,h]$, $x/0,j/$, $j \in [0,k]$

and every vector $x_f \in R^n$ there exists a sequence of inputs $u/i,j/$, $/0,0/ \leq /i,j/ < /h,k/$ such that $x/h,k/ = x_f$.

## Theorem 3.23

F-MM I is locally controllable in the rectangle $[/0,0/, /h,k/]$ if and only if

$$\text{rank} \begin{bmatrix} B & A_{21}^{1,0} B & A_{21}^{0,1} B & \ldots & A_{21}^{h-2,k} B & A_{21}^{h-1,k} B \end{bmatrix} = n \qquad /3.86/$$

where

$$\begin{bmatrix} A_{11}^{i,j} & A_{12}^{i,j} \\ A_{21}^{i,j} & A_{22}^{i,j} \end{bmatrix} = \begin{bmatrix} A_2 & A_0 + A_2 A_1 \\ I_n & A_1 \end{bmatrix}^{i,j} \qquad /3.87/$$

**Proof:** Using the formula /1.80/ and taking into account that $x/h,k/ = x_f$ we can write

$$x_f - \sum_{i=0}^{h} A^{h-i,k} x/i,0/ - \sum_{j=0}^{k} A_{21}^{h,k-j} x/0,j+1/ - A_1 x/0,j/ =$$

$$= \sum_{/0,0/ \leq /i,j/ < /h,k/} A_{21}^{h-i-1,k-j} Bu/i,j/ =$$

$$= \begin{bmatrix} B & A^{1,0} B & A^{0,1} B & \ldots & A^{h-2,k} B & A^{h-1,k} B \end{bmatrix} \begin{bmatrix} u/h-1,k/ \\ u/h-2,k/ \\ u/h-1,k-1/ \\ \vdots \\ u/1,0/ \\ u/0,0/ \end{bmatrix} \qquad /3.88/$$

From /3.88/ it follows that F-MM I is locally controllable in the rectangle $[/0,0/, /h,k/]$ if and only if the condition /3.86/ holds. □

Consider the second Fornasini-Marchesini's model /F-MM II/ described by the equations

$$x/i+1,j+1/ = A_1 x/i,j+1/ + A_2 x/i+1,j/ + B_2 u/i+1,j/ + B_1 u/i,j+1/ \qquad /3.89/$$

$$y/i,j/ = Cx/i,j/ + Du/i,j/ \qquad /3.90/$$

where

$i,j$ are integer-valued vertical and horizontal coordinates, respectively,

$x/i,j/ \in R^n$ is the local state vector at $/i,j/$,

$u/i,j/ \in R^m$ is the input vector,

$y/i,j/ \in R^l$ is the output vector,

$A_1$, $A_2$, $B_1$, $B_2$, C, D are real matrices of appropriate dimensions.

The local controllability of F-MM II is defined in the same way as for F-MM I /Definition 3.12/.

## Theorem 3.24

F-MM II is locally controllable in the rectangle $[/0,0/, /h,k/]$ if and only if

$$\operatorname{rank}\left[B_1 \; B_2 \; A^{1,0}B_1 \; A^{0,1}B_2 \; \ldots \; A^{h-1,k}B_1 + A^{h,k-1}B_2\right] = n \qquad /3.91/$$

where $A^{i,j}$ is defined by /1.84/.

Proof: Using the formula /1.86/ and taking into account that $x/h,k/ = x_f$ we can write

$$x_f - \sum_{i=0}^{h-1} A^{i,k} x/h-i,0/ - \sum_{j=0}^{k-1} A^{h,j} x/0,k-j/ =$$

$$= \sum_{/0,0/ \leqslant /i,j/ < /h,k/} \left[A^{i-1,j}B_1 + A^{i,j-1}B_2\right] u/h-i,k-j/ = \qquad /3.92/$$

$$= \left[B_1 \; B_2 \; A^{1,0}B_1 \; A^{0,1}B_2 \; \ldots \; A^{h-1,k}B_1 + A^{h,k-1}B_2\right] \begin{bmatrix} u/h-1,k/ \\ u/h,k-1/ \\ u/h-2,k/ \\ u/h,k-2/ \\ \cdots \\ u/0,0/ \end{bmatrix}$$

From /3.92/ it follows that F-MM II is locally controllable in the rectangle $[/0,0/, /h,k/]$ if and only if the condition /3.91/ holds. □

In a similar way as for RM it is easy to show that for F-MM II /and, respectively, for F-MM I/ $A^{i,j}$ with $i+j \geqslant n$ are linear combination of $A^{i,j}$ with $i+j < n$ /see Problem 5 in Chapter 2/. Therefore F-MM II is locally controllable in the rectangle $[/0,0/, /h,k/]$ for $h+k \geqslant n$ only if it is locally controllable in the rectangle $[/0,0/, /\bar{h},\bar{k}/]$ for $\bar{h}+\bar{k} < n$.

## 3.6.2 Local observability.

### Definition 3.13

F-MM II is said to be locally observable in the rectangle $\left[/0,0/,\ /h,k/\right]$ iff there is no local initial state $x/0,0/ \neq 0$ such that for zero inputs $u/i,j/$, $/0,0/ \leqslant /i,j/ \leqslant /h,k/$ and zero boundary conditions $x/i,0/ = 0$, $i \in [1,h]$, $x/0,j/ = 0$, $j \in [1,k]$ the output is also zero $y/i,j/ = 0$, $/0,0/ \leqslant /i,j/ \leqslant /h,k/$.

### Theorem 3.25

F-MM II is locally observable in the rectangle $\left[/0,0/,\ /h,k/\right]$ if and only if

$$\operatorname{rank} \begin{bmatrix} C \\ CA^{1,0} \\ CA^{0,1} \\ \vdots \\ CA^{h,k} \end{bmatrix} = n \qquad /3.93/$$

Proof: From Theorem 1.7 for $u/i,j/ = 0$, $/0,0/ \leqslant /i,j/ \leqslant /h,k/$ and $x/i,0/ = 0$, $i \in [1,h]$, $x/0,j/ = 0$, $j \in [1,k]$ it follows that

$$x/i,j/ = A^{i,j} x/0,0/ \qquad /3.94/$$

Substitution of /3.94/ into /3.90/ yields

$$y/i,j/ = CA^{i,j} x/0,0/ \quad \text{for} \quad /0,0/ \leqslant /i,j/ \leqslant /h,k/ \qquad /3.95/$$

Taking into account that $y/i,j/ = 0$, $/0,0/ \leqslant /i,j/ \leqslant /h,k/$ we obtain

$$\begin{bmatrix} C \\ CA^{1,0} \\ CA^{0,1} \\ \vdots \\ CA^{h,k} \end{bmatrix} x/0,0/ = 0 \qquad /3.96/$$

Note that if /3.93/ holds, then /3.96/ implies $x/0,0/ = 0$ and if /3.93/ is not satisfied, then there exists $x/0,0/ \neq 0$ such that /3.96/ holds.

□

Since $A^{i,j}$ with $i+j \geq n$ are linear combinations of $A^{i,j}$ with $i+j < n$, F-MM II is locally observable in the rectangle $[/0,0/, /h,k/]$ for $h+k \geq n$ only if it is locally observable in the rectangle $[/0,0/, /\bar{h},\bar{k}/]$ for $\bar{h}+\bar{k} < n$.

PROBLEMS

1. Show that the matrix $F = \begin{bmatrix} F_1 & F_2 \end{bmatrix}$ can be chosen so that

$$A_c = A + BF = \begin{bmatrix} A_{c1} & A_{c2} \\ 0 & A_{c4} \end{bmatrix}$$

if and only if

$$\text{rank } B_2 = \text{rank}\begin{bmatrix} B_2 & A_3 \end{bmatrix}$$

where

$$A = \begin{bmatrix} A_1 & A_2 \\ A_3 & A_4 \end{bmatrix}, \quad B = \begin{bmatrix} B_1 \\ B_2 \end{bmatrix}$$

Hint: Use the Kronecker-Cappeli's theorem.

2. Show that the matrix /3.75/ is non-singular if and only if RM is locally controllable in the rectangle $\begin{bmatrix} /0,0/, /h,k/ \end{bmatrix}$.

Hint: Show that the nonsingularity of /3.75/ is equivalent to the nonsingularity of /3.7/.

3. Show that T-PM with

$$A = \begin{bmatrix} A_{11} & A_{12} & A_{13} \\ A_{21} & A_{22} & A_{23} \\ A_{31} & A_{32} & A_{33} \end{bmatrix}, \quad B = \begin{bmatrix} B_1 \\ B_2 \\ B_3 \end{bmatrix}, \quad A_{ij} \in R^{n_i \times n_j}, \quad B_i \in R^{n_i \times m}$$

$$i, j = 1, 2, 3$$

is locally controllable in the cube $\begin{bmatrix} /0,0,0/, /r,p,q/ \end{bmatrix}$ if and only if

$$\text{rank}\begin{bmatrix} M/1,0,0/ & M/0,1,0/ & M/0,0,1/ & \cdots & M/i,j,k/ & \cdots & M/r,p,q/ \end{bmatrix} = n_1 + n_2 + n_3$$

where

$$M/i,j,k/ = A^{i-1,j,k}\begin{bmatrix} B_1 \\ 0 \\ 0 \end{bmatrix} + A^{i,j-1,k}\begin{bmatrix} 0 \\ B_2 \\ 0 \end{bmatrix} + A^{i,j,k-1}\begin{bmatrix} 0 \\ 0 \\ B_3 \end{bmatrix}$$

Hint: Use a similar procedure as in the **proof** of Theorem 3.1.

4. Show that if for RM

$$\begin{bmatrix} \bar{x}^h/i,j/ \\ \bar{x}^v/i,j/ \end{bmatrix} = \begin{bmatrix} T_1 & 0 \\ 0 & T_2 \end{bmatrix}\begin{bmatrix} x^h/i,j/ \\ x^v/i,j/ \end{bmatrix}, \quad \det T_1 \neq 0, \quad \det T_2 \neq 0$$

then

$$\bar{A}_{11} = T_1 A_{11} T_1^{-1}, \quad \bar{A}_{12} = T_1 A_{12} T_2^{-1}, \quad \bar{A}_{21} = T_2 A_{21} T_1^{-1}, \quad \bar{A}_{22} = T_2 A_{22} T_2^{-1},$$

$$\bar{B}_1 = T_1 B_1, \quad \bar{B}_2 = T_2 B_2, \quad \bar{C}_1 = C_1 T_1^{-1}, \quad \bar{C}_2 = C_2 T_2^{-1}, \quad \bar{D} = D$$

and

$$G/z_1, z_2/ = \begin{bmatrix} C_1 & C_2 \end{bmatrix} \begin{bmatrix} I_{n_1} z_1 - A_{11} & -A_{12} \\ -A_{21} & I_{n_2} z_2 - A_{22} \end{bmatrix}^{-1} \begin{bmatrix} B_1 \\ B_2 \end{bmatrix} + D =$$

$$= \begin{bmatrix} \bar{C}_1 & \bar{C}_2 \end{bmatrix} \begin{bmatrix} I_{n_1} z_1 - \bar{A}_{11} & -\bar{A}_{12} \\ -\bar{A}_{21} & I_{n_2} z_2 - \bar{A}_{22} \end{bmatrix}^{-1} \begin{bmatrix} \bar{B}_1 \\ \bar{B}_2 \end{bmatrix} + \bar{D}$$

<u>Hint</u>: Note that

$$\begin{bmatrix} I_{n_1} z_1 - \bar{A}_{11} & -\bar{A}_{12} \\ -\bar{A}_{21} & I_{n_2} z_2 - \bar{A}_{22} \end{bmatrix} = \begin{bmatrix} T_1 & 0 \\ 0 & T_2 \end{bmatrix} \begin{bmatrix} I_{n_1} z_1 - A_{11} & -A_{12} \\ -A_{21} & I_{n_2} z_2 - A_{22} \end{bmatrix} \begin{bmatrix} T_1^{-1} & 0 \\ 0 & T_2^{-1} \end{bmatrix}$$

5. Show that the controllability /observability/ of the pair A,B /C,A/ in a 1-D sense implies the local controllability /observability/ of RM.

   <u>Hint</u>: Show that

   $$\sum_{i=0}^{k} M/i, k-i/ = A^{k-1} B \quad \text{and} \quad \sum_{i=0}^{k} A^{i,k-i} = A^k \quad \text{for } k > 0.$$

6. Show that if RM is modally controllable, then it is separately locally controllable.

   <u>Hint</u>: Use Theorem 3.16 and Theorem 3.5.

7. Show that for

   $$A = \begin{bmatrix} A_{11} & 0 \\ A_{21} & A_{22} \end{bmatrix}$$

   the following relations hold:

   $$A^{i,j} = \begin{bmatrix} 0 & 0 \\ A_{22}^{j-1} A_{21} A_{11}^{i} & 0 \end{bmatrix} \quad i, j \geq 1$$

$$M/i,j/ = A^{i-1,j}\begin{bmatrix} B_1 \\ 0 \end{bmatrix} + A^{i,j-1}\begin{bmatrix} 0 \\ B_2 \end{bmatrix} = \begin{bmatrix} 0 \\ A_{22}^{j-1} A_{21} A^{i-1} B_1 \end{bmatrix} \qquad i,j \geq 1.$$

Hint: Use the definition of $A^{i,j}$.

8. Solve the minimum energy control problem for RM with

$$A = \begin{bmatrix} A_{11} & A_{12} \\ A_{21} & A_{22} \end{bmatrix} = \left[\begin{array}{cc|c} 0 & 0 & 1 \\ 0 & 0 & 1 \\ \hline 1 & 0 & 0 \end{array}\right], \quad B = \begin{bmatrix} B_1 \\ B_2 \end{bmatrix} = \begin{bmatrix} 1 \\ 1 \\ 0 \end{bmatrix}, \quad Q = 1$$

zero boundary conditions and

$$x_o = \begin{bmatrix} x^h/0,0/ \\ x^v/0,0/ \end{bmatrix} = \begin{bmatrix} 1 \\ 5 \\ 3 \end{bmatrix}, \quad x_f = \begin{bmatrix} x^h/1,2/ \\ x^v/1,2/ \end{bmatrix} = \begin{bmatrix} 3 \\ 2 \\ 1 \end{bmatrix}$$

Answer:   $\hat{u}/0,0/ = 0$,    $\hat{u}/1,1/ = 3$
          $\hat{u}/1,0/ = 0$,    $\hat{u}/0,2/ = 1$
          $\hat{u}/0,1/ = 1$,    $\hat{u}/i,j/ = 0$    otherwise

$I/\hat{u}/ = 11$.

# REFERENCES

[1] Eising R.: Controllability and observability of 2-D systems. IEEE Trans. Autom. Control vol. AC-24, no.1, Febr. 1979, pp.132-133.

[2] Eising R.: Realization and stabilization of 2-D systems. IEEE Trans. Autom. Control vol. AC-23, no.5, Oct. 1978, pp.793-799.

[3] Eising R.: 2-D Systems; An Algebraic Approach. Mathematisch Centrum, Amsterdam 1979.

[4] Fornasini E., Marchesini G.: Global properties and duality in 2-D systems. Systems and Control Letters vol.2, no.1, July 1982, pp.30-38.

[5] Fornasini E., MarchesiniG.: State-space realization theory of two-dimensional filters. IEEE Trans. Autom. Control vol. AC-21, no.4, Aug. 1976, pp.484-491.

[6] Heymann M.: Comments on "pole assignment in multi-input controllable linear systems". IEEE Trans. Autom. Control vol. AC-13, 1968, pp.748-749.

[7] Hinamoto T.: Realization of a state-space model from two-dimensional input-output map. IEEE Trans. Circuits and Systems vol. CAS-27, no.1, Jan. 1980, pp.36-44.

[8] Hinamoto T., Maekawa S.: Canonic form state space realization of two-dimensional transfer functions having separable denominator. Int.J.Systems Sci. vol.13, no.10, 1982, pp.1083-1091.

[9] Kaczorek T.: Minimum energy control for 3-D systems. Control and cybernetics vol.12, no.3-4, 1983, pp.53-58.

[10] Kaczorek T.: Separability of transfer function matrices of 2-D linear systems by state feedbacks. Int.J.Systems Sci. vol.13, no.9, 1982, pp.1013-1018.

[11] Klamka J.: Controllability of M-dimensional linear systems. Foundations of Control Engineering, vol.8, no.2, 1983, pp.65-74.

[12] Klamka J.: Controllability and optimal control of 2-D linear systems. Foundations of Control Engineering vol.9, 1984, pp.

[13] Klamka J.: Minimum energy control for 2-D systems. Systems Science, vol.9, 1983 /in press/

[14] Kung S.Y., Lévy B.C., Morf M., Kailath T.: New results in 2-D systems theory; part II: 2-D state-space models - realization and the notions of controllability, observability and minimality. Proc. IEEE vol.65, no.6, June 1977, pp.945-961.

[15] Morf M., Lévy B.C., Kung S.Y.: New results in 2-D systems theory, Part I: 2-D polynomial matrices, Factorization, and coprimeness. Proc. IEEE vol.65, no.6, June 1977, pp.861-872.

[16] Paraskevopoulos P.N.: Characteristic polynomial assignment and determination of the residual polynomial in 2-D systems. IEEE Trans. Autom. Control vol. AC-26, no.2, 1981, pp.541-543.

[17] Paraskevopoulos P.N., Mertzios B.G.: Transfer function factorization of 2-D systems using state feedback. Int.J.Systems Sci. vol. 12, 1981, pp.1135-1147.

[18] Roesser R.P.: A discrete state-space model for linear image processing. IEEE Trans. Autom. Control vol. AC-20, no.1, Febr. 1975, pp.1-10.

[19] Turhan Çiftçibaşi, Önder Yüksel: Sufficient or necessary conditions

for modal controllability and observability of Roesser's 2-D system model. IEEE Trans. Autom. Control vol. AC-28, no.4, 1983.

# 4. STABILITY AND STABILIZATION

## 4.1 STABILITY OF 2-D LINEAR INPUT - OUTPUT SYSTEMS

### 4.1.1 Definition of BIBO stability and Shank's theorem.

Consider an input - output system described by the equation

$$y/h,k/ = \sum_{i=0}^{h} \sum_{j=0}^{k} g/h-i,k-j/\, u/i,j/ \qquad h,k \geq 0 \qquad /4.1/$$

where $u/i,j/ \in R^m$ is the input vector at $/i,j/$,
$y/i,j/ \in R^l$ is the output vector,
$g/i,j/ \in R^{l \times m}$ is the impulse response of the system.

The system is causal if $g/i,j/ = 0$ for $i < 0$ or $j < 0$ /Definition 2.1/. In what follows we assume that the system is causal.

### Definition 4.1

The system /4.1/ is said to be bounded-input bounded-output /BIBO/ stable iff for every $M > 0$ there exists an $N > 0$ such that, if $\|u/i,j/\| \leq M$ for all $/i,j/$, then $\|y/h,k/\| \leq N$ for all $/h,k/$, where $\|v\|$ is a norm of $v$.

### Theorem 4.1

The system /4.1/ is BIBO stable if and only if

$$\sum_{i=0}^{\infty} \sum_{j=0}^{\infty} \|g/i,j/\| < \infty \qquad /4.2/$$

Proof: From /4.1/ and the well known properties of the norm we have

$$\|y/h,k/\| = \left\| \sum_{i=0}^{h} \sum_{j=0}^{k} g/h-i,k-j/\, u/i,j/ \right\| =$$

$$= \left\| \sum_{i=0}^{\infty} \sum_{j=0}^{\infty} g/h-i,k-j/\, u/i,j/ \right\| \leq$$

$$\leq \sum_{i=0}^{\infty} \sum_{j=0}^{\infty} \|g/h-i,k-j/\|\, \|u/i,j/\| \leq M \sum_{i=0}^{\infty} \sum_{j=0}^{\infty} \|g/h-i,k-j/\| \qquad /4.3/$$

where $M \geq \|u/i,j/\|$.

If /4.2/ holds then $\|y/h,k/\| \leq N$ for all /h,k/ and the system is BIBO stable. To prove the necessity let us assume $\|u/i,j/\| = 1$ for all /i,j/. Then from /4.3/ it follows that the system is BIBO stable only if /4.2/ holds. □

Let $d/z_1,z_2/$ be the least common multiple of the denominators of the entries of a transfer function matrix $G/z_1,z_2/$.
Then

$$G/z_1,z_2/ = \frac{N/z_1,z_2/}{d/z_1,z_2/} \qquad /4.4/$$

where $N/z_1,z_2/$ is a polynomial matrix in $z_1,z_2$ and

$$d/z_1,z_2/ = \sum_{i=0}^{n_1} \sum_{j=0}^{n_2} d_{ij} z_1^i z_2^j \qquad /4.5/$$

We assume that $N/z_1,z_2/$ and $d/z_1,z_2/$ are factor coprime, i.e. a common divisor of $d/z_1,z_2/$ and all entries of $N/z_1,z_2/$ is a constant.

**Theorem 4.2**

The system with transfer function matrix /4.4/ is BIBO stable if

$$d/z_1,z_2/ \neq 0 \quad \text{for} \quad |z_1| \geq 1, \ |z_2| \geq 1. \qquad /4.6/$$

**Proof:** If /4.6/ then /4.4/ is analytic for $|z_1| \geq 1$, $|z_2| \geq 1$ and the series expansion

$$G/z_1,z_2/ = \sum_{i=0}^{\infty} \sum_{j=0}^{\infty} g/i,j/ \ z_1^{-i} z_2^{-j} \qquad /4.7/$$

is absolutely convergent for $|z_1| \geq 1$, $|z_2| \geq 1$. This implies /4.2/ and the BIBO stability of the system. □

The condition /4.6/ is not a necessary condition for BIBO stability of the system. This is because of the possible occourance of nonessential singularities of the second kind on $|z_1| = 1$, $|z_2| = 1$ for /4.4/.
A point $/z_1,z_2/$ such that $d/z_1,z_2/ = 0$ and $N/z_1,z_2/ = 0$ is called a nonessential singularity of the second kind /the value of /4.4/ is undefined/.

### Theorem 4.3

If the system with transfer function matrix /4.4/ is BIBO stable, then

$$d/z_1,z_2/ \neq 0 \quad \text{for} \quad |z_1| \geq 1, \; |z_2| \geq 1$$

and /4.4/ has no nonessential singularities of the second kind for $|z_1| \geq 1$, $|z_2| \geq 1$ except possibly on $|z_1| = 1$, $|z_2| = 1$.

**Proof:** If the system is BIBO stable, then /4.2/ holds and /4.7/ is absolutely convergent for $|z_1| \geq 1$, $|z_2| \geq 1$. Therefore /4.4/ is analytic for $|z_1| \geq 1$, $|z_2| \geq 1$ and /4.6/ holds. Note that if /4.4/ is analytic then it has no poles and no nonessential singularities of the second kind for $|z_1| \geq 1$, $|z_2| \geq 1$ except possibly on $|z_1| = 1$, $|z_2| = 1$. □

From Theorem 4.2 and Theorem 4.3 the following theorem follows [23]

### Shank's theorem.

The system with transfer function matrix /4.4/, which has no nonessential singularities of the second kind for $|z_1| = 1$, $|z_2| = 1$, is BIBO stable if and only if /4.6/ holds.

Multiplying the numerator and the denominator of /4.4/ by $z_1^{-n_1} z_2^{-n_2}$ we obtain

$$d/z_1,z_2/ = \sum_{i=0}^{n_1} \sum_{j=0}^{n_2} d_{ij} z_1^{i-n_1} z_2^{j-n_2} = \sum_{i=0}^{n_1} \sum_{j=0}^{n_2} \bar{d}_{ij} z_1^{-i} z_2^{-j} \qquad /4.8/$$

where $\bar{d}_{ij} = d_{n_1-i, n_2-j}$

Without loss of generality we can assume $d_{n_1 n_2} = \bar{d}_{00} = 1$ and $d/z_1^{-1}, z_2^{-1}/$ can be written in the form

$$d/z_1^{-1}, z_2^{-1}/ = 1 + \sum_{/0,0/ < /i,j/ < /n_1,n_2/} \bar{d}_{ij} z_1^{-i} z_2^{-j} \qquad /4.9/$$

### Theorem 4.4

The system with proper transfer function matrix /4.4/ which has no nonessential singularities of the second kind for $|z_1| = 1$, $|z_2| = 1$ is BIBO stable if

$$\sum_{/0,0/ < /i,j/ < /n_1,n_2/} |\bar{d}_{ij}| < 1 \qquad /4.10/$$

Proof: Using /4.9/ we can write

$$\left|d/z_1^{-1},z_2^{-1}/\right| = \left|1 + \sum_{/0,0/</i,j/\leq/n_1,n_2/} \bar{d}_{ij}z_1^{-i}z_2^{-j}\right| \geq$$

$$\geq 1 - \left|\sum_{/0,0/</i,j/\leq/n_1,n_2/} \bar{d}_{ij}z_1^{-i}z_2^{-j}\right| \geq$$

$$\geq 1 - \sum_{/0,0/</i,j/\leq/n_1,n_2/} |\bar{d}_{ij}||z_1^{-i}||z_2^{-j}|$$

For $|z_1^{-1}| \leq 1$, $|z_2^{-1}| \leq 1$ we have

$$\left|d/z_1^{-1},z_2^{-1}/\right| \geq 1 - \sum_{/0,0/</i,j/\leq/n_1,n_2/} |\bar{d}_{ij}|$$

and $d/z_1^{-1},z_2^{-1}/ \neq 0$ for $|z_1^{-1}| \leq 1$, $|z_2^{-1}| \leq 1$ if /4.10/ holds. Therefore, by Shank's theorem the system is BIBO stable. □

Example 4.1

Test the BIBO stability of the system with

$$d/z_1,z_2/ = z_1 z_2 + \frac{1}{3}z_1 + \frac{1}{4}z_2 - \frac{1}{3}z_1 z_2$$

In this case $n_1 = n_2 = 1$ and

$$d/z_1^{-1},z_2^{-1}/ = 1 + \frac{1}{3}z_2^{-1} + \frac{1}{4}z_1^{-1} - \frac{1}{3}z_1^{-1}z_2^{-1}$$

Hence $\bar{d}_{00}=1$, $\bar{d}_{01}=\frac{1}{3}$, $\bar{d}_{10}=\frac{1}{4}$, $\bar{d}_{11}=\frac{-1}{3}$

and

$$\sum_{/0,0/</i,j/\leq/n_1,n_2/} |\bar{d}_{ij}| = \frac{1}{3} + \frac{1}{4} + \frac{1}{3} = \frac{11}{12}$$

Therefore the condition /4.10/ is satisfied and by Theorem 4.4 the system is BIBO stable.

### 4.1.2 Huang's theorem and Strinzis' theorems.

Consider a transfer function of the form

$$G/z_1^{-1},z_2^{-1}/ = \frac{n/z_1^{-1},z_2^{-1}/}{d/z_1^{-1},z_2^{-1}/} \qquad /4.11/$$

where $n/z_1^{-1},z_2^{-1}/$, $d/z_1^{-1},z_2^{-1}/$ are polynomials in $z_1^{-1}$ and $z_2^{-1}$. In what follows we assume that /4.11/ has no nonessential singularities of the second kind for $|z_1^{-1}| = 1$, $|z_2^{-1}| = 1$.

**Huang's theorem** [15]

The system with transfer function /4.11/ is BIBO stable if and only if

$1^o \quad d/z_1^{-1},0/ \neq 0 \quad \text{for} \quad |z_1^{-1}| \leq 1 \qquad /4.12a/$

$2^o \quad d/z_1^{-1},z_2^{-1}/ \neq 0 \quad \text{for} \quad |z_1^{-1}| = 1, \quad |z_2^{-1}| \leq 1 \qquad /4.12b/$

**Proof:** From Shank's theorem it follows that to prove the theorem we have to show that the conditions /4.12/ are equivalent to the condition

$$d/z_1^{-1},z_2^{-1}/ \neq 0 \quad \text{for} \quad |z_1^{-1}| \leq 1, \quad |z_2^{-1}| \leq 1 \qquad /4.13/$$

Note that the condition /4.13/ implies the conditions /4.12/. Let $z_2^{-1} = f/z_1^{-1}/$ be the algebraic function determined by $d/z_1^{-1},z_2^{-1}/ = 0$. The condition /4.12a/ implies that $f/z_1^{-1}/ \neq 0$ for $|z_1^{-1}| = 1$ and the condition /4.12b/ implies that $|f/z_1^{-1}/| > 1$ for $|z_1^{-1}| = 1$. From minimum modulus theorem it follows that the minimum modulus $|f/z_1^{-1}/|$ for $|z_1^{-1}| \leq 1$ cannot occur for such that $|z_1^{-1}| < 1$. Therefore $|f/z_1^{-1}/| > 1$ for $|z_1^{-1}| \leq 1$ and $d/z_1^{-1},z_2^{-1}/ \neq 0$ for $|z_1^{-1}| \leq 1$, $|z_2^{-1}| \leq 1$. □

The polynomial /4.5/ can be written in the form

$$d/z_1,z_2/ = \sum_{j=0}^{n_2} d_j/z_1/ \, z_2^j \qquad /4.14/$$

We assume that $d_{n_2}/z_1/$ is not the zero polynomial.

**Theorem 4.5**

The system with proper transfer function matrix /4.4/ is BIBO stable if and only if

$1°$ $d_{n_2}/z_1/ \neq 0$ for $|z_1| \geq 1$ /4.15a/

$2°$ $d/z_1,z_2/ \neq 0$ for $|z_1|=1$, $|z_2| \geq 1$ /4.15b/

**Proof:** Multiplying the numerator and the denominator of /4.4/ by $z_2^{-n_2}$ and taking into account /4.14/ we obtain

$$d/z_1,z_2^{-1}/ = d_{n_2}/z_1/ + d_{n_2-1}/z_1/ z_2^{-1} + \ldots + d_0/z_1/ z_2^{-n_2} \qquad /4.16/$$

From Huang's theorem and /4.16/ it follows that the condition /4.15a/ is equivalent to /4.12a/ and the condition /4.15b/ is equivalent to /4.12b/. □

Note that the possible occurrence of nonessential singularities of the second kind can only violate the condition /4.15b/. Therefore BIBO stability of the system implies /4.15a/.

## Example 4.2

Using theorem 4.5 test the BIBO stability of the system with

$$d/z_1,z_2/ = z_1 z_2 + \frac{2}{3} z_1 + \frac{1}{3} z_2 + \frac{1}{3}$$

In this case $n_1 = n_2 = 1$ and

$$d_1/z_1/ = z_1 + \frac{1}{3} \neq 0 \quad \text{for} \quad |z_1| \geq 1$$

So the condition /4.15a/ is satisfied.
It is easy to see that

$$d/e^{j\theta},z_2/ = z_2/e^{j\theta} + \frac{1}{3}/ + \frac{2}{3} e^{j\theta} + \frac{1}{3} \neq 0 \quad \text{for} \quad |z_2| \geq 1 \text{ and } \theta \in [0,2\pi].$$

Therefore the condition /4.15b/ is satisfied and the system is BIBO stable.

Note that in this case the condition /4.10/ is not satisfied.

## Strintzis' theorem 1 [24]

The system with transfer function /4.11/ is BIBO stable if and only if

$1°$ $d/z_1^{-1},a/ \neq 0$ for $|z_1^{-1}| \leq 1$ and some $a$, $|a| \leq 1$ /4.17a/

$2°$ $d/z_1^{-1},z_2^{-1}/ \neq 0$ for $|z_1^{-1}|=1$, $|z_2^{-1}| \leq 1$ /4.17b/

**Proof:** To prove the theorem we have to show that the conditions /4.17/

are equivalent to the condition /4.13/.
Note that the condition /4.13/ implies the conditions /4.17/.
To show that /4.17/ imply /4.13/ we denote by $n/z_2^{-1}/$ the number of zeros in $z_1^{-1}$ of $d/z_1^{-1}, z_2^{-1}/$ that lie in the region $|z_1^{-1}| \leqslant 1$. It is well known that $n/z_2^{-1}/$ is given by the Cauchy principal value formula [22]:

$$n/z_2^{-1}/ = \frac{1}{2\pi j} \oint_{|z_1^{-1}|=1} \frac{\partial d/z_1^{-1}, z_2^{-1}/}{\partial z_1^{-1}} \frac{1}{d/z_1^{-1}, z_2^{-1}/} dz_1^{-1} \qquad /4.18/$$

From /4.17b/ it follows that the integrand in /4.18/ is a continuous function of $z_1^{-1}$ and $z_2^{-1}$ in the region $\{|z_1^{-1}| = 1, |z_2^{-1}| \leqslant 1\}$. Since the integral is over a set of finite measure, the Bounded Convergence Theorem [22] immediately implies that $n/z_2^{-1}/$ is a continuous function of $z_2^{-1}$ when $|z_2^{-1}| \leqslant 1$. It is also integer-valued and hence a constant $n/z_2/ = n$ for all $|z_2^{-1}| \leqslant 1$. By /4.17a/ this constant is zero and hence $d/z_1^{-1}, z_2^{-1}/ \neq 0$ for $|z_1^{-1}| \leqslant 1$, $|z_2^{-1}| \leqslant 1$. □

Note that Huang's theorem is a particular case of Strintzis' theorem 1 for $a = 0$.

### Strintzis' theorem 2 [24]

The system with transfer function /4.11/ is BIBO stable if and only if

$1°\quad d/a, z_2^{-1}/ \neq 0$ for some $a$, $|a| \leqslant 1$, $|z_2^{-1}| \leqslant 1$ \qquad /4.19a/

$2°\quad d/z_1^{-1}, b/ \neq 0$ for $|z_1^{-1}| \leqslant 1$, some $b$, $|b| = 1$ \qquad /4.19b/

$3°\quad d/z_1^{-1}, z_2^{-1}/ \neq 0$ for $|z_1^{-1}| = |z_2^{-1}| = 1$ \qquad /4.19c/

Proof: To prove the theorem we have to show that the conditions /4.19/ are equivalent to the condition /4.13/.
Note that the condition /4.13/ implies the conditions /4.19/.
To prove the sufficiency of /4.19/ we shall use the same line as in the proof of Strintzis' theorem 1. From /4.19c/ it follows that the integrand in /4.18/ is continuous in $z_1^{-1}$ and $z_2^{-1}$ for $|z_2^{-1}| = 1$, and the integration is over a set of finite measure. This implies that $n/z_2^{-1}/$ is a continuous function of $z_2^{-1}$ when $|z_2^{-1}| = 1$. Further, for each fixed $z_2^{-1}$, $n/z_2^{-1}/$ is equal to the corresponding number of zeros in $z_1^{-1}$ of $d/z_1^{-1}, z_2^{-1}/$ that lie in the region $|z_1^{-1}| \leqslant 1$, and hence is integer valued. Thus $n/z_2^{-1}/ = n$ must be constant for all $z_2^{-1}$, $|z_2^{-1}| = 1$. By /4.19b/ this constant is zero and hence $d/z_1^{-1}, z_2^{-1}/ \neq 0$ for $|z_1^{-1}| \leqslant 1$, $|z_2^{-1}| \leqslant 1$.

In particular case for $a = b = 1$ from Strintzis' theorem 2 we have

Corollary 4.1

The system with transfer function /4.11/ is BIBO stable if and only if

1° $d/1, z_2^{-1}/ \neq 0$ for $|z_2^{-1}| \leq 1$ /4.20a/

2° $d/z_1^{-1}, 1/ \neq 0$ for $|z_1^{-1}| \leq 1$ /4.20b/

3° $d/z_1^{-1}, z_2^{-1}/ \neq 0$ for $|z_1^{-1}| = |z_2^{-1}| = 1$ /4.20c/

Example 4.3

Using Corollary 4.1 find the range of real values of h for which the system with

$$d/z_1, z_2/ = h z_1 z_2 + \frac{2}{3} z_1 + \frac{1}{3} z_2 + 1 \qquad /4.21/$$

is BIBO stable.
For
$$d/z_1^{-1}, z_2^{-1}/ = h + \frac{2}{3} z_2^{-1} + \frac{1}{3} z_1^{-1} + z_1^{-1} z_2^{-1} \qquad /4.21/$$

from /4.20/ we obtain

1. $d/1, z_2^{-1}/ = h + \frac{2}{3} z_2^{-1} + \frac{1}{3} + z_2^{-1} \neq 0$ for $|z_2^{-1}| \leq 1$ /4.22a/

2. $d/z_1^{-1}, 1/ = h + \frac{2}{3} + \frac{1}{3} z_1^{-1} + z_1^{-1} \neq 0$ for $|z_1^{-1}| \leq 1$ /4.22b/

3. $d/e^{j\theta_1}, e^{j\theta_2}/ = h + \frac{2}{3} e^{j\theta_2} + \frac{1}{3} e^{j\theta_1} + e^{j/\theta_1+\theta_2/} \neq 0$ for all real $\theta_1$ and $\theta_2$. /4.22c/

It is easy to prove that /4.22a/ is satisfied for

$$h \in \left[(-\infty, -2) \cup \left(\frac{4}{3}, +\infty\right)\right] \qquad /4.23a/$$

and /4.22b/ is satisfied for

$$h \in \left[(-\infty, -2) \cup \left(\frac{2}{3}, +\infty\right)\right] \qquad /4.23b/$$

The condition /4.22c/ will be satisfied if and only if the equations

$$h + \frac{2}{3} \cos \theta_2 + \frac{1}{3} \cos \theta_1 + \cos/\theta_1+\theta_2/ = 0 \qquad /4.24a/$$

$$\frac{2}{3} \sin \theta_2 + \frac{1}{3} \sin \theta_1 + \sin/\theta_1+\theta_2/ = 0 \qquad /4.24b/$$

have no real solutions for $\theta_1$ and $\theta_2$.

From /4.24a/ we have

$$h \in [(-\infty, -2) \cup (2, +\infty)] \qquad /4.23c/$$

Therefore, from /4.23/ it follows that the system is BIBO stable for

$$h \in [(-\infty, -2) \cup (2, +\infty)]$$

Corollary 4.2

The system with transfer function /4.11/ is BIBO stable if and only if

$$1° \quad d/z^{-1}, z^{-1}/ \neq 0 \quad \text{for } |z^{-1}| \leq 1 \qquad /4.25a/$$

$$2° \quad d/z_1^{-1}, z_2^{-1}/ \neq 0 \quad \text{for } |z_1^{-1}| = |z_2^{-1}| = 1 \qquad /4.25b/$$

Proof: We shall show that the conditions /4.25/ are equivalent to the conditions /4.17/. Note that for $a = z_1^{-1} = z^{-1}$ from /4.17a/ we obtain /4.25a/ and /4.17b/ implies /4.25b/. It can be also easily shown that /4.25/ imply /4.17/. □

Example 4.4

Using Corollary 4.2 find the range of real values of h for which the system with /4.21/ is BIBO stable.
For /4.21a/ using /4.25/ we obtain

1. $d/z^{-1}, z^{-1}/ = 1 + z^{-1} + hz^{-2} \neq 0 \quad \text{for } |z^{-1}| \leq 1 \qquad /4.26a/$

2. $d/e^{j\theta_1}, e^{j\theta_2}/ = 1 + \frac{2}{3} e^{j\theta_2} + \frac{1}{3} e^{j\theta_1} + h e^{j/\theta_1 + \theta_2/} \neq 0 \qquad /4.26b/$
   for all real $\theta_1, \theta_2$.

It is easy to verify that /4.26a/ is satisfied for

$$h \in [(-\infty, -2) \cup (1, +\infty)] \qquad /4.27a/$$

In the same way as in Example 4.3 it can be shown that /4.26b/ is satisfied for

$$h \in [(-\infty, -2) \cup (2, +\infty)] \qquad /4.27b/$$

Therefore, the system is BIBO stable for h given by /4.27b/

The above considerations can be extended for n-D systems [24, 6].

## 4.2 STABILITY OF ROESSER'S MODEL

### 4.2.1 BIBO stability of Roesser's model.

Consider Roesser's model /RM/ described by the equations /3.1/ and /3.2/ with $D = 0$. The transfer function matrix of RM is given by

$$G/z_1^{-1}, z_2^{-1}/ = C \begin{bmatrix} I_{n_1} - z_1^{-1}A_{11} & -z_1^{-1}A_{12} \\ -z_2^{-1}A_{21} & I_{n_2} - z_2^{-1}A_{22} \end{bmatrix}^{-1} B = \frac{N/z_1^{-1}, z_2^{-1}/}{d/z_1^{-1}, z_2^{-1}/} \qquad /4.28/$$

where

$$N/z_1^{-1}, z_2^{-1}/ = C \text{ adj} \begin{bmatrix} I_{n_1} - z_1^{-1}A_{11} & -z_1^{-1}A_{12} \\ -z_2^{-1}A_{21} & I_{n_2} - z_2^{-1}A_{22} \end{bmatrix} B \qquad /4.29a/$$

$$d/z_1^{-1}, z_2^{-1}/ = \det \begin{bmatrix} I_{n_1} - z_1^{-1}A_{11} & -z_1^{-1}A_{12} \\ -z_2^{-1}A_{21} & I_{n_2} - z_2^{-1}A_{22} \end{bmatrix} \qquad /4.29b/$$

We assume that $N/z_1^{-1}, z_2^{-1}/$ and $d/z_1^{-1}, z_2^{-1}/$ are factor coprime and have no nonessential singularities of the second kind for $|z_1^{-1}| = |z_2^{-1}| = 1$. Taking into account that

$$\begin{bmatrix} I_{n_1} & 0 \\ z_2^{-1}A_{21}\left[I_{n_1} - z_1^{-1}A_{11}\right]^{-1} & I_{n_2} \end{bmatrix} \begin{bmatrix} I_{n_1} - z_1^{-1}A_{11} & -z_1^{-1}A_{12} \\ -z_2^{-1}A_{21} & I_{n_2} - z_2^{-1}A_{22} \end{bmatrix} =$$

$$= \begin{bmatrix} I_{n_1} - z_1^{-1}A_{11} & -z_1^{-1}A_{12} \\ 0 & I_{n_2} - z_2^{-1}\left[A_{22} + A_{21}\left[I_{n_1}z_1 - A_{11}\right]^{-1}A_{12}\right] \end{bmatrix}$$

and

$$\begin{bmatrix} I_{n_1} & z_1^{-1}A_{12}\left[I_{n_2} - z_2^{-1}A_{22}\right]^{-1} \\ 0 & I_{n_2} \end{bmatrix} \begin{bmatrix} I_{n_1} - z_1^{-1}A_{11} & -z_1^{-1}A_{12} \\ -z_2^{-1}A_{21} & I_{n_2} - z_2^{-1}A_{22} \end{bmatrix} =$$

$$\begin{bmatrix} I_{n_1} - z_1^{-1}\left[A_{11} + A_{12}\left[I_{n_2}z_2 - A_{22}\right]^{-1}A_{21}\right] & 0 \\ -z_2^{-1}A_{21} & I_{n_2} - z_2^{-1}A_{22} \end{bmatrix}$$

we can write the polynomial /4.29b/ in the form

$$d/z_1^{-1},z_2^{-1}/ = \det\left[I_{n_1} - z_1^{-1}A_{11}\right] \cdot \det\left\{I_{n_2} - z_2^{-1}\left[A_{22} + A_{21}\left[I_{n_1}z_1 - A_{11}\right]^{-1}A_{12}\right]\right\}$$
/4.30a/

or

$$d/z_1^{-1},z_2^{-1}/ = \det\left[I_{n_2} - z_2^{-1}A_{22}\right] \cdot \det\left\{I_{n_1} - z_1^{-1}\left[A_{11} + A_{12}\left[I_{n_2}z_2 - A_{22}\right]^{-1}A_{21}\right]\right\}$$
/4.30b/

## Theorem 4.6 [26]

RM is BIBO stable if and only if one group of the following equivalent conditions is satisfied

$1^o$ i. $A_{11}$ is stable,

ii. $A_{22} + A_{21}\left[I_{n_1}z_1 - A_{11}\right]^{-1}A_{12}$ with $|z_1| = 1$ is stable;

$2^o$ i. $A_{22}$ is stable,

ii. $A_{11} + A_{12}\left[I_{n_2}z_2 - A_{22}\right]^{-1}A_{21}$ with $|z_2| = 1$ is stable;

$3^o$ i.
$$A = \begin{bmatrix} A_{11} & A_{12} \\ A_{21} & A_{22} \end{bmatrix} \text{ is stable,}$$

ii. $A_{11}$ has no eigenvalues on the unit circle,

iii. $A_{22} + A_{21}\left[I_{n_1}z_1 - A_{11}\right]^{-1}A_{12}$ with $|z_1| = 1$ has no eigenvalues on the unit circle;

$4^o$ i.
$$A = \begin{bmatrix} A_{11} & A_{12} \\ A_{21} & A_{22} \end{bmatrix} \text{ is stable,}$$

ii. $A_{22}$ has no eigenvalues on the unit circle,

iii. $A_{11} + A_{12}\left[I_{n_2}z_2 - A_{22}\right]^{-1}A_{21}$ with $|z_2| = 1$ has no eigenvalues on the unit circle.

Proof: Using Huang's theorem and /4.30/ it is easy to see that the condition $1^o$i. ($2^o$i.) is equivalent to /4.12a/ and the condition $1^o$ii. ($2^o$ii.) is equivalent to /4.12b/. Next note that the condition $3^o$i. ($4^o$i.) is equivalent to /4.25a/ and the conditions $3^o$ii. and $3^o$iii. ($4^o$ii. and $4^o$iii.) are equivalent to /4.25b/.

## Example 4.4

Test the BIBO stability of RM with

$$A = \begin{bmatrix} A_{11} & A_{12} \\ A_{21} & A_{22} \end{bmatrix} = \frac{1}{4}\begin{bmatrix} 2 & 1 \\ -1 & 2 \end{bmatrix} \qquad /4.31/$$

To test the stability we shall use the condition $1^o$ of Theorem 4.6. Hence

i. $A_{11} = \frac{1}{2} < 1$

and

ii. $\left| A_{22} + A_{21}\left[I_{n_1}z_1 - A_{11}\right]^{-1} A_{12} \right| = \left| \frac{1}{2} - \frac{1}{8/2z_1 - 1/} \right| < 1$ for $|z_1| = 1$.

Therefore, the conditions $1^o$ are satisfied and RM with /4.31/ is BIBO stable.

From Theorem 4.6 we have the following sufficient conditions for BIBO unstability of RM

## Corollary 4.3

RM is BIBO unstable if one of the following conditions is satisfied

1/ A is unstable

2/ $A_{11}$ is unstable

3/ $A_{22}$ is unstable

For example, RM with

$$A = \begin{bmatrix} A_{11} & A_{12} \\ A_{21} & A_{22} \end{bmatrix} = \begin{bmatrix} 1 & 0.5 \\ 0.5 & 0.25 \end{bmatrix}$$

is BIBO unstable because $A_{11}$ is unstable. Similarly, it is easy to check that RM with

$$A = \begin{bmatrix} A_{11} & A_{12} \\ A_{21} & A_{22} \end{bmatrix} = \begin{bmatrix} 0.5 & -2 \\ 1 & 0.5 \end{bmatrix}$$

is BIBO unstable because A is unstable even though $A_{11}$ and $A_{22}$ are stable.

Consider a non-singular matrix

$$T = \begin{bmatrix} T_1 & 0 \\ 0 & T_2 \end{bmatrix}, \quad \det T_i \neq 0 \quad /i=1,2/ \qquad /4.32/$$

and

$$\bar{A} = TAT^{-1} = \begin{bmatrix} T_1 A_{11} T_1^{-1} & T_1 A_{12} T_2^{-1} \\ T_2 A_{21} T_1^{-1} & T_2 A_{22} T_2^{-1} \end{bmatrix} = \begin{bmatrix} \bar{A}_{11} & \bar{A}_{12} \\ \bar{A}_{21} & \bar{A}_{22} \end{bmatrix} \qquad /4.33/$$

Note that

$$d/z_1^{-1}, z_2^{-1}/ = \det \begin{bmatrix} I_{n_1} - z_1^{-1} A_{11} & -z_1^{-1} A_{12} \\ -z_2^{-1} A_{21} & I_{n_2} - z_2^{-1} A_{22} \end{bmatrix} =$$

$$= \det \begin{bmatrix} I_{n_1} - z_1^{-1} \bar{A}_{11} & -z_1^{-1} \bar{A}_{12} \\ -z_2^{-1} \bar{A}_{21} & I_{n_2} - z_2^{-1} \bar{A}_{22} \end{bmatrix} \qquad /4.34/$$

Also each term in /4.30/ is invariant under the transformation.

<u>Theorem 4.7</u> [26]

RM is BIBO stable if

i. $A_{11}$ and $A_{22}$ are stable

ii. $A_{11}$ and $A_{22}$ are diagonalizable, i.e. there exist non-singular matrices $T_1, T_2$ such that $\bar{A}_{11} = T_1 A_{11} T_1^{-1}$ and $\bar{A}_{22} = T_2 A_{22} T_2^{-1}$ are diagonal

iii. $\|A_{12}\| \|A_{21}\| < (1-\lambda_{11})(1-\lambda_{22})$  /4.35/

where $\|A\|$ is the induced norm of A defined as

$$\|A\| = \max_i \left[ \lambda_i /A^T A/ \right]^{\frac{1}{2}} \qquad /4.36/$$

$$\lambda_{11} = \max_i |\lambda_i /A_{11}/|, \quad \lambda_{22} = \max_i |\lambda_i /A_{22}/| \qquad /4.37/$$

and $\lambda_i/A_{11}/ \; (\lambda_i/A_{22}/)$ is the i-th eigenvalue of $A_{11} \; (A_{22})$.

**Proof:** By condition 1° of Theorem 4.6 it is sufficient to show that the norm of $\overline{A}_{22} + \overline{A}_{21}\left[I_{n_1}z_1 - \overline{A}_{11}\right]^{-1}\overline{A}_{12}$ with $|z|=1$ is less than one.
We choose $T_1$ and $T_2$ so that

$$\overline{A}_{11} = T_1 A_{11} T_1^{-1} = \text{diag}\left[\lambda_1/A_{11}/ \quad \lambda_2/A_{11}/ \quad \cdots \quad \lambda_{n_1}/A_{11}/\right]$$

$$\overline{A}_{22} = T_2 A_{22} T_2^{-1} = \text{diag}\left[\lambda_1/A_{22}/ \quad \lambda_2/A_{22}/ \quad \cdots \quad \lambda_{n_2}/A_{22}/\right]$$

Note that we can take $\|\overline{A}_{22}\| = \lambda_{22}$ and

$$\left\|\left[I_{n_1}z_1 - \overline{A}_{11}\right]^{-1}\right\| = \left\|\text{diag}\left\{\left[z_1 - \lambda_1/A_{11}/\right]^{-1} \quad \cdots \quad \left[z_1 - \lambda_{n_1}/A_{11}/\right]^{-1}\right\}\right\| < \frac{1}{1 - \lambda_{11}}$$

Thus we have

$$\left\|\overline{A}_{22} + \overline{A}_{21}\left[I_{n_1}z_1 - \overline{A}_{11}\right]^{-1}\overline{A}_{12}\right\| \leq \lambda_{22} + \frac{\|\overline{A}_{12}\|\|\overline{A}_{21}\|}{1 - \lambda_{11}} < 1$$

if /4.35/ is satisfied.

□

### 4.2.2 Asymptotic stability of Roesser's model.

**Definition 4.2**
RM is said to be asymptotically stable iff for zero input $u/i,j/ = 0$ and finite

$$X^h = \sup_j \|x^h/0,j/\| \quad \text{and} \quad X^v = \sup_i \|x^v/i,0/\|,$$

$\sup_{h,k} \|x/h,k/\|$ is finite and $\lim_{\substack{h \to \infty \\ k \to \infty}} x/h,k/ = 0$.

**Theorem 4.8**
RM is asymptotically stable if and only if

$$d/z_1^{-1}, z_2^{-1}/ = \det\begin{bmatrix} I_{n_1} - z_1^{-1}A_{11} & -z_1^{-1}A_{12} \\ -z_2^{-1}A_{21} & I_{n_2} - z_2^{-1}A_{22} \end{bmatrix} \neq 0 \quad \text{for } |z_1^{-1}| \leq 1,\ |z_2^{-1}| \leq 1.$$

/4.38/

Proof: For $u/i,j/ = 0$ from /1.28/ we have

$$x/h,k/ = \sum_{i=0}^{h} A^{h-i,k} \begin{bmatrix} 0 \\ x^v/i,0/ \end{bmatrix} + \sum_{j=0}^{k} A^{h,k-j} \begin{bmatrix} x^h/0,j/ \\ 0 \end{bmatrix} \qquad /4.39/$$

In Appendix it is shown that

$$\begin{bmatrix} I_{n_1} - z_1^{-1} A_{11} & -z_1^{-1} A_{12} \\ -z_2^{-1} A_{21} & I_{n_2} - z_2^{-1} A_{22} \end{bmatrix}^{-1} = \sum_{i=0}^{\infty} \sum_{j=0}^{\infty} A^{i,j} z_1^{-i} z_2^{-j} \qquad /4.40/$$

From Definition 4.2 and /4.39/ it follows that RM is asymptotically stable if

$$\sum_{i=0}^{\infty} \sum_{j=0}^{\infty} \|A^{i,j}\| z_1^{-i} z_2^{-j} \qquad /4.41/$$

converges for $|z_1^{-1}| \leq 1$, $|z_2^{-1}| \leq 1$.
Conversely, if /4.41/ converges, then /4.38/ holds. Therefore RM is asymptotically stable if and only if /4.38/ is satisfied. □

From comparison of Shank's theorem and Theorem 4.8 it follows that BIBO stability test can also be used for testing asymptotic stability of RM.

Theorem 4.9

RM is asymptotically /BIBO/ stable if

$$\|A_{11}\| < 1 \quad \text{and} \quad \|A_{22}\| + \|A_{21}\|(1 - \|A_{11}\|)^{-1} \|A_{12}\| < 1$$

or

$$\|A_{22}\| < 1 \quad \text{and} \quad \|A_{11}\| + \|A_{12}\|(1 - \|A_{22}\|)^{-1} \|A_{21}\| < 1$$

Proof: Taking into account /4.30a/

$$d/z_1^{-1}, z_2^{-1}/ = \det\left[I_{n_1} - z_1^{-1} A_{11}\right] \cdot \det\left\{I_{n_2} - z_2^{-1}\left[A_{22} + A_{21}\left[I_{n_1} z_1 - A_{11}\right]^{-1} A_{12}\right]\right\}$$

it is easy to see that /4.38/ is satisfied if $\|A_{11}\| < 1$ and

$$\left\|A_{22} + A_{21}\left[I_{n_1} z_1 - A_{11}\right]^{-1} A_{12}\right\| < 1 \quad \text{with } |z_1| = 1 \qquad /4.42/$$

Using the property $\|A - B\| \geq \|A\| - \|B\|$ to the relation

$$z_1 \left[I_{n_1} z_1 - A_{11}\right]^{-1} - \left[I_{n_1} z_1 - A_{11}\right]^{-1} A_{11} = I_{n_1} \quad \text{with } |z_1| = 1$$

we obtain

$$\left\| \left[ I_{n_1} z_1 - A_{11} \right]^{-1} \right\| - \left\| \left[ I_{n_1} z_1 - A_{11} \right]^{-1} \right\| \cdot \|A_{11}\| \leq 1$$

and

$$\left\| \left[ I_{n_1} z_1 - A_{11} \right]^{-1} \right\| \leq \left[ 1 - \|A_{11}\| \right]^{-1} \quad \text{with } |z_1| = 1$$

Therefore

$$\left\| A_{22} + A_{21} \left[ I_{n_1} z_1 - A_{11} \right]^{-1} A_{12} \right\| \leq \|A_{22}\| + \|A_{21}\| \cdot \left[ 1 - \|A_{11}\| \right]^{-1} \cdot \|A_{12}\|$$

and /4.42/ is satisfied if

$$\|A_{22}\| + \|A_{21}\| \cdot \left[ 1 - \|A_{11}\| \right]^{-1} \cdot \|A_{12}\| < 1.$$

The proof of the second part is similar. □

### Example 4.5

Using Theorem 4.9 show that RM with /4.21/ is asymptotically stable. In this case we have $A_{11} = A_{22} = 0.5$, $A_{12} = -A_{21} = 0.25$ and $\|A_{11}\| = \|A_{22}\| = 0.5$, $\|A_{12}\| = \|A_{21}\| = 0.25$. Hence $\|A_{11}\| < 1$ and

$$\|A_{22}\| + \|A_{21}\|[1 - \|A_{11}\|]^{-1} \|A_{12}\| = 0.625 < 1.$$

Therefore the conditions of Theorem 4.9 are satisfied and RM with /4.31/ is asymptotically stable.

### Theorem 4.10 [20]

RM with $A \geq 0$ /i.e. with nonnegative entries $a_{ij} \geq 0$/ is asymptotically stable if and only if

$$A_{11} \quad \text{and} \quad A_{22} + A_{21} \left[ I_{n_1} - A_{11} \right]^{-1} A_{12} \quad \text{are stable}$$

or

$$A_{22} \quad \text{and} \quad A_{11} + A_{12} \left[ I_{n_2} - A_{22} \right]^{-1} A_{21} \quad \text{are stable.}$$

**Proof:** From Theorem 4.6 it follows that it is sufficient to show that for $A \geq 0$ the matrix

$$A_{22} + A_{21} \left[ I_{n_1} - A_{11} \right]^{-1} A_{12}$$

is stable if and only if

$$A_{22} + A_{21} \left[ I_{n_1} z_1 - A_{11} \right]^{-1} A_{12} \quad \text{with } |z_1| = 1 \quad \text{is stable.}$$

Note that stability of
$$A_{22} + A_{21}\left[I_{n_1}z_1 - A_{11}\right]^{-1} A_{12} \text{ with } |z_1| = 1$$
implies stability of
$$A_{22} + A_{21}\left[I_{n_1} - A_{11}\right]^{-1} A_{12}.$$
From the relation
$$\left[I_{n_1}z_1 - A_{11}\right]^{-1} = \sum_{i=0}^{\infty} \frac{A_{11}^i}{z_1^{i+1}}$$
it follows that / for $A_{11} \geq 0$ and $z_1 = 1$ /
$$\left[I_{n_1} - A_{11}\right]^{-1} \geq 0.$$
Therefore, the stability of
$$A_{22} + A_{21}\left[I_{n_1} - A_{11}\right]^{-1} A_{12}$$
implies stability of
$$A_{22} + A_{21}\left[I_{n_1}z_1 - A_{11}\right]^{-1} A_{12} \text{ with } |z_1| = 1.$$
□

In a similar way the following theorem can be proved:

**Theorem 4.11** [20]
RM with $A^{1,0} \leq 0$ and $A^{0,1} \geq 0$ is asymptotically /BIBO/ stable if and only if

$A_{11}$ and $A_{22} - A_{21}\left[I_{n_1} + A_{11}\right]^{-1} A_{12}$ are stable

or

$A_{22}$ and $A_{11} + A_{12}\left[I_{n_2} - A_{22}\right]^{-1} A_{21}$ are stable.

**Remark 4.1**

By choosing suitable nonsingular state transformations $\bar{x} = Tx$ /for example, $T = $ block diag $[-I \; T]$ / it is possible to extend the class of RM to which Theorem 4.10 and Theorem 4.11 can be applied.

**Example 4.6**

Test the asymptotic /BIBO/ stability of RM with

$$A = \begin{bmatrix} A_{11} & A_{12} \\ A_{21} & A_{22} \end{bmatrix} = \begin{bmatrix} \frac{1}{3} & \frac{1}{2} & -\frac{1}{2} \\ 0 & \frac{1}{4} & -\frac{1}{4} \\ -\frac{1}{3} & 0 & \frac{1}{4} \end{bmatrix} \qquad /4.43/$$

To obtain $\overline{A} \geq 0$ we take

$$T = \begin{bmatrix} 1 & 0 & 0 \\ 0 & 1 & 0 \\ 0 & 0 & -1 \end{bmatrix}$$

and

$$\overline{A} = TAT^{-1} = \begin{bmatrix} \frac{1}{3} & \frac{1}{2} & \frac{1}{2} \\ 0 & \frac{1}{4} & \frac{1}{4} \\ \frac{1}{3} & 0 & \frac{1}{4} \end{bmatrix} = \begin{bmatrix} \overline{A}_{11} & \overline{A}_{12} \\ \overline{A}_{21} & \overline{A}_{22} \end{bmatrix}$$

Hence

$$\overline{A}_{22} + \overline{A}_{21}\left[I_{n_1} - \overline{A}_{11}\right]^{-1}\overline{A}_{12} = \begin{bmatrix} 0.25 & 0.25 \\ 0.25 & 0.5 \end{bmatrix} \qquad /4.44/$$

It is easy to check that /4.44/ is stable.
Therefore the conditions of Theorem 4.10 are satisfied and RM with /4.43/ is asymptotically stable.


## 4.3 ASYMPTOTIC AND EXPONENTIAL STABILITY OF FORNASINI-MARCHESINI´S MODELS

### 4.3.1 Asymptotic stability.

Consider the second Fornasini-Marchesini´s model /F-MM II/ described by the equations /3.89/ and /3.90/. Following [9, 10] we introduce the notation

$$X_r = \left\{ x/h,k/ : \; x/h,k/ \in X, \; h+k = r \right\} \qquad h,k \in Z$$

where $x/h,k/ \in X = R^n$ /local state space/ and $Z$ is the set of nonnegative integers.

Let $\|x\|$ denote the Euclidean norm of x in X and let

$$\|X_r\| = \sup_{n \in Z} \|x/r-n,n/\|$$

### Definition 4.3 [9, 10]
F-MM II is asymptotically stable iff, assuming zero input $an/i,j/ = 0$ and $\|X_0\|$ finite, $\|X_i\| \to 0$ as $i \to +\infty$.

### Theorem 4.12
F-MM II is asymptotically stable if and only if

$$d/z_1^{-1},z_2^{-1}/ = \det\left[I_n - A_1 z_1^{-1} - A_2 z_2^{-1}\right] \neq 0 \quad \text{for } |z_1^{-1}| \leq 1, |z_2^{-1}| \leq 1 \quad /4.45/$$

The proof is similar to the proof of Theorem 4.8.

Note that for testing asymptotic stability of F-MM II all BIBO stability tests can be used.

### Example 4.7
Test the asymptotic stability of F-MM II with

$$A_1 = \begin{bmatrix} 0.1 & 0 \\ -0.2 & 0 \end{bmatrix}, \quad A_2 = \begin{bmatrix} 0 & 0.2 \\ 0 & 0.1 \end{bmatrix} \qquad /4.46/$$

In this case we have

$$d/z_1^{-1},z_2^{-1}/ = \det\left[I_n - A_1 z_1^{-1} - A_2 z_2^{-1}\right] = \begin{vmatrix} 1 - 0.1 z_1^{-1} & -0.2 z_2^{-1} \\ 0.2 z_1^{-1} & 1 - 0.1 z_2^{-1} \end{vmatrix}$$

$$= 0.05 z_1^{-1} z_2^{-1} - 0.1 z_1^{-1} - 0.1 z_2^{-1} + 1$$

Using /4.10/ for $\bar{d}_{10} = -0.1$, $\bar{d}_{01} = -0.1$, $\bar{d}_{11} = 0.05$ we obtain

$$\sum_{/0,0/ < /i,j/ \leq /n_1,n_2/} |\bar{d}_{ij}| = 0.25 < 1$$

Therefore the condition of Theorem 4.4 is satisfied and F-MM II with /4.46/ is asymptotically stable.

Since F-MM I is a particular case of RM, therefore for testing stability of F-MM I all stability tests for RM can be used.

## 4.3.2 Exponential stability.

Consider F-MM II with zero input $u/i,j/=0$ described by the equation

$$x/i+1,j+1/ = A_1 x/i,j+1/ + A_2 x/i+1,j/ \qquad /4.47/$$

with boundary conditions

$$x/i,0/ \;,\; x/0,j/ \quad \text{for} \quad i,j \geqslant 0 \qquad /4.47a/$$

where $x/i,j/ \in R^n$ is the local state vector at $/i,j/$ and $A_1$, $A_2$ are $n \times n$ real matrices.

### Definition 4.4 [21]
F-MM II is said to be exponentially stable iff it is possible to find numbers $M > 0$, $\alpha \in [0,1)$ such that

$$\|x/i,j/\| \leqslant M \left\{ \alpha^j \max_{0 \leqslant h \leqslant j} \|x/0,h/\| + \alpha^i \max_{0 \leqslant k \leqslant i} \|x/k,0/\| \right\} \quad \text{for} \quad i,j > 0 \qquad /4.48/$$

From comparison of Definitions 4.3 and 4.4 it follows that the exponential stability implies asymptotic stability.

### Theorem 4.13
F-MM II is exponentially stable if and only if

$$d/z_1^{-1}, z_2^{-1}/ = \det\left[I_n - z_1^{-1} A_1 - z_2^{-1} A_2\right] \neq 0 \quad \text{for} \quad |z_1^{-1}| \leqslant 1 \;,\; |z_2^{-1}| \leqslant 1 \qquad /4.49/$$

Proof: The necessity follows from Theorem 4.12.
The solution to /4.47/ with /4.47a/ is given by

$$x/i,j/ = \sum_{h=0}^{i} A^{i-h,j} \bar{x}/h,0/ + \sum_{k=0}^{j} A^{i,j-k} \bar{x}/0,k/ \qquad /4.50/$$

where

$$Z\left[A^{i,j}\right] = \sum_{i=0}^{\infty} \sum_{j=0}^{\infty} A^{i,j} z_1^{-i} z_2^{-j} = \left[I - z_1^{-1} A_1 - z_2^{-1} A_2\right]^{-1} \qquad /4.51a/$$

$$\bar{x}/h,0/ = x/h,0/ - A_1 x/h-1,0/ \;,\; \bar{x}/0,k/ = x/0,k/ - A_2 x/0,k-1/ \qquad /4.51b/$$

Hence

$$\|x/i,j/\| \leq \sum_{h=0}^{i} \|A^{i-h,j}\| \|\bar{x}/h,0/\| + \sum_{k=0}^{j} \|A^{i,j-k}\| \|\bar{x}/0,k/\| \leq$$

$$\leq \sum_{h=0}^{i} \|A^{i-h,j}\| \cdot \max_{0\leq h\leq i} \|\bar{x}/h,0/\| + \sum_{k=0}^{j} \|A^{i,j-k}\| \cdot \max_{0\leq k\leq j} \|\bar{x}/0,k/\| \quad /4.52/$$

From /4.51b/ we have

$$\max_{0\leq h\leq i} \|\bar{x}/h,0/\| \leq \beta \max_{0\leq h\leq i} \|x/h,0/\|$$

$$\max_{0\leq k\leq j} \|\bar{x}/0,k/\| \leq \beta \max_{0\leq k\leq j} \|x/0,k/\| \quad /4.53/$$

where $\beta > 1 + \|A_1\| + \|A_2\|$

Note that to prove the sufficiency we have to show the existance of numbers $M > 0$, $\alpha \in [0, 1)$ such that

$$\sum_{h=0}^{i} \|A^{h,j}\| \leq M \alpha^j \quad \text{and} \quad \sum_{k=0}^{j} \|A^{i,k}\| \leq M \alpha^i \quad /4.54/$$

The condition /4.49/ implies that the series /4.51a/ is absolutely convergent for $|z_1^{-1}| < 1$ and $|z_2^{-1}| < 1$. Therefore, there exist numbers $M < 0$ and $\alpha \in [0, 1)$ such that

$$\sum_{h=0}^{\infty} \sum_{k=0}^{\infty} \alpha^{-h} \alpha^{-k} \|A^{h,k}\| \leq M$$

and

$$\alpha^{-j} \sum_{h=0}^{i} \|A^{h,j}\| \leq \alpha^{-j} \sum_{h=0}^{i} \alpha^{-h} \|A^{h,j}\| \leq$$

$$\leq \sum_{h=0}^{\infty} \sum_{k=0}^{\infty} \alpha^{-h} \alpha^{-k} \|A^{h,k}\| \leq M$$

In a similar way we can prove the second relation of /4.54/.

The relation /4.48/ follows from /4.52/, /4.53/ and /4.54/.

□

## 4.4 MARGIN OF STABILITY

Consider a 2-D linear system with the transfer function

$$G/z_1^{-1},z_2^{-1}/ = \frac{n/z_1^{-1},z_2^{-1}/}{d/z_1^{-1},z_2^{-1}/} \qquad /4.55/$$

where $n/z_1^{-1},z_2^{-1}/$, $d/z_1^{-1},z_2^{-1}/$ are factor coprime polynomials in $z_1^{-1}$ and $z_2^{-1}$.

The margin of stability for 2-D linear system can be defined in different ways [1].

### Definition 4.5a

We call $\sigma_1$ the margin of stability /MS/ for the system with $G/z_1^{-1},z_2^{-1}/$ iff

$$U_{\sigma_1}^2 = \left\{ /z_1^{-1},z_2^{-1}/ : |z_1^{-1}| < 1 + \sigma_1, \ |z_2^{-1}| < 1 \right\} \qquad /4.56a/$$

is a largest analytic bidisk of $G/z_1^{-1},z_2^{-1}/$ with its center at $/0,0/$.

### Definition 4.5b

We call $\sigma_2$ the margin of stability /MS/ for the system with $G/z_1^{-1},z_2^{-1}/$ iff

$$U_{\sigma_2}^2 = \left\{ /z_1^{-1},z_2^{-1}/ : |z_1^{-1}| < 1, \ |z_2^{-1}| < 1 + \sigma_2 \right\} \qquad /4.56b/$$

is a largest analytic bidisk of $G/z_1^{-1},z_2^{-1}/$ with its center at $/0,0/$.

The following theorem shows the relation between $\sigma_1 / \sigma_2 /$ and the impulse response $g/i,j/$ of the system:

### Theorem 4.14

$\sigma_1 / \sigma_2 /$ gives a lower bound for MS of any sequence $g/i,j/$ for all i and fixed j / $g/i,j/$ for all j and fixed i /.

Proof: Consider $G/z_1^{-1},z_2^{-1}/$ with the largest analytic bidisk $U_{\sigma_1}^2$. Then we have that

$$G/z_1^{-1},z_2^{-1}/ = \sum_{i=0}^{\infty}\sum_{j=0}^{\infty} g/i,j/ z_1^{-i} z_2^{-j} = \sum_{j=0}^{\infty} G_j /z_1^{-1}/ z_2^{-j}$$

is analytic for $|z_1^{-1}| < 1+\delta_1$, $|z_2^{-1}| < 1$.
Therefore

$$G_j/z_1^{-1}/ = \sum_{i=0}^{\infty} g/i,j/ z_1^{-i}$$

is analytic for every j and at least for $|z_1^{-1}| < 1+\delta_1$.

This implies that $\delta_1$ is the lower bound for MS of any sequence $g/i,j/$ for all i and fixed j.
The second part can be proved in a similar way.

□

In Definition 4.5 we defined MS using an extension of the analytic region of $G/z_1^{-1}, z_2^{-1}/$ with respect to $z_1^{-1}$ or $z_2^{-1}$. In the following definition the analytic region will be extended in both $z_1^{-1}$ and $z_2^{-1}$:

## Definition 4.6 [1]

We call $\delta$ the margin of stability /MS/ for the system with $G/z_1^{-1}, z_2^{-1}/$ iff

$$U_\delta^2 = \left\{ /z_1^{-1}, z_2^{-1}/ : |z_1^{-1}| < 1+\delta, |z_2^{-1}| < 1+\delta \right\}$$

is a largest analytic bidisk of $G/z_1^{-1}, z_2^{-1}/$ wwith its center at /0,0/.

## Theorem 4.15

$\delta$ gives a lower bound for MS of the sequences $\{g/i,j-i/,\ i-\text{fixed}\}$ and $\{g/j-i,i/,\ i-\text{fixed}\}$ for all i.

**Proof:** Consider $G/z_1^{-1}, z_2^{-1}/$ with the largest analytic bidisk $U_\delta^2$.
Note that $G/z_1^{-1}, z_2^{-1}/$ can be written as

$$G/z_1^{-1}, z_2^{-1}/ = \sum_{i=0}^{\infty} \sum_{j=1}^{\infty} g/i,j-i/ z_1^{-i} z_2^{i-j} = \sum_{i=0}^{\infty} \left[ \sum_{j=1}^{\infty} g/i,j-i/ z_2^{-j} \right] \left[ \frac{z_1^{-1}}{z_2^{-1}} \right]^i =$$

$$= \sum_{i=0}^{\infty} G_i/z_2^{-1}/ z_2^{-i} \left[ \frac{z_1^{-1}}{z_2^{-1}} \right]^i$$

where

$$G_i/z_2^{-1}/ = \sum_{j=0}^{\infty} g/i,j/ z_2^{-j}.$$

$G/z_1^{-1}, z_2^{-1}/$ is analytic for $|z_1^{-1}| < 1+\delta$ and $|z_2^{-1}| < 1+\delta$. This implies that for every i

$$G_i/z_2^{-1}/ z_2^{-i} = \sum_{j=i}^{\infty} g/i, j-i/ z_2^{-j}$$

is analytic at least for $|z_2^{-1}| < 1+\delta$. Therefore $\delta$ is the lower bound for MS of the sequences $\{g/i, j-i/, \; i\text{-fixed}\}$ and $\{g/j-i, i/, \; i\text{-fixed}\}$ for all i. □

Note that the conditions $\delta_1 > 0$ and $\delta_2 > 0$ or $\delta > 0$ are necessary and sufficient for asymptotic /BIBO/ stability.
From Huang's theorem it follows that $\delta_1$ can be found as the minimal value $\delta_1$ which satisfies the condition

$$d/z_1^{-1}, z_2^{-1}/ = 0 \quad \text{for} \quad |z_1^{-1}| = 1+\delta_1 \quad \text{and} \quad |z_2^{-1}| = 1.$$

In [1] it was shown that computation of $\delta_1$ can be reduced to an optimization problem. The above considerations can be extended for n-D systems [25].

## 4.5 STABILIZATION OF 2-D SYSTEMS BY STATE FEEDBACK OR OUTPUT FEEDBACK

Consider RM described by the equations /3.1/ and /3.2/ with the state feedback

$$u = Kx \qquad /4.57/$$

where $K = [K_1 \; K_2]$ and $K_1$, $K_2$ are $m \times n_1$, $m \times n_2$ real matrices, respectively.
Substitution of /4.57/ into /3.1/ yields

$$x' = A_c x \qquad /4.58/$$

where

$$A_c = A + BK = \begin{bmatrix} A_{11} + B_1 K_1 & A_{12} + B_1 K_2 \\ A_{21} + B_2 K_1 & A_{22} + B_2 K_2 \end{bmatrix} \qquad /4.59/$$

## Definition 4.6

We call RM stabilizable by state feedback /4.57/ if there exists a matrix K such that the resulting closed-loop system is BIBO /asymptotically/ stable.

Similarly, let us consider RM with the output feedback

$$u = Fy \qquad /4.60/$$

where $F \in R^{m \times l}$ is an output feedback matrix.

Substitution of $u = FCx$ into /3.1/ yields /4.58/ with

$$A_c = A + BFC = \begin{bmatrix} A_{11} + B_1 FC_1 & A_{12} + B_1 FC_2 \\ A_{21} + B_2 FC_1 & A_{22} + B_2 FC_2 \end{bmatrix} \qquad /4.61/$$

## Definition 4.7

We call RM stabilizable by output feedback /4.60/ iff there exists a matrix F such that the resulting closed-loop system is BIBO /asymptotically/ stable.

## Theorem 4.16 [26]

RM is stabilizable by state feedback only if

1. $/A,B/$ as an 1-D system is stabilizable by state feedback,
2. $/A_{11},B_1/$ and $/A_{22},B_2/$ are stabilizable by state feedbacks.

Proof: From Theorem 4.6 it follows that RM is BIBO stable only if $A$, $A_{11}$ and $A_{22}$ are stable. Therefore, RM is stabilizable by state feedback only if $/A,B/$, $/A_{11},B_1/$ and $/A_{22},B_2/$ are stabilizable by state feedbacks. □

In a similar way we can prove the following

## Theorem 4.17 [26]

RM is stabilizable by output feedback only if

1. $/A,B,C/$ as an 1-D system is stabilizable by output feedback,
2. $/A_{11},B_1,C_1/$, $/A_{22},B_2,C_2/$ are stabilizable by output feedbacks.

Those theorems provide some possibilities to verify the unstabilizability by using the corresponding results from 1-D systems theory.

## Theorem 4.18 [26]

RM is stabilizable by state feedback /4.57/ if either there exists a matrix $K_1$ such that $A_{21} + B_2K_1 = 0$ with $A_{11} + B_1K_1$ stable and $/A_{22},B_2/$ is stabilizable by state feedback, or there exists a matrix $K_2$ such that $A_{12} + B_1K_2 = 0$ with $A_{22} + B_2K_2$ stable and $/A_{11},B_1/$ is stabilizable by state feedback.

Proof: If there exists a matrix $K_1$ such that $A_{21} + B_2K_1 = 0$ with $A_{11} + B_1K_1$ stable, then RM is stabilizable by state feedback /4.57/ whenever $/A_{22},B_2/$ is stabilizable, since the closed-loop matrix /4.59/ is upper triangular. The proof of the second part is similar. □

In a similar way we can prove the following

## Theorem 4.19

RM is stabilizable by output feedback /4.60/ if either there exists a matrix F such that $A_{12} + B_1FC_2 = 0$ with $A_{11} + B_1FC_1$ and $A_{22} + B_2FC_2$ stable, or there exists a matrix F such that $A_{21} + B_2FC_1 = 0$ with $A_{11} + B_1FC_1$ and $A_{22} + B_2FC_2$ stable.

It is easy to prove that the equation

$$B_1FC_2 = -A_{12} \quad /4.62/$$

has a solution if and only if

$$B_1B_1^+A_{12}C_2^+C_2 = A_{12} \quad /4.63/$$

where $B_1^+$ $/C_2^+/$ is the generalized inverse of $B_1$ $/C_2/$ satisfying the condition

$$B_1B_1^+B_1 = B_1 .$$

The general solution to /4.62/ is

$$F = -B_1^+A_{12}C_2^+ + F_1 - B_1^+B_1F_1C_2C_2^+ \quad /4.64/$$

where $F_1$ is an arbitrary matrix of appropriate size.

Substitution of /4.64/ into $A_{11} + B_1FC_1$ and $A_{22} + B_2FC_2$ yields

$$A_{11} + B_1FC_1 = A_{11} - B_1B_1^+A_{12}C_2^+C_1 + B_1F_1\left[I_1 - C_2C_2^+\right]C_1 \quad /4.65a/$$

and

$$A_{22} + B_2FC_2 = A_{22} - B_2B_1^+A_{12}C_2^+C_2 + B_2[I_m - B_1^+B_1]F_1C_2 \qquad /4.65b/$$

A similar analysis may be carried out for the equation

$$B_2FC_1 = -A_{21} \qquad /4.66/$$

which has a solution if and only if

$$B_2B_2^+A_{21}C_1^+C_1 = A_{21} \qquad /4.67/$$

The general solution to /4.66/ has the form

$$F = -B_2^+A_{21}C_1^+ + F_2 - B_2^+B_2F_2C_1C_1^+ \qquad /4.68/$$

where $F_2$ is an arbitrary matrix of appropriate size, and

$$A_{11} + B_1FC_1 = A_{11} - B_1B_2^+A_{21}C_1^+C_1 + B_1[I_m - B_2^+B_2]F_2C_1 \qquad /4.69a/$$

$$A_{22} + B_2FC_2 = A_{22} - B_2B_2^+A_{21}C_1^+C_1 + B_2F_2[I_1 - C_1C_1^+]C_2 \qquad /4.69b/$$

Thus we have

### Corollary 4.4 [26]

RM is stabilizable by output feedback /4.60/ if either /4.63/ holds and the systems $\{A_{11} - B_1B_1^+A_{12}C_2^+C_1\,,\ B_1\,,\ [I_1 - C_2C_2^+]C_1\}$, $\{A_{22} - B_2B_1^+A_{12}C_2^+C_2\,,\ B_2[I_m - B_1^+B_1]\,,\ C_2\}$ can be stabilized by the same output feedback matrix $F_1$, then the desired feedback matrix F is given by /4.64/; or /4.67/ holds and the systems $\{A_{11} - B_1B_2^+A_{21}C_1^+C_1\,,\ B_1[I_m - B_2^+B_2]\,,\ C_1\}$ and $\{A_{22} - B_2B_2^+A_{21}C_1^+C_2\,,\ B_2\,,\ [I_1 - C_1C_1^+]C_2\}$ can be stabilized by the same output feedback $F_2$, then the desired feedback matrix F is given by /4.68/.

Using Theorem 4.6 we can reduce the stabilizing of RM to stabilizing a 1-D constant system and then stabilizing another 1-D system with a complex parameter.

### Theorem 4.20 [26]

RM is stabilizable by state feedback if and only if there exist two matrices $K_1$ and $K_2$ such that either

$1^o$    i. $A_{11} + B_1K_1$ is stable

ii. $A_{22} + B_2K_2 + [A_{21} + B_2K_1][z_1 I_{n_1} - [A_{11} + B_1K_1]]^{-1}[A_{12} + B_1K_2]$
with $|z_1^{-1}| = 1$ is stable

or

$2°$  i. $A_{22} + B_2K_2$ is stable

ii. $A_{11} + B_1K_1 + [A_{12} + B_1K_2][z_2 I_{n_2} - [A_{22} + B_2K_2]]^{-1}[A_{21} + B_2K_1]$
with $|z_2^{-1}| = 1$ is stable.

**Proof:** The theorem follows from conditions $1°$ and $2°$ of Theorem 4.6 applied to the closed-loop system with /4.59/.  □

Note that the condition $1°$ ii. of Theorem 4.20 can be rewritten in the form

$$A_{22} + B_2K_2 + [A_{21} + B_2K_1][z_1 I_{n_1} - [A_{11} + B_1K_1]]^{-1}[A_{12} + B_1K_2] =$$
$$= A_2/z_1^{-1}/ + B_2/z_1^{-1}/ K_2 \qquad /4.70/$$
with $|z_1^{-1}| = 1$

where

$$A_2/z_1^{-1}/ = A_{22} + z_1^{-1}[A_{21} + B_2K_1][I_{n_1} - z_1^{-1}[A_{11} + B_1K_1]]^{-1} A_{12} \qquad /4.71a/$$

$$B_2/z_1^{-1}/ = B_2 + z_1^{-1}[A_{21} + B_2K_1][I_{n_1} - z_1^{-1}[A_{11} + B_1K_1]]^{-1} B_1 \qquad /4.71b/$$

Let

$$\mathbb{R} = \left\{ \frac{b/z_1^{-1}/}{a/z_1^{-1}/} \; : \; a/z^{-1}/ \neq 0 \text{ for } |z^{-1}| \leq 1 \right\} \qquad /4.72/$$

where $a/z^{-1}/$ and $b/z^{-1}/$ are polynomials in $z^{-1}$ with real coefficients.

It is well known that $\mathbb{R}$ is a principal integral domain [8, 26]. It is also known [26] that if the pair $A_1/z_1^{-1}/$, $B_1/z_1^{-1}/$ is $\mathbb{R}$-reachable, i.e. every $x/z_1^{-1}/ \in \mathbb{R}^{n_2}$ is an $\mathbb{R}$-linear combination of the columns of

$$[B_2/z_1^{-1}/ \quad A_2/z_1^{-1}/ B_2/z_1^{-1}/ \quad \ldots \quad A_2/z_1^{-1}/^{n_2-1} B_2/z_1^{-1}/],$$

then for every $p_1/z_1^{-1}/, \ldots, p_{n_2}/z_1^{-1}/ \in \mathbb{R}$ there exists $K_2/z_1^{-1}/ \in \mathbb{R}^{m \times n_2}$ such that

$$\det[z_2 I_{n_2} - [A_2/z_1^{-1}/ + B_2/z_1^{-1}/ K_2/z_1^{-1}/]] = \prod_{i=1}^{n_2}(z_2 - p_i/z_1^{-1}/) \qquad /4.73/$$

Similarly, the condition $2^\circ$ ii. of Theorem 4.20 can be rewritten in the form

$$A_{11} + B_1K_1 + \left[A_{12} + B_1K_2\right]\left[z_2 I_{n_2} - \left[A_{22} + B_2K_2\right]\right]^{-1}\left[A_{21} + B_2K_1\right] =$$
$$= A_1/z_2^{-1}/ + B_1/z_2^{-1}/ K_1 \qquad /4.74/$$

with $|z_2^{-1}| = 1$

where

$$A_1/z_2^{-1}/ = A_{11} + z_2^{-1}\left[A_{12} + B_1K_2\right]\left[I_{n_2} - z_2^{-1}\left[A_{22} + B_2K_2\right]\right]^{-1} A_{21} \qquad /4.75a/$$

$$B_1/z_2^{-1}/ = B_1 + z_2^{-1}\left[A_{12} + B_1K_2\right]\left[I_{n_2} - z_2^{-1}\left[A_{22} + B_2K_2\right]\right]^{-1} B_2 \qquad /4.75b/$$

From the above considerations and Theorem 4.20 the following theorem follows:

### Theorem 4.21 [26]

RM is stabilizable by state feedback with the matrix

$$K = \left[K_1 \quad K_2/z_1^{-1}/\right]$$

where $K_1$ is a constant $m \times n_1$ matrix and $K_2/z_1^{-1}/ \in \mathbb{R}^{m \times n_2}$ if

i. $A_{11}, B_1$ is stabilizable by $K_1$,

ii. for some $K_1$ stabilizing $A_{11}, B_1$, the pair $A_2/z_1^{-1}/, B_2/z_1^{-1}/$ defined by /4.71/ is $\mathbb{R}$-reachable.

Similarly, RM is stabilizable by state feedback with the matrix

$$K = \left[K_1/z_2^{-1}/ \quad K_2\right]$$

where $K_1/z_2^{-1}/ \in \mathbb{R}^{m \times n_1}$ and $K_2$ is a constant $m \times n_2$ matrix if

i. $A_{22}, B_2$ is stabilizable by $K_2$,

ii. for some $K_2$ stabilizing $A_{22}, B_2$, the pair $A_1/z_2^{-1}/, B_1/z_2^{-1}/$ defined by /4.75/ is $\mathbb{R}$-reachable.

## 4.6. THE LYAPUNOV EQUATION FOR 2-D SYSTEMS

Following [27] let us consider RM described by equations /3.1/ and /3.4/
The characteristic polynomial of RM is given by

$$d(z_1^{-1}, z_2^{-1}) = \det[I_n - z_1^{-1} A_2 - z_2^{-1} A_1] \qquad /4.76/$$

where

$$A = \begin{bmatrix} 0 & 0 \\ A_{21} & A_{22} \end{bmatrix}, \quad A_2 = \begin{bmatrix} A_{11} & A_{12} \\ 0 & 0 \end{bmatrix}$$

By Theorem 4.8 RM is asymptotically stable if

$$d(z_1^{-1}, z_2^{-1}) \neq 0 \quad \text{for} \quad |z_1^{-1}| \leq 1, \quad |z_2^{-1}| \leq 1 \qquad /4.77/$$

The Lyapunow equation for RM has the form

$$-Q = [A_1 + A_2]^T P [A_1 + A_2] - P = A^T PA - P =$$

$$= \begin{bmatrix} A_{11} & A_{12} \\ A_{21} & A_{22} \end{bmatrix}^T \begin{bmatrix} P_1 & 0 \\ 0 & P_2 \end{bmatrix} \begin{bmatrix} A_{11} & A_{12} \\ A_{21} & A_{22} \end{bmatrix} - \begin{bmatrix} P_1 & 0 \\ 0 & P_2 \end{bmatrix} \qquad /4.78/$$

where

$$P = \begin{bmatrix} P_1 & 0 \\ 0 & P_2 \end{bmatrix} \in R^{n \times n}, \quad Q \in R^{n \times n}$$

**Theorem 4.22**

RM is asymptotically stable if there exists a block diagonal positive definite matrix P such that the matrix Q, given by /4.78/, is positive definite.

**Proof**

Consider a positive definite Lyapunov function of the quadratic form

$$V(x) = x^* P x$$

Taking into account that $x' = Ax$ ($u = 0$) we may write

$$\Delta Vx = V(x') - V(x) = x^*[A^T PA - P]x \qquad /4.79/$$

Thus, RM is asymptotically stable if there exists a block diagonal positive definite matrix P such that the matrix Q, given by /4.78/ is positive definite. □

## Example 4.8

Test the asymptotic stability of RM with

$$A = \begin{bmatrix} A_{11} & A_{12} \\ A_{21} & A_{22} \end{bmatrix} = \begin{bmatrix} \frac{1}{2} & 0 \\ -1 & \frac{1}{3} \end{bmatrix} \qquad /4.80/$$

using the Theorem 4.22.

For /4.80/ from /4.78/ we obtain

$$A^T PA - P = \begin{bmatrix} \frac{1}{2} & -1 \\ 0 & \frac{1}{3} \end{bmatrix} \begin{bmatrix} P_1 & 0 \\ 0 & P_2 \end{bmatrix} \begin{bmatrix} \frac{1}{2} & 0 \\ -1 & \frac{1}{3} \end{bmatrix} - \begin{bmatrix} P_1 & 0 \\ 0 & P_2 \end{bmatrix} =$$

$$= \begin{bmatrix} \frac{3}{4}P_1 - P_2 & \frac{1}{3}P_2 \\ \frac{1}{3}P_2 & \frac{8}{9}P_2 \end{bmatrix} = -Q$$

It is easy to check for $\frac{3}{4}P_1 > P_2 > 0$ the matrices

$$P = \begin{bmatrix} P_1 & 0 \\ 0 & P_2 \end{bmatrix}, \quad Q = \begin{bmatrix} \frac{3}{4}P_1 - P_2, & \frac{1}{3}P_2 \\ \frac{1}{3}P_2 & \frac{8}{9}P_2 \end{bmatrix}$$

are positive definite. Therefore, by Theorem 4.22, RM with /4.80/ is asymptotically stable.

### Theorem 4.23

Given Q positive definite and P block diagonal such that /4.78/ is satisfied, then /4.77/ holds if and only if P is positive definite.

### Proof

To show the necessity we assume that Q is positive definite and /4.77/ holds. We shall show that the block diagonal matrix P is positive definite.

From

$$\left[ I - z_1^{-1} A_2 - z_2^{-1} A_1 \right] x = 0$$

we obtain a nontrivial solution

$$\left[ A_1 + A_2 \right] x_o = \left[ z_1 I^{1,0} + z_2 I^{1,0} \right] x_o \qquad /4.81/$$

for some $(z_1, z_2)$ such that $|z_1| \leq 1$, $|z_2| \leq 1$.

Using /4.78/ and /4.81/ we obtain

$$x_o^*([A_1+A_2]^T P[A_1+A_2]-P)x_o =$$
$$= x_o^*([z_1^* I^{1,0}+z_2^* I^{0,1}]P[z_1 I^{1,0}+z_2 I^{0,1}]-P)x_o = -x_o^* Q x_o$$

and

$$x_o^*\left[(|z_1|^2-1)P^{1,0}+(|z_2|^2-1)P^{0,1}+Q\right]x_o = 0 \qquad /4.82/$$

since

$$I^{1,0}PI^{0,1} = P^{1,0}, \quad I^{0,1}PI^{0,1} = P^{0,1}, \quad I^{1,0}PI^{0,1} = 0$$

and

$$I^{0,1}PI^{1,0} = 0$$

Due to the fact that $x_o \neq 0$ it follows that

$$\det\left[(|z_1|^2-1)P^{1,0}+(|z_2|^2-1)P^{0,1}+Q\right] = 0$$

for some $(z_1, z_2)$ such that $|z_1| \leq 1$, $|z_2| \leq 1$.

This can be considered as the characteristic equation of the following pencil

$$P^{1,0} - \lambda_1\left[(|z_2|^2-1)P^{0,1}+Q\right] \qquad /4.83/$$

where

$$\lambda_1 = \frac{1}{1-|z_1|^2} \qquad /4.84/$$

From /4.77/ we have

$$\det\left[I-z_1^{-1}A_2-z_2^{-1}A_1\right] = 0 \text{ for } |z_2^{-1}| = 1, \quad |z_1^{-1}| > 1 \qquad /4.85/$$

For $|z_2^{-1}| = 1$ the pencil /4.83/ becomes

$$P^{1,0} - \lambda_1 Q \qquad /4.86/$$

which is a regular pencil due to $Q > 0$.

Using the extremal properties of regular pencils we obtain

$$\lambda_{1\min} \leq \frac{x^T P^{1,0} x}{x^T Q x} \leq \lambda_{1\max} \qquad /4.87/$$

for all $x \neq 0$

From /4.85/ and /4.84/ it follows that $\lambda_1 > 0$. Therefore /4.87/ gives

$$0 < \lambda_{1min} \leq \frac{x_1^T P_1 x_1}{x^T Q x} \leq \lambda_{1max} \quad /4.88/$$

for all $x \neq 0$, $x_1 \neq 0$, where $x^T = [x_1^T, x_1^T]$, since $x^T P^{1,0} x = x_1^T P_1 x_1$. Using the fact that Q is positive definite /4.88/ implies that $P_1$ is positine definite too. The positive definiteness of $P_2$ can be proved using a similar approach for the pencil given by

$$P^{0,1} - \lambda_2 [(|z_1|^2 - 1) P^{0,1} + Q]$$

where

$$\lambda_2 = \frac{1}{1 - |z_2|^2}$$

instead of the pencil given by /4.83/.

The sufficiency follows from Theorem 4.22.

Note that Theorem 4.23 assumes the existence of a Q positive definite and P block diagonal such that /4.78/ is satisfied. It is possible that for asymptotically stable RM no such matrices P and Q exist. In such case RM must have a characteristic polynomial which is of higher order than 1 in both variables $z_1$ and $z_2$.

### Theorem 4.24

RM is asymptotically stable if the matrix $I - |A|$ has all the principle minors positive, where $|A|$ denotes the matrix with elements $a_{ij} = |a_{ij}|$

### Proof

In [28] it was shown that if the matrix $I - |A|$ has all the principle minors positive, then there exists a diagonal positive definite matrix P such thst the matrix $P - A^T P A$ is positive definite. Therefore by Theorem 4.22 RM is asymptotically stable if $I - |A|$ has all the principle minors positive. □

## Example 4.9

Test the asymptotic stability of RM with

$$A = \begin{bmatrix} A_{11} & A_{12} \\ A_{21} & A_{22} \end{bmatrix} = \left[\begin{array}{cc|c} -\frac{1}{2} & 0 & \frac{1}{3} \\ 0 & -\frac{1}{2} & -\frac{1}{4} \\ \hline -\frac{1}{4} & -\frac{1}{3} & -\frac{1}{3} \end{array}\right] \qquad /4.81/$$

using Theorem 4.24

In this case

$$|A| = \begin{bmatrix} \frac{1}{2} & 0 & \frac{1}{3} \\ 0 & \frac{1}{2} & \frac{1}{4} \\ \frac{1}{4} & \frac{1}{3} & \frac{1}{3} \end{bmatrix}$$

and

$$I - |A| = \begin{bmatrix} \frac{1}{2} & 0 & -\frac{1}{3} \\ 0 & \frac{1}{2} & -\frac{1}{4} \\ -\frac{1}{4} & -\frac{1}{3} & -\frac{2}{3} \end{bmatrix} \qquad /4.90/$$

It is easy to check that all the principal minors of the matrix /4.90/ are positive. Therefore, by Theorem 4.24 RM wirh /4.89/ is asymptotically stable. Note that the characteristic polynomial /4.76/ of /4.89/ is

$$d(z_1^{-1}, z_2^{-1}) = \begin{bmatrix} 1+\frac{1}{2}z_1^{-1} & 0 & \frac{1}{3}z_1^{-1} \\ 0 & 1+\frac{1}{2}z_1^{-1} & \frac{1}{4}z_1^{-1} \\ \frac{1}{4}z_2^{-1} & \frac{1}{3}z_2^{-1} & 1+\frac{1}{3}z_2^{-1} \end{bmatrix} = \left(z_1^{-1}+\frac{1}{2}\right)^2 \left(z_2^{-1}+\frac{1}{3}\right)$$

and has no zeros in the unit bidisk.

## PROBLEMS

1. Show that if $n_1 = n_2 = 1$ then RM is BIBO stable if and only if

$$|A_{11}| < 1$$

and

$$\max\left\{\left|A_{22} + \frac{A_{12}A_{21}}{1 - A_{11}}\right|, \left|A_{22} - \frac{A_{12}A_{21}}{1 + A_{11}}\right|\right\} < 1$$

Hint: Use the conditions $1°$ of Theorem 4.6.

2. Show that RM is BIBO stable if

   i. $A_{11}$ and $A_{22}$ are stable
   ii. $A_{11}^T A_{11} = A_{11} A_{11}^T$ and $A_{22}^T A_{22} = A_{22} A_{22}^T$
   iii. $\|A_{12}\| \|A_{21}\| < (1 - \lambda_{11})(1 - \lambda_{22})$

   where $\|\cdot\|$ is the induced norm defined by /4.36/ and $\lambda_{11}$, $\lambda_{22}$ are given by /4.37/.

   Hint: Note that $\|\bar{A}_{12}\| = \|T_1 A_{12} T_2^T\| \leq \|A_{12}\|$ and

   $$\|\bar{A}_{21}\| = \|T_2 A_{21} T_1^T\| \leq \|A_{21}\|$$

   $T_1, T_2$ are orthonormal matrices.

3. Show that RM with $n_2 = 1$ is BIBO stable if and only if

   i. $A$ and $A_{11}$ are stable
   ii. $$\left| z - \frac{\prod_{i=1}^{n_1+1}(z - \alpha_i)}{\prod_{i=1}^{n_1}(z - \beta_i)} \right| \neq 1 \quad \text{for } |z| = 1$$

   where $\alpha_i$ /$i = 1, 2, \ldots, n_1+n_2$/ and $\beta_i$ /$i = 1, 2, \ldots, n_1$/ are the eigenvalues of $A$ and $A_1$, respectively.

   Hint: Use the conditions $3°$ of Theorem 4.6 and note that

   $$\det\left[zI_{n_2} - \left[A_{22} + A_{21}\left[zI_{n_1} - A_{11}\right]^{-1} A_{12}\right]\right] =$$

$$= \frac{\det\begin{bmatrix} zI_{n_1} - A_{11} & -A_{12} \\ -A_{21} & zI_{n_2} - A_{22} \end{bmatrix}}{\det\begin{bmatrix} zI_{n_1} - A_{11} \end{bmatrix}} =$$

$$= \frac{\prod_{i=1}^{n_1+n_2}(z-\alpha_i)}{\prod_{i=1}^{n_1}(z-\beta_i)}$$

4. Show that RM is asymptotically /BIBO/ stable only if the matrices

$A_{11}$, $A_{22}$, $A_{11} + A_{12}[I - A_{22}]^{-1}A_{21}$, $A_{22} + A_{21}[I - A_{11}]^{-1}A_{12}$,

$A_{11} + A_{12}[I + A_{22}]^{-1}A_{21}$ and $A_{22} + A_{21}[I + A_{11}]^{-1}A_{12}$ are stable.

Hint: Use

$$A^{i,j} = \sum_{h=0}^{i} A^{h,0} A^{0,1} A^{i-h,j-1}$$

and

$$\sum_{i=0}^{\infty} A^{i,j} = \sum_{i=0}^{\infty} A^{i,0} \left[ A^{0,1} \sum_{i=0}^{\infty} A^{i,0} \right]^j$$

5. Show that RM is asymptotically /BIBO/ stable only if

$\sigma/A_{11}/ < 1$ and $\sigma/A_{22}/ < 1$

where $\sigma/A_{11}/ = \max_i |\lambda_i/A_{11}/|$

Hint: Use Theorem 4.6.

6. Show that the condition

rank $A$ = rank $[A \vdots B]$

is equivalent to the condition $AA^+B = B$, where $A^+$ is the generalized inverse of $A$.

Hint: Consider the equation $AK = B$.

# REFERENCES

[1] Agathoklis P., Jury E.I., **Mansour M.**: The margin of stability of 2-D linear discrete systems. IEEE Trans. Acoustics, Speech, Signal Processing vol. ASSP-30, no.6, 1982, pp.869-873.

[2] Ahmed A.R.E.: On the stability of two-dimensional discrete systems. IEEE Trans. Autom. Control vol. AC-25, no.3, 1980, pp.551-552.

[3] Basu S., Bose N.K.: Stability of 2-D matrix rational approximants from input data. IEEE Trans. Autom. Control vol. AC-26, no.2, 1981, pp.540-541.

[4] Chiasson J.N., Brierly S.D.: New stability tests for 2-D polynomials. **Transactions CDC 1984 /in press/**

[5] Davis D.L.: A correct proof of Huang's theorem. IEEE Trans. Acoustics, Speech, Signal Processing, October 1976, pp.425-426.

[6] DeCarlo R.A., Murray J., Sacks R.: Multivariable Nyquist theory. Int. J. Control vol.25, no.5, 1977, pp.657-675.

[7] Delsarte Ph., Genin Y.V., Kamp Y.G.: A simple proof of Rudin's multivariable stability theorem. IEEE Acoustics, Speech, Signal Processing vol. ASSP-28, no.6, 1980, pp.701-705.

[8] Eising R.: 2-D Systems; An Algebraic Approach. Mathematish Centrum, Amsterdam 1979.

[9] Fornasini E., Marchesini G.: Doubly - indexed dynamical systems;: State - space models and structural properties. Mathematical Systems Theory vol.12, 1978, pp.59-72.

[10] Fornasini E., Marchesini G.: On the internal stability of two-dimensional filters. IEEE Trans. Autom. Control vol. AC-24, 1979, pp.129-130.

[11] Fornasini E., Marchesini G.: Stability analysis od 2-D systems. IEEE Trans. Circuits, Systems vol. CAS-27, 1980, pp.1210-1217.

[12] Goodman D.: An alternate proof of Huang's theorem. IEEE Trans. Acoustics, Speech, Signal Processing, Oct. 1976, pp.426-427.

[13] Goodman D.: Some stability properties of two-dimensional linear shift - invariant digital filters. IEEE Trans. Circuits Syst. vol. CAS-24, no.4, 1977, pp.201-208.

[14] Hormander L.: An Introduction to Complex Analysis in Several Variables. North - Holland 1973.

[15] Huang T.S.: Stability of two-dimensional recursive filters. IEEE Trans. Audio Electroacoust. vol. AU-20, no.2, 1972, pp.158-163.

[16] Jury E.I., Mansour M.: A set of necessary and sufficient stability conditions for low order two-dimensional polynomials. IEEE Trans. Autom. Control vol. AC-27, no.1, 1982, pp.192-194.

[17] Justice J.H., Shanks J.L.: Stability criterion for n-dimensional digital filters. IEEE Trans. Autom. Control, June 1973, pp.284-286.

[18] Kamen E.W.: Asymptotic stability of linear shift invariant two-dimensional digital filters. IEEE Trans. Circuits, Systems vol. CAS-27, 1980, pp.1234-1240.

[19] Kamen E.W.: On the relationship between zero criteria for two-variable polynomials and asymptotic stability of delay differential equations. IEEE Trans. Autom. Control vol. AC-25, no.5, 1980.

[20] Kurek J.E. Stabilities of 2-D linear discrete-time systems. Preprints IFAC 9th World Congress, Budapest July 2-6,1984,pp.150-154

[21] Pandolfi L. Exponential stability of 2-D systems. Systems and Control Letters, vol.4, 1984, pp. 381-385

[22] Rudin W.: Principles of Mathematical Analysis. McGraw-Hill, New York 1964.

[23] Shanks J.L., Treitel S., Justice J.H.: Stability and synthesis of two-dimensional recursive filters. IEEE Trans. Audio Electroacoust. vol. AU-20, no.2, 1972, pp.115-128.

[24] Strintzis M.G.: Tests of stability of multidimensional filters. IEEE Trans. Circuits Systems vol. CAS-24, no.8, 1977, pp.432-437.

[25] Walach E., Zeheb E.: N-dimensional stability margins computation and a variable transformation. IEEE Trans. Acoustics, Speech, Signal Processing vol. ASSP-30, no.6, 1982, pp.887-893.

[26] Wu-Sheng Lu, Lee E.B.: Stability analysis for two-dimensional systems. IEEE Trans. Circuits, Systems vol. CAS-30, no.7, 1983, pp.455-461.

[27] Agathoklis P., Jury E, Mansour M.: Asymptotic stability and the Lyaponov equation for two-dimensional discrete systems. Preprints IFAC 9th World Congress, Budapest,July 2-6, 1984, vol. VIII, pp. 155-158

[28] Moylan P.J.: Matrices with positive principal minors. Linear Algebra and its Applications, vol. 17, 1977, pp. 53-58.

# 5. CHARACTERISTIC POLYNOMIAL AND EIGENVALUE ASSIGNMENT

## 5.1 PARASKEVOPOULOS´ METHOD OF COEFFICIENT ASSIGNMENT

### 5.1.1 Problem formulation.

Consider Roesser´s model /RM/ described by the equations

$$x´ = Ax + Bu \qquad /5.1/$$
$$y = Cx \qquad /5.2/$$

where

$$x´ = \begin{bmatrix} x^h/i+1,j/ \\ x^v/i,j+1/ \end{bmatrix}, \quad x = \begin{bmatrix} x^h/i,j/ \\ x^v/i,j/ \end{bmatrix},$$

$$A = \begin{bmatrix} A_{11} & A_{12} \\ A_{21} & A_{22} \end{bmatrix}, \quad B = \begin{bmatrix} B_1 \\ B_2 \end{bmatrix},$$

$$C = \begin{bmatrix} C_1 & C_2 \end{bmatrix},$$

$x^h/i,j/ \in R^{n_1}$ is the horizontal state vector,

$x^v/i,j/ \in R^{n_2}$ is the vertical state vector,

$u = u/i,j/ \in R^m$ is the input vector,

$y = y/i,j/ \in R^l$ is the output vector

and $A_{ij}, B_j, C_i$ /i,j = 1,2/ are real constant matrices of appropriate dimensions.

Let the state feedback law applied have the form [14]

$$u = Kx + v \qquad /5.3/$$

where $K \in R^{m \times n}$ /$n = n_1+n_2$/ is a real constant matrix,

$v = v/i,j/ \in R^m$ is a new input vector.

Substitution of /5.3/ into /5.1/ yields the closed-loop system equation

$$x´ = [A + BK]x + Bv \qquad /5.4/$$

Let $d_c/z_1,z_2/$ be the desirable 2-D characteristic polynomial of the closed-loop system. Then, the coefficient assignment problem via the state feedback can be stated as follows

**Problem 1.**

Given $A,B$ and $d_c/z_1,z_2/$, find the state feedback matrix $K$ such that
$$d_c/z_1,z_2/ = \det[Z - A - BK] \qquad /5.5/$$
where
$$Z = \begin{bmatrix} z_1 I_{n_1} & 0 \\ 0 & z_2 I_{n_2} \end{bmatrix}$$

Let the output feedback law applied have the form
$$u = Fy + v \qquad /5.6/$$
where $F \in R^{m \times l}$ is a real constant matrix.

Substitution of $u = FCx + v$ into /5.1/ yields the closed-loop system equation
$$x' = [A + BFC]x + Bv \qquad /5.7/$$

The coefficient assignment problem via the output feedback can be stated as follows

**Problem 2.**

Given $A,B,C$ and $d_c/z_1,z_2/$, find the output feedback matrix $F$ such that
$$d_c/z_1,z_2/ = \det[Z - A - BFC] \qquad /5.8/$$

### 5.1.2 Problem solutions for single-input systems.

**Problem 1.**

For a single-input system /5.5/ has the form [13]
$$d_c/z_1,z_2/ = \det[Z - A - bk^T] \qquad /5.9/$$
where
$$b = \begin{bmatrix} b_1 \\ b_2 \end{bmatrix} \in R^{n \times 1}, \quad k^T = [k_1 \; k_2 \; \ldots \; k_n]$$
and superscript $T$ denotes the transposition of a matrix /or a vector/.

Using the following well-known relationship
$$\det[Z - A - bk^T] = \det[Z - A] - k^T\left(\mathrm{adj}[Z - A]\right)b \qquad /5.10/$$

we can write /5.9/ in the form

$$d_c/z_1,z_2/ = d_o/z_1,z_2/ - k^T M/z_1,z_2/ \qquad /5.11/$$

where

$$d_o/z_1,z_2/ = \det[Z - A] \qquad /5.12/$$

is the open-loop characteristic polynomial and

$$M/z_1,z_2/ = (\text{adj}[Z - A])b \qquad /5.13/$$

The equation /5.11/ may further be written as

$$M^T/z_1,z_2/\,k = d/z_1,z_2/ \qquad /5.14/$$

where

$$d/z_1,z_2/ = d_o/z_1,z_2/ - d_c/z_1,z_2/ = \sum_{i=0}^{n_1} \sum_{j=0}^{n_2} d_{ij} z_1^i z_2^j \qquad /5.15/$$

with $d_{n_1 n_2} = 0$.

Substituting

$$M/z_1,z_2/ = \sum_{i=0}^{n_1} \sum_{j=0}^{n_2} m_{ij} z_1^i z_2^j \quad \text{with} \quad m_{n_1 n_2} = 0 \qquad /5.16/$$

and /5.15/ into /5.14/ we obtain

$$\sum_{i=0}^{n_1} \sum_{j=0}^{n_2} m_{ij}^T k\, z_1^i z_2^j = \sum_{i=0}^{n_1} \sum_{j=0}^{n_2} d_{ij} z_1^i z_2^j \qquad /5.17/$$

Comparison of the coefficients of like powers of $z_1$ and $z_2$ on both sides of /5.17/ yields

$$M k = d \qquad /5.18/$$

where

$$M = \begin{bmatrix} m_{00}^T \\ m_{10}^T \\ m_{01}^T \\ \cdots \\ m_{n_1-1,n_2}^T \\ m_{n_1,n_2-1}^T \end{bmatrix}, \quad d = \begin{bmatrix} d_{00} \\ d_{10} \\ d_{01} \\ \cdots \\ d_{n_1-1,n_2} \\ d_{n_1,n_2-1} \end{bmatrix} \qquad /5.18a/$$

Note that /5.18/ is a linear system of $N = (n_1+1)(n_2+1) - 1 = n_1 n_2 + n_1 + n_2$ algebraic equations with $n = n_1 + n_2$ unknowns.

Thus, Problem 1 has a solution if and only if there exists a solution k to /5.18/.

Let us assume that

$$\text{rank } M = r \leq n$$

Then, there exists a nonsingular matrix $P \in R^{N \times N}$ of elementary row operations such that

$$PM = \begin{bmatrix} M_1 \\ 0 \end{bmatrix} \qquad /5.19/$$

where $M_1 \in R^{r \times n}$.

Premultiplying /5.18/ by P and taking into account /5.19/ we obtain

$$\begin{bmatrix} M_1 \\ 0 \end{bmatrix} k = \begin{bmatrix} d_1 \\ d_2 \end{bmatrix} = Pd$$

or

$$M_1 k = d_1 \qquad /5.20/$$

and

$$d_2 = 0 \qquad /5.21/$$

where $d_1 \in R^{r \times 1}$, $d_2 \in R^{(N-r) \times 1}$

We have thus established the following

Theorem 5.1

Problem 1 for single-input system has a solution if and only if the condition /5.21/ is satisfied.

If /5.21/ is satisfied, then the desired k is given by

$$k = M_1^T [M_1 M_1^T]^{-1} d_1 \qquad /5.22/$$

and in a particular case when r=n

$$k = M_1^{-1} d_1 \qquad /5.22'/$$

Example 5.1

Find conditions under which Problem 1 has a solution for

$$A = \begin{bmatrix} 0 & 1 \\ -1 & -2 \end{bmatrix}, \quad b = \begin{bmatrix} 1 \\ 1 \end{bmatrix} \quad \text{and} \quad d_c/z_1,z_2/ = z_1 z_2 + a_1 z_1 + a_2 z_2 + a_0 \qquad /5.23/$$

In this case $n_1 = n_2 = 1$,

$$d_0/z_1,z_2/ = \det[Z - A] = \begin{bmatrix} z_1 & -1 \\ 1 & z_2+2 \end{bmatrix} = z_1 z_2 + 2z_1 + 1$$

$$M/z_1,z_2/ = (\text{adj}[Z - A])b = \left(\text{adj}\begin{bmatrix} z_1 & -1 \\ 1 & z_2+2 \end{bmatrix}\right)\begin{bmatrix} 1 \\ 1 \end{bmatrix} = \begin{bmatrix} z_2+3 \\ z_1-1 \end{bmatrix}$$

and

$$d/z_1,z_2/ = d_0/z_1,z_2/ - d_c/z_1,z_2/ = 2 - a_1 z_1 - a_2 z_2 + 1 - a_0$$

Hence

$$M = \begin{bmatrix} 3 & -1 \\ 0 & 1 \\ 1 & 0 \end{bmatrix} \quad \text{and} \quad d = \begin{bmatrix} 1-a_0 \\ 2-a_1 \\ -a_2 \end{bmatrix}$$

For

$$P = \begin{bmatrix} 0 & 0 & 1 \\ 0 & 1 & 0 \\ 1 & 1 & -3 \end{bmatrix}$$

we obtain

$$PM = \begin{bmatrix} 1 & 0 \\ 0 & 1 \\ 0 & 0 \end{bmatrix}, \quad Pd = \begin{bmatrix} -a_2 \\ 2 - a_1 \\ 3(1+a_2) - a_1 - a_0 \end{bmatrix}$$

and

$$M_1 = \begin{bmatrix} 1 & 0 \\ 0 & 1 \end{bmatrix}, \quad d_1 = \begin{bmatrix} -a_2 \\ 2 - a_1 \end{bmatrix}, \quad d_2 = 3(1+a_2) - a_1 - a_0$$

The problem has a solution if and only if $d_2 = 0$, i.e. iff

$$a_0 = 3(1+a_2) - a_1 \qquad /5.24/$$

From /5.22'/ we have

$$k = M_1^{-1} d_1 = \begin{bmatrix} -a_2 \\ 2-a_1 \end{bmatrix}$$

and

$$A_c = A + bk^T = \begin{bmatrix} 0 & 1 \\ -1 & -2 \end{bmatrix} + \begin{bmatrix} 1 \\ 1 \end{bmatrix}[-a_2 \quad 2-a_1] = \begin{bmatrix} -a_2 & 3-a_1 \\ 1-a_2 & -a_1 \end{bmatrix}$$

Therefore

$$d_c/z_1,z_2/ = \begin{vmatrix} z_1+a_2 & a_1-3 \\ 1+a_2 & z_2+a_1 \end{vmatrix} = z_1 z_2 + a_1 z_1 + a_2 z_2 + 3(1+a_2) - a_1$$

and the closed-loop characteristic polynomial has the given form /5.23/ iff /5.24/ is satisfied.

Problem 2.

For a single-input system /5.8/ has the form

$$d_c/z_1,z_2/ = \det\left[Z - A - b f^T C\right] \qquad /5.25/$$

where $f^T = \begin{bmatrix} f_1 & f_2 & \cdots & f_1 \end{bmatrix}$

In a similar way as in the case of state feedback we obtain

$$\bar{M} f = d \qquad /5.26/$$

where $\bar{M} = M C^T$.

Note that /5.26/ is a linear system of N equations with 1 unknowns. Let us assume that $\text{rank}\,\bar{M} = \bar{r} \leq 1$. Then, there exists a nonsingular matrix $\bar{P} \in R^{N \times N}$ of elementary row operations such that

$$\bar{P}\bar{M} = \begin{bmatrix} \bar{M}_1 \\ 0 \end{bmatrix} \qquad /5.27/$$

where $\bar{M}_1 \in R^{\bar{r} \times 1}$.

Premultiplying /5.26/ by $\bar{P}$ and taking into account /5.27/ we obtain

$$\begin{bmatrix} \bar{M}_1 \\ 0 \end{bmatrix} f = \begin{bmatrix} \bar{d}_1 \\ \bar{d}_2 \end{bmatrix} = \bar{P} d$$

or

$$\bar{M}_1 f = \bar{d}_1 \qquad /5.28/$$

and

$$\bar{d}_2 = 0 \qquad /5.29/$$

where $\bar{d}_1 \in R^{\bar{r} \times 1}$, $\bar{d}_2 \in R^{(N-\bar{r}) \times 1}$

We have thus established the following

Theorem 5.2

Problem 2 for single-input system has a solution if and only if the condition /5.29/ is satisfied.

If /5.29/ is satisfied, then the desired f is given by

$$f = \bar{M}_1^T \left[\bar{M}_1 \bar{M}_1^T\right]^{-1} \bar{d}_1 \qquad /5.30/$$

and in a particular case when $\bar{r} = 1$

$$f = \bar{M}_1^{-1} \bar{d}_1 \qquad /5.30'/$$

Example 5.2

Find conditions under which Problem 2 has a solution for /5.23/ and

$$C = \begin{bmatrix} -1 & 1 \end{bmatrix}$$

Using the results obtained in Example 5.1 we have

$$\bar{M} = M C^T = \begin{bmatrix} 3 & -1 \\ 0 & 1 \\ 1 & 0 \end{bmatrix} \begin{bmatrix} -1 \\ 1 \end{bmatrix} = \begin{bmatrix} -4 \\ 1 \\ -1 \end{bmatrix}$$

For

$$\bar{P} = \begin{bmatrix} 0 & 1 & 0 \\ 0 & 1 & 1 \\ 1 & 4 & 0 \end{bmatrix}$$

we obtain

$$\bar{P}\bar{M} = \begin{bmatrix} 1 \\ 0 \\ 0 \end{bmatrix}, \quad \bar{P}d = \begin{bmatrix} 0 & 1 & 0 \\ 0 & 1 & 1 \\ 1 & 4 & 0 \end{bmatrix} \begin{bmatrix} 1-a_0 \\ 2-a_1 \\ -a_2 \end{bmatrix} = \begin{bmatrix} 2 - a_1 \\ 2 - a_1 - a_2 \\ 9 - a_0 - 4a_1 \end{bmatrix}$$

and

$$\bar{M}_1 = 1, \quad \bar{d}_1 = 2 - a_1, \quad \bar{d}_2 = \begin{bmatrix} 2 - a_1 - a_2 \\ 9 - a_0 - 4a_1 \end{bmatrix}$$

The problem has a solution if and only if $\bar{d}_2 = 0$; i.e., iff

$$a_0 = 9 - 4a_1, \quad a_2 = 2 - a_1 \qquad /5.31/$$

From /5.30'/ we have

$$f = \bar{M}_1^{-1} \bar{d}_1 = 2 - a_1$$

and

$$A_c = A + b f^T C = \begin{bmatrix} 0 & 1 \\ -1 & -2 \end{bmatrix} + \begin{bmatrix} 1 \\ 1 \end{bmatrix}(2-a_1)\begin{bmatrix} -1 & 1 \end{bmatrix} = \begin{bmatrix} a_1 - 2 & 3 - a_1 \\ a_1 - 3 & -a_1 \end{bmatrix}$$

Therefore

$$d_c/z_1, z_2/ = \begin{vmatrix} z_1 + 2 - a_1 & a_1 - 3 \\ 3 - a_1 & z_2 + a_1 \end{vmatrix} = z_1 z_2 + a_1 z_1 + (2-a_1) z_2 + 9 - 4a_1$$

and the closed-loop characteristic polynomial has the given form /5.23/ if /5.31/ are satisfied.

### 5.1.3 Problem solutions for multi-input systems.

#### Problem 1.

We assume that K has the dyadic structure

$$K = q k^T \qquad /5.32/$$

where $q \in R^{m \times 1}$ and $k \in R^{n \times 1}$.

Note that the closed-loop system matrix $A_c = A + BK$ has the same form $A_c = A + bk^T$ as the single-input system if $b = Bq$. Therefore the method developed for single-input systems can be applied for finding k for $b = Bq$. If the problem has a solution the following algorithm can be used for evaluating K:

#### Algorithm 5.1

Step 1. Write K in the dyadic form /5.32/.

Step 2. Assume a nonzero vector for q /one of the elements in q may be taken 1/.

Step 3. Using /5.12/, /5.13/ and /5.15/ find $d_o/z_1, z_2/$, $M/z_1, z_2/$ and $d/z_1, z_2/$.

Step 4. Using /5.18a/ find M and d.

Step 5. Find P satisfying /5.19/ and evaluate $M_1$, $d_1$ and $d_2$.

Step 6. Using /5.22/ (or /5.22'/) and /5.32/ calculate k and K.

#### Example 5.3

Find conditions under which Problem 1 has a solution for

$$A = \begin{bmatrix} A_{11} & A_{12} \\ A_{21} & A_{22} \end{bmatrix} = \begin{bmatrix} 0 & 1 & 0 \\ -1 & -2 & -1 \\ \hline 1 & 0 & -2 \end{bmatrix}, \quad B = \begin{bmatrix} B_1 \\ B_2 \end{bmatrix} = \begin{bmatrix} 1 & 0 \\ 0 & 1 \\ \hline 0 & -1 \end{bmatrix} \qquad /5.33/$$

and

$$d_c/z_1, z_2/ = z_1^2 z_2 + a_{20} z_1^2 + a_{11} z_1 z_2 + a_{10} z_1 + a_{01} z_2 + a_{00} \qquad /5.34/$$

Step 1.

$$K = \begin{bmatrix} q_1 \\ q_2 \end{bmatrix} \begin{bmatrix} k_1 & k_2 & k_3 \end{bmatrix}$$

Step 2.

$$\text{We assume} \quad q = \begin{bmatrix} q_1 \\ q_2 \end{bmatrix} = \begin{bmatrix} 1 \\ 1 \end{bmatrix}$$

Step 3.

$$d_0/z_1,z_2/ = \det[Z-A] = \begin{vmatrix} z_1 & -1 & 0 \\ 1 & z_1+2 & 1 \\ -1 & 0 & z_2+2 \end{vmatrix} =$$

$$= z_1^2 z_2 + 2z_1^2 + 2z_1 z_2 + 4z_1 + z_2 + 3$$

$$M/z_1,z_2/ = \text{adj}[Z-A]\cdot Bq = \text{adj}\begin{bmatrix} z_1 & -1 & 0 \\ 1 & z_1+2 & 1 \\ -1 & 0 & z_2+2 \end{bmatrix} \cdot \begin{bmatrix} 1 \\ 1 \\ -1 \end{bmatrix} =$$

$$= \begin{bmatrix} (z_1+3)(z_2+2)+1 \\ (z_1-1)(z_2+3) \\ (z_1+2)(1-z_1) \end{bmatrix}$$

and

$$d/z_1,z_2/ = d_0/z_1,z_2/ - d_c/z_1,z_2/ =$$

$$= (2-a_{20})z_1^2 + (2-a_{11})z_1 z_2 + (4-a_{10})z_1 + (1-a_{01})z_2 +$$

$$+ 3 - a_{00}$$

Step 4.

Using /5.18a/ we obtain

$$M = \begin{bmatrix} 7 & -3 & 22 \\ 2 & 3 & -1 \\ 3 & -1 & 0 \\ 1 & 1 & 0 \\ 0 & 0 & -1 \end{bmatrix}, \quad d = \begin{bmatrix} 3 - a_{00} \\ 4 - a_{10} \\ 1 - a_{01} \\ 2 - a_{11} \\ 2 - a_{20} \end{bmatrix}$$

Step 5.

It is easy to check that for

$$P = \begin{bmatrix} 0 & -1 & 0 & 3 & 1 \\ 0 & 1 & 0 & -2 & -1 \\ 0 & 0 & 0 & 0 & -1 \\ 1 & 10 & 0 & -27 & -8 \\ 0 & 4 & 1 & -11 & -4 \end{bmatrix} \quad \text{we have}$$

$$PM = \begin{bmatrix} 1 & 0 & 0 \\ 0 & 1 & 0 \\ 0 & 0 & 1 \\ 0 & 0 & 0 \\ 0 & 0 & 0 \end{bmatrix}, \quad Pd = \begin{bmatrix} a_{10} - a_{20} - 3a_{11} + 4 \\ a_{20} + 2a_{11} - a_{10} - 2 \\ a_{20} - 2 \\ 8a_{20} + 27a_{11} - 10a_{10} - a_{00} - 27 \\ 4a_{20} + 11a_{11} - 4a_{10} - a_{01} - 13 \end{bmatrix}$$

and

$$M_1 = \begin{bmatrix} 1 & 0 & 0 \\ 0 & 1 & 0 \\ 0 & 0 & 1 \end{bmatrix}, \quad d_1 = \begin{bmatrix} a_{10} - a_{20} - 3a_{11} + 4 \\ a_{20} + 2a_{11} - a_{10} - 2 \\ a_{20} - 2 \end{bmatrix},$$

$$d_2 = \begin{bmatrix} 8a_{20} + 27a_{11} - 10a_{10} - a_{00} - 27 \\ 4a_{20} + 11a_{11} - 4a_{10} - a_{01} - 13 \end{bmatrix}.$$

The problem has a solution if and only if $d_2 = 0$; that is, iff

$$\begin{aligned} a_{00} &= 8a_{20} + 27a_{11} - 10a_{10} - 27 \\ a_{01} &= 4a_{20} + 11a_{11} - 4a_{10} - 13 \end{aligned} \qquad /5.35/$$

Step 6.

From /5.22´/ we obtain

$$k = M_1^{-1} d_1 = \begin{bmatrix} a_{10} - a_{20} - 3a_{11} + 4 \\ a_{20} + 2a_{11} - a_{10} - 2 \\ a_{20} - 2 \end{bmatrix}$$

and

$$K = q k^T = \begin{bmatrix} a_{10} - a_{20} - 3a_{11} + 4 & a_{20} + 2a_{11} - a_{10} - 2 & a_{20} - 2 \\ a_{10} - a_{20} - 3a_{11} + 4 & a_{20} + 2a_{11} - a_{10} - 2 & a_{20} - 2 \end{bmatrix}$$

Therefore

$$A_c = A + BK = \begin{bmatrix} a_{10} - a_{20} - 3a_{11} + 4 & a_{20} + 2a_{11} - a_{10} - 1 & a_{20} - 2 \\ a_{10} - a_{20} - 3a_{11} + 3 & a_{20} + 2a_{11} - a_{10} - 4 & a_{20} - 3 \\ -a_{10} + a_{20} + 3a_{11} - 3 & -a_{20} - 2a_{11} + a_{10} + 2 & -a_{20} \end{bmatrix}$$

and

$$d_c/z_1, z_2/ = \begin{vmatrix} z_1 - a_{10} + a_{20} + 3a_{11} - 4 & -a_{20} - 2a_{11} + a_{10} + 1 & -a_{20} + 2 \\ -a_{10} + a_{20} + 3a_{11} - 3 & z_1 - a_{20} - 2a_{11} + a_{10} + 4 & -a_{20} + 3 \\ a_{10} - a_{20} - 3a_{11} + 3 & a_{20} + 2a_{11} - a_{10} - 2 & z_2 + a_{20} \end{vmatrix}$$

$$= z_1^2 z_2 + a_{20} z_1^2 + a_{11} z_1 z_2 + a_{10} z_1 + \left( 4a_{20} + 11a_{11} + \right.$$
$$\left. - 4a_{10} - 13 \right) z_2 + 8a_{20} + 27a_{11} - 10a_{10} - 27$$

Thus the closed-loop characteristic polynomial has the preassigned form /5.34/ if /5.35/ are satisfied.

Problem 2.

We assume that F has the dyadic structure

$$F = q f^T \qquad /5.36/$$

where $q \in R^{m \times 1}$, $f \in R^{l \times 1}$.

Note that the closed-loop system matrix $A_c = A + BFC$ has the same form $A_c = A + b f^T C$ as the single-input system if $b = Bq$. Therefore the method developed for single-input system can be applied for finding f for $b = Bq$. From comparison of /5.5/ and /5.8/ it follows that

$$FC = K \qquad /5.37/$$

Therefore, another method of determining F is to solve /5.37/ for the known C and K. The equation /5.37/ may be solved in various ways [9] to yield F. Let us assume that $\text{rank } C = 1$ and let $C_1$ /$C_2$/ be a matrix which consists of the linearly independent /dependent/ columns of C. From /5.37/ it follows that

$$FC_1 = KI_1 \qquad /5.38a/$$
$$FC_2 = KI_2 \qquad /5.38b/$$

where $I_1$ /$I_2$/ is $n \times 1$ matrix / $n \times (n-1)$ matrix / which consists of the columns of unit matrix $I_n$ corresponding to the suitable columns of $C_1$ /$C_2$/.
From /5.38a/ we have

$$F = K I_1 C_1^{-1} \qquad /5.39/$$

Substitution of /5.39/ into /5.38b/ yields

$$K I_1 C_1^{-1} C_2 = K I_2$$

and

$$K \left[ I_2 - I_1 C_1^{-1} C_2 \right] = 0 \qquad /5.40/$$

The matrix /5.39/ is a solution to /5.37/ if and only if /5.40/ is satisfied.

Exapmle 5.4

Find conditions under which Problem 2 has a solution for /5.33/, /5.34/ and

$$C = \begin{bmatrix} 1 & -1 & 0 \\ 0 & 1 & 1 \end{bmatrix}$$

To solve the problem we shall apply the last method.
In this case

$$C_1 = \begin{bmatrix} 1 & -1 \\ 0 & 1 \end{bmatrix}, \quad C_2 = \begin{bmatrix} 0 \\ 1 \end{bmatrix}$$

and using the results of Example 5.3 we obtain

$$K\left[I_2 - I_1 C_1^{-1} C_2\right] = \begin{bmatrix} a_{10}-a_{20}-3a_{11}+4 & a_{20}+2a_{11}-a_{10}-2 & a_{20}-2 \\ a_{10}-a_{20}-3a_{11}+4 & a_{20}+2a_{11}-a_{10}-2 & a_{20}-2 \end{bmatrix} \begin{bmatrix} -1 \\ -1 \\ 1 \end{bmatrix} =$$

$$= \begin{bmatrix} a_{20}+a_{11}-4 \\ a_{20}+a_{11}-4 \end{bmatrix} = \begin{bmatrix} 0 \\ 0 \end{bmatrix}$$

The matrix

$$F = KI_1 C_1^{-1} = \begin{bmatrix} a_{10}-a_{20}-3a_{11}+4 & -a_{11}+2 \\ a_{10}-a_{20}-3a_{11}+4 & -a_{11}+2 \end{bmatrix}$$

is a solution to the equation

$$F \begin{bmatrix} 1 & -1 & 0 \\ 0 & 1 & 1 \end{bmatrix} = \begin{bmatrix} a_{10}-a_{20}-3a_{11}+4 & a_{20}+2a_{11}-a_{10}-2 & a_{20}-2 \\ a_{10}-a_{20}-3a_{11}+4 & a_{20}+2a_{11}-a_{10}-2 & a_{20}-2 \end{bmatrix}$$

if $a_{20}+a_{11} = 4$.

Therefore, the problem has a solution if /5.35/ and $a_{20}+a_{11}=4$ are satisfied.

## 5.2  CHARACTERISTIC POLYNOMIAL ASSIGNMENT AND DETERMINATION OF THE RESIDUAL POLYNOMIAL

### 5.2.1 Problem formulation.

Consider a single-input RM described by /5.1/ and /5.2/ with the state feedback /5.3/. The closed-loop characteristic polynomial $d_c/z_1,z_2/$ has the form /5.9/. We assume that $d_c/z_1,z_2/$ can be written as

$$d_c/z_1,z_2/ = d_a/z_1,z_2/ \cdot d_r/z_1,z_2/ \qquad /5.41/$$

where $d_a/z_1,z_2/$ is the assignable /or controllable/ and $d_r/z_1,z_2/$ is the residual /or unassignable/ part of $d_c/z_1,z_2/$ having the forms

$$d_a/z_1,z_2/ = \sum_{i=0}^{q_1} \sum_{j=0}^{q_2} \bar{d}_{ij} z_1^i z_2^j \, , \quad \bar{d}_{q_1 q_1} = 1 \qquad /5.42a/$$

$$d_r/z_1,z_2/ = \sum_{i=0}^{r_1} \sum_{j=0}^{r_2} d_{ij} z_1^i z_2^j \, , \quad d_{r_1 r_2} = 1 \qquad /5.42b/$$

where $r_1 = n_1 - q_1$ and $r_2 = n_2 - q_2$.

We assume that the number of coefficients of $d_a/z_1,z_2/$ is equal to the number of elements of k; i.e., $q_1 q_2 + q_1 + q_2 = n_1 + n_2$. Note that the number of the elements of k is $n_1 n_2$ less than the number $N = n_1 n_2 + n_1 + n_2$ of the coefficients of $d_c/z_1,z_2/$.

The residual polynomial $d_r/z_1,z_2/$ may be written as

$$d_r/z_1,z_2/ = z_1^{r_1} z_2^{r_2} + h/z_1,z_2/ d_r^T \qquad /5.43/$$

where

$$h/z_1,z_2/ = \begin{bmatrix} z_1^{r_1-1} z_2^{r_2} \\ z_1^{r_1} z_2^{r_2-1} \\ \cdots \\ z_1^0 z_2 \\ z_1 z_2^0 \\ z_1^0 z_2^0 \end{bmatrix}, \quad d_r = \begin{bmatrix} d_{r_1-1,r_2} \\ d_{r_1,r_2-1} \\ \cdots \\ d_{01} \\ d_{10} \\ d_{00} \end{bmatrix}$$

Substitution of /5.41/ and /5.43/ into /5.14/ yields

$$M^T/z_1,z_2/ k = d_0/z_1,z_2/ - d_a/z_1,z_2/ \left[ z_1^{r_1} z_2^{r_2} + h/z_1,z_2/ d_r^T \right]$$

or

$$v/z_1,z_2/ x = e/z_1,z_2/ \qquad /5.44/$$

where

$$v/z_1,z_2/ = \begin{bmatrix} M^T/z_1,z_2/ & \vdots & d_a/z_1,z_2/ h^T/z_1,z_2/ \end{bmatrix}, \quad x = \begin{bmatrix} k \\ d_r \end{bmatrix} \in R^{M \times 1}$$

$$e/z_1,z_2/ = d_0/z_1,z_2/ - z_1^{r_1} z_2^{r_2} d_a/z_1,z_2/, \quad M = n_1 + n_2 + r_1 r_2 + r_1 + r_2$$

and $M/z_1,z_2/$ is defined by /5.13/.


Problem 3.

Given A, B and $d_a/z_1,z_2/$, find the state feedback matrix k and the residual polynomial $d_r/z_1,z_2/$ such that the assignable part of $d_c/z_1,z_2/$ is $d_a/z_1,z_2/$.

### 5.2.2 Problem solution.

Two procedures for solving Problem 3 will be presented [13].

**Procedure 1.**

Evaluating /5.44/ in M different points $/z_{1i}, z_{2j}/$ we obtain the set of equations

$$V x = E \qquad /5.45/$$

where

$$V = \begin{bmatrix} v/z_{11}, z_{21}/ \\ v/z_{12}, z_{22}/ \\ \vdots \\ v/z_{1M}, z_{2M}/ \end{bmatrix}, \quad E = \begin{bmatrix} e/z_{11}, z_{21}/ \\ e/z_{12}, z_{22}/ \\ \vdots \\ e/z_{1M}, z_{2M}/ \end{bmatrix}$$

The equation /5.45/ is a linear system of M equations with M unknowns and may be solved to yield x, provided that $\det V \neq 0$. If $\det V = 0$, then choose another set of M points $/z_{1i}, z_{2j}/$ such that $\det V \neq 0$.

**Procedure 2.**

Let

$$v/z_1, z_2/ = \sum_{i=0}^{n_1} \sum_{j=0}^{n_2} v_{ij} z_1^i z_2^j, \quad v_{n_1 n_2} = 0 \qquad /5.46a/$$

and

$$e/z_1, z_2/ = \sum_{i=0}^{n_1} \sum_{j=0}^{n_2} e_{ij} z_1^i z_2^j, \quad e_{n_1 n_2} = 0 \qquad /5.46b/$$

Substitution of /5.46/ into /5.44/ yields

$$\sum_{i=0}^{n_1} \sum_{j=0}^{n_2} v_{ij} \times z_1^i z_2^j = \sum_{i=0}^{n_1} \sum_{j=0}^{n_2} e_{ij} z_1^i z_2^j \qquad /5.47/$$

Comparison of the coefficients of like $z_1^i z_2^j$ terms on both sides of /5.47/ yields

$$\overline{V} x = \overline{E} \qquad /5.48/$$

where $\overline{V}$ and $\overline{E}$ are given by the following relations:

$$V = \begin{bmatrix} v_{n_1-1,n_2} \\ v_{n_1,n_2-1} \\ \cdots \\ v_{01} \\ v_{10} \\ v_{00} \end{bmatrix}, \quad E = \begin{bmatrix} e_{n_1-1,n_2} \\ e_{n_1,n_2-1} \\ \cdots \\ e_{01} \\ e_{10} \\ e_{00} \end{bmatrix}$$

The equation /5.48/ is a linear system of $N = n_1 n_2 + n_1 + n_2$ equations with $M = n_1 + n_2 + r_1 r_2 + r_1 + r_2$ unknowns. Thus, /5.48/ is an overdetermined system of equations involving N-M more equations than unknowns. If rank $\overline{V} = M$, then solving /5.48/ we obtain

$$x = \left[V^T V\right]^{-1} V^T E \qquad /5.49/$$

Note that /5.49/ is an exact solution to /5.48/ if and only if

$$V\left[V^T V\right]^{-1} V^T E = E \qquad /5.50/$$

If /5.49/ is not an exact solution to /5.48/, then $d_a/z_1, z_2/$ is not an assignable part of $d_c/z_1, z_2/$.

Example 5.5

Given

$$A = \begin{bmatrix} 0 & 1 \\ -1 & -2 \end{bmatrix}, \quad b = \begin{bmatrix} 1 \\ 1 \end{bmatrix} \quad \text{and} \quad d_a/z_1, z_2/ = z_1 + 1$$

find $k = \begin{bmatrix} k_1 \\ k_2 \end{bmatrix}$ and $d_r/z_1, z_2/ = z_2 + a$.

In Example 5.1 we have found

$$d_o/z_1, z_2/ = z_1 z_2 + 2z_1 + 1 \quad \text{and} \quad M/z_1, z_2/ = \begin{bmatrix} z_2 + 3 \\ z_1 - 1 \end{bmatrix}$$

Taking into account that in this case $r_1 = 0$, $r_2 = 1$, $h/z_1, z_2/ = 1$, $d_r = a$ we obtain /5.44/ in the form

$$\begin{bmatrix} z_2 + 3 & z_1 - 1 & z_1 + 1 \end{bmatrix} \begin{bmatrix} k_1 \\ k_2 \\ a \end{bmatrix} = \begin{bmatrix} 2z_1 - z_2 + 1 \end{bmatrix} \qquad /5.51/$$

Procedure 1.

Evaluating /5.51/ in the points /0,1/, /1,0/, /1,1/ we obtain the equation

$$\begin{bmatrix} 4 & -1 & 1 \\ 3 & 0 & 2 \\ 4 & 0 & 2 \end{bmatrix} \begin{bmatrix} k_1 \\ k_2 \\ a \end{bmatrix} = \begin{bmatrix} 0 \\ 3 \\ 2 \end{bmatrix}$$

The solution reads $k_1 = k_2 = -1$, $a = 3$. Therefore $k = \begin{bmatrix} -1 \\ -1 \end{bmatrix}$ and $d_r/z_1,z_2/ = z_2 + 3$.

Hence

$$A_c = A + bk^T = \begin{bmatrix} 0 & 1 \\ -1 & -2 \end{bmatrix} + \begin{bmatrix} 1 \\ 1 \end{bmatrix} \begin{bmatrix} -1 & -1 \end{bmatrix} = \begin{bmatrix} -1 & 0 \\ -2 & -3 \end{bmatrix}$$

and

$$d_c/z_1,z_2/ = (z_1+1)(z_2+3) = d_a/z_1,z_2/\, d_r/z_1,z_2/$$

Procedure 2.

Comparison of the coefficients of like $z_1^i z_2^j$ terms of both sides of /5.51/ yields the equation

$$\begin{bmatrix} 3 & -1 & 1 \\ 1 & 0 & 0 \\ 0 & 1 & 1 \end{bmatrix} \begin{bmatrix} k_1 \\ k_2 \\ a \end{bmatrix} = \begin{bmatrix} 1 \\ -1 \\ 2 \end{bmatrix}$$

which has the solution $k_1 = k_2 = -1$, $a = 3$. So we obtained the same result.

The above considerations can be easily extended for multi-input systems.

## 5.3 CHARACTERISTIC POLYNOMIAL ASSIGNMENT BY DYNAMIC OUTPUT FEEDBACK

### 5.3.1 Problem formulation.

The ring of real polynomials in one indeterminate $z$ and the ring of real polynomials in two indeterminates $z_1, z_2$ will be denoted by $R[z]$ and $R[z_1,z_2]$, respectively. The ring of real polynomials in $z_2$ with coefficients in $R[z_1]$ and the ring of real polynomials in $z_2$ with coefficients in the field of rational functions $R(z_1)$ will be denoted by $R[z_1][z_2]$ and $R(z_1)[z_2]$, respectively. Any two polynomials are factor coprime iff they have no common nonconstant factor and they are zero coprime iff they have no zero in common.

Consider a 2-D linear single-input single-output system /plant/ with the transfer function

$$T = T/z_1, z_2/ = \frac{b}{a} \qquad /5.52/$$

where $a, b \in R[z_1, z_2]$ are factor coprime.

Let us consider the closed-loop system with a linear output feedback dynamic compensator having the transfer function

$$T_c = T_c/z_1, z_2/ = \frac{y}{x} \qquad /5.53/$$

where $x, y \in R[z_1, z_2]$ are factor coprime.

The transfer function of closed-loop system is

$$\frac{T}{1 + T_c T} = \frac{bx}{ax + by} \qquad /5.54/$$

The problem can be stated as follows [17]. Given the transfer function /5.52/ and a desired characteristic polynomial $c \in R[z_1, z_2]$, find the transfer function /5.53/ such that the closed-loop characteristic polynomial is equal to c.

## 5.3.2 Problem solution.

From /5.54/ it follows that

$$ax + by = c \qquad /5.55/$$

Therefore, the problem is reduced to solving the equation /5.55/, i.e. to finding x,y for the given a,b,c. We assume that a,b,c are factor coprime, i.e. that the common factor has already been cancelled from /5.55/. Note that zeros of a common factor of a and b can not be shifted.

### Theorem 5.3

The problem has a solution for every c if and only if a and b are zero coprime.

Proof. In Appendix it is shown that there exist two polynomials $p, q \in R[z_1, z_2]$ such that

$$pa + qb = 1 \qquad /5.56/$$

if and only if a,b are zero coprime.

Multiplying /5.56/ by c and comparing with /5.55/ we obtain $x = pc$ and $y = qc$.

Therefore /5.55/ and the problem has a solution for every c if and only if a and b are zero coprime.

□

To solve /5.55/ the methods given in Appendix can be used.

## Example 5.6

Given

$$T = \frac{2 - z_2}{1 + z_1 + z_1 z_2} \qquad /5.57/$$

and

$$c = 1 + z_1 + 3z_1 z_2 - z_1 z_2^2 + z_1^2 z_2 + z_1^3 z_2 + z_1^3 z_2^2 \qquad /5.58/$$

find /5.53/ such that the closed-loop characteristic polynomial is equal to /5.58/.

Let

$$x = x_{00} + x_{10} z_1 + x_{01} z_2 + x_{11} z_1 z_2 + x_{21} z_1^2 z_2$$

and

$$y = y_{00} + y_{10} z_1 + y_{01} z_2 + y_{11} z_1 z_2$$

Then equation /5.55/ takes the form

$$(1+z_1+z_1 z_2)(x_{00}+x_{10} z_1+x_{01} z_2+x_{11} z_1 z_2+x_{21} z_1^2 z_2) + (2-z_2)(y_{00}+y_{01} z_1+$$
$$+y_{01} z_2+y_{11} z_1 z_2) = 1+z_1+3z_1 z_1-z_1 z_2+z_1^2 z_2+z_1^3 z_2+z_1^3 z_2^2 \qquad /5.59/$$

Comparison of coefficients of like $z_1^i z_2^j$ terms of both sides of /5.59/ yields

$$\begin{aligned}
&x_{00}+2y_{00}=1, & &x_{00}+x_{10}-2y_{10}=1, \\
&x_{01}+2y_{01}-y_{00}=0, & &x_{00}+x_{01}+x_{11}+2y_{11}-y_{10}=3, \\
&x_{10}=0, & &x_{11}+x_{10}+x_{21}=1, \\
&x_{01}+y_{11}=-1, & &x_{11}=0, & &x_{21}=1.
\end{aligned} \qquad /5.60/$$

Assuming $x_{01} = 0$ from /5.60/ we obtain

$$x_{00}=1, \quad x_{10}=x_{01}=x_{11}=0, \quad x_{21}=1,$$
$$y_{00}=y_{01}=y_{10}=0, \quad y_{11}=1$$

and

$$T_c = \frac{z_1 z_2}{1 + z_1^2 z_2} \qquad /5.61/$$

A /system having/ transfer function /5.52/ is called causal (strictly causal) iff $a/0,0/ \neq 0$ ($a/0,0/ \neq 0$ and $b/0,0/ = 0$).

### Theorem 5.4

If $a/0,0/ \neq 0$, $c/0,0/ \neq 0$ and the equation has a solution, then it also has a solution for which $x/0,0/ \neq 0$.

<u>Proof</u>. If $b/0,0/ = 0$, then from /5.55/ we have $a/0,0/ \, x/0,0/ = c/0,0/$. Therefore, if $a/0,0/ \neq 0$ and $c/0,0/ \neq 0$, then we obtain $x/0,0/ \neq 0$. If $b/0,0/ \neq 0$ /5.55/ may have a non-causal solution $x_o$ such that $x_o/0,0/ = 0$. In this case a causal solution can be found from general solution in the form

$$x = x_o + bt$$
$$y = y_o - at$$

by taking t /which is arbitrary/ such that $t/0,0/ \neq 0$, i.e.

$$x/0,0/ = b/0,0/ \, t/0,0/ \neq 0$$

□

Note that in Example 5.6 for the causal system transfer function /5.57/ we obtained the strictly causal compensator transfer function /5.61/.

In general case let us consider a 2-D linear multi-input multi-output system /plant/ with the transfer function matrix

$$T = A^{-1} B \qquad \qquad /5.62/$$

where $A \in R^{l \times l}[z_1, z_2]$, $B \in R^{l \times m}[z_1, z_2]$.

Let the transfer function matrix of output-feedback dynamic compensator be of the form

$$T_c = Y X^{-1} \qquad \qquad /5.63/$$

where $X \in R^{l \times l}[z_1, z_2]$, $Y \in R^{m \times l}[z_1, z_2]$.

In general case the problem can be stated as follows. Given /5.62/ and a desired characteristic polynomial $c \in R[z_1, z_2]$, find /5.63/ such that the closed-loop characteristic polynomial is equal to c.

The transfer function matrix of the closed-loop system is

$$[I_1 + T T_c]^{-1} T = [I_1 + A^{-1} B Y X^{-1}]^{-1} A^{-1} B = X[AX + BY]^{-1} B \qquad /5.64/$$

Note that it is always possible to choose a matrix C so that $\det C = c$; for example, $C = \text{diag}[1 \; 1 \; \dots \; c]$.

If we assume that

$$AX + BY = C \qquad /5.65/$$

then the closed-loop characteristic polynomial is equal to $c$. Therefore, the problem is reduced to solving the equation /5.65/, i.e. to finding $X, Y$ for the given $A, B, C$. To find $X, Y$ the methods given in Appendix can be used.

Example 5.7

Given

$$A = \begin{bmatrix} z_1 & 0 \\ 1 & z_2 \end{bmatrix}, \quad B = \begin{bmatrix} 1 \\ z_2 \end{bmatrix} \quad \text{and} \quad c = z_1^2 z_2 + z_1 z_2 + z_2 + 1 \qquad /5.66/$$

find $X, Y$ such that the closed-loop characteristic polynomial equals $c$.

Let us assume that

$$C = \begin{bmatrix} 1 & 0 \\ 0 & z_1^2 z_2 + z_1 z_2 + z_2 + 1 \end{bmatrix}$$

In this case equation /5.65/ takes the form

$$\begin{bmatrix} z_1 & 0 \\ 1 & z_2 \end{bmatrix} X + \begin{bmatrix} 1 \\ z_2 \end{bmatrix} Y = \begin{bmatrix} 1 & 0 \\ 0 & z_1^2 z_2 + z_1 z_2 + z_2 + 1 \end{bmatrix} \qquad /5.67/$$

To solve /5.67/ we carry out the reduction

$$\begin{bmatrix} A & B & C \\ I_1 & 0 & 0 \\ 0 & I_m & 0 \end{bmatrix} = \left[\begin{array}{cc|c|cc} z_1 & 0 & 1 & 1 & 0 \\ 1 & z_2 & z_2 & 0 & z_1^2 z_2 + z_1 z_2 + z_2 + 1 \\ \hline 1 & 0 & 0 & 0 & 0 \\ 0 & 1 & 0 & 0 & 0 \\ \hline 0 & 0 & 1 & 0 & 0 \end{array}\right]$$

$$\longrightarrow \left[\begin{array}{cc|c|cc} z_1 & 0 & 1 & 0 & 0 \\ 1 & z_2 & z_2 & 0 & 0 \\ \hline 1 & 0 & 0 & 0 & -(z_1^2 z_2 + z_1 z_2 + z_2 + 1) \\ 0 & 1 & 0 & 1 & -z_1(z_1^2 z_2 + z_1 z_2 + z_2 + 1) \\ \hline 0 & 0 & 1 & -1 & z_1(z_1^2 z_2 + z_1 z_2 + z_2 + 1) \end{array}\right]$$

Hence
$$X = \begin{bmatrix} 0 & z_1^2 z_2 + z_1 z_2 + z_2 + 1 \\ -1 & z_1(z_1^2 z_2 + z_1 z_2 + z_2 + 1) \end{bmatrix}, \quad Y = \begin{bmatrix} 1. & -z_1(z_1^2 z_2 + z_1 z_2 + z_2 + 1) \end{bmatrix}$$

The above considerations can be extemded for 3-D systems [8].

## 5.4 CHARACTERISTIC POLYNOMIAL ASSIGNMENT USING PID CONTROLLERS

### 5.4.1 Problem formulations.

Consider RM described by the equations /5.1/ and /5.2/ with the state feedback control law of the form [15]

$$U/z_1, z_2/ = K/z_1, z_2/ \; X/z_1, z_2/ + U_n/z_1, z_2/ \qquad /5.68/$$

where

$$K/z_1, z_2/ = \frac{K_1}{z_1} + \frac{K_2}{z_2} + K_3 + z_1 K_4 + z_2 K_5 \qquad /5.69/$$

$K_i \in R^{m \times n}$ /i=1,2,...,5/ and $U/z_1,z_2/$, $X/z_1,z_2/$, $U_n/z_1,z_2/$ are the 2-D Z transforms of $u/i,j/$, $x/i,j/$ and $u_n/i,j/$ /the new input vector/, respectively.

The closed-loop characteristic polynomial is

$$d_c/z_1, z_2/ = \det\left[Z - A - BK/z_1, z_2/\right] \qquad /5.70/$$

**Problem 4.**

Given A,B and the desirable $d_c/z_1,z_2/$, find $K_i$ for i=1,...,5 of the state feedback PID controller such that $d_c/z_1,z_2/$ is as prespecified one.

Let the output feedback law applied have the form

$$U/z_1, z_2/ = F/z_1, z_2/ \; Y/z_1, z_2/ + U_n/z_1, z_2/ \qquad /5.71/$$

where

$$F/z_1, z_2/ = \frac{F_1}{z_1} + \frac{F_2}{z_2} + F_3 + z_1 F_4 + z_2 F_5 \qquad /5.72/$$

$F_i \in R^{m \times 1}$ /i=1,2,...,5/ and $Y/z_1,z_2/$ is the 2-D Z transform of $y/i,j/$.
The closed-loop characteristic polynomial is

$$d_c/z_1,z_2/ = \det\left[Z - A - BF/z_1,z_2/C\right] \qquad /5.73/$$

**Problem 5.**

Given A,B,C and the desirable $d_c/z_1,z_2/$, find $F_i$ for i=1,...,5 of the output feedback PID controller such that $d_c/z_1,z_2/$ is as prespecified one.

### 5.4.2 Problem solutions for single-input systems.

**Problem 4.**

For a single-input system /5.70/ has the form

$$d_c/z_1,z_2/ = \det\left[Z - A - b k^T/z_1,z_2/\right] \qquad /5.74/$$

where

$$b = \begin{bmatrix} b_1 \\ b_2 \end{bmatrix} \in R^{n \times 1}, \quad k/z_1,z_2/ = \frac{k_1}{z_1} + \frac{k_2}{z_2} + k_3 + z_1 k_4 + z_2 k_5$$

$k_i \in R^{n \times 1}$ for i=1,2,...,5.

In a similar way as for Problem 1 we obtain

$$M^T/z_1,z_2/\left[\frac{k_1}{z_1} + \frac{k_2}{z_2} + k_3 + z_1 k_4 + z_2 k_5\right] = d/z_1,z_2/ \qquad /5.75/$$

or

$$M^T/z_1,z_2/\left[k_1 z_2 + k_2 z_1 + k_3 z_1 z_2 + k_4 z_1^2 z_2 + k_5 z_1 z_2^2\right] =$$
$$= z_1 z_2 d/z_1,z_2/ \qquad /5.76/$$

where $M/z_1,z_2/$ is defined by /5.13/ and

$$d/z_1,z_2/ = d_o/z_1,z_2/ - d_c/z_1,z_2/ = \sum_{i=0}^{n_1+1} \sum_{j=0}^{n_2+1} d_{ij} z_1^i z_2^j \qquad /5.77/$$

Substitution of /5.13/ and /5.77/ into /5.76/ yields

$$\sum_{i=0}^{n_1} \sum_{j=0}^{n_2} m_{ij}^T \left[k_1 z_1^i z_2^{j+1} + k_2 z_1^{i+1} z_2^j + k_3 z_1^{i+1} z_2^{j+1} + k_4 z_1^{i+2} z_2^{j+1} + \right.$$

$$+ k_5 z_1^{i+1} z_2^{j+2}\Big] = \sum_{i=0}^{n_1+1} \sum_{j=0}^{n_2+1} d_{ij} z_1^{i+1} z_2^{j+1} \qquad /5.78/$$

Comparison of the coefficients of like $z_1^i z_2^j$ terms of both sides of /5.78/ yields

$$\overline{M}\,\overline{k} = \overline{d} \qquad /5.79/$$

where

$$\overline{M} = \begin{bmatrix} m_{00}^T & 0 & 0 & 0 & 0 \\ 0 & m_{00}^T & 0 & 0 & 0 \\ m_{10}^T & m_{01}^T & m_{00}^T & 0 & 0 \\ m_{20}^T & m_{11}^T & m_{10}^T & m_{00}^T & 0 \\ m_{11}^T & m_{02}^T & m_{01}^T & 0 & m_{00}^T \\ \cdot & \cdot & \cdot & \cdot & \cdot \\ 0 & 0 & 0 & m_{n_1,n_2-1}^T & 0 \\ 0 & 0 & 0 & 0 & m_{n_1-1,n_2}^T \end{bmatrix}, \quad \overline{d} = \begin{bmatrix} 0 \\ 0 \\ d_{00} \\ d_{10} \\ d_{01} \\ \cdot \\ d_{n_1+1,n_2-1} \\ d_{n_1-1,n_2+1} \end{bmatrix}$$

$$\overline{k} = \begin{bmatrix} k_1 \\ k_2 \\ \cdots \\ k_5 \end{bmatrix}$$

The equation /5.78/ is a linear system of $(n_1+2)\cdot(n_2+2)-1$ equations with $5(n_1+n_2)$ unknowns.

From Kronecker-Capelli's theorem it follows that /5.78/ has a solution if and only if

$$\text{rank } \overline{M} = \text{rank}\begin{bmatrix} \overline{M} & \overline{d} \end{bmatrix} \qquad /5.80/$$

We have thus established the following

<u>Theorem 5.5</u>

Problem 4 for single-input system has a solution if and only if the condition /5.80/ is satisfied.

## Example 5.8

Given

$$A = \begin{bmatrix} 0 & 1 \\ -1 & -1 \end{bmatrix}, \quad b = \begin{bmatrix} 0 \\ 1 \end{bmatrix} \qquad /5.81/$$

and

$$d_c/z_1,z_2/ = -2z_1^2 + z_1 z_2 + z_1 - z_2 - 1 \qquad /5.82/$$

find $k/z_1,z_2/$ of the PID controller such that the closed-loop characteristic polynomial is equal to /5.82/.

In this case

$$d_0/z_1,z_2/ = \det[Z - A] = \begin{vmatrix} z_1 & -1 \\ 1 & z_2+1 \end{vmatrix} = z_1 z_2 + z_1 + 1$$

$$M/z_1,z_2/ = (\text{adj}[Z - A])b = \begin{bmatrix} z_2+1 & 1 \\ -1 & z_1 \end{bmatrix} \begin{bmatrix} 0 \\ 1 \end{bmatrix} = \begin{bmatrix} 1 \\ z_1 \end{bmatrix}$$

and

$$d/z_1,z_2/ = d_0/z_1,z_2/ - d_c/z_1,z_2/ = 2z_1^2 + z_2 + 2$$

The equation /5.76/ takes the form

$$\begin{bmatrix} 1 & z_1 \end{bmatrix} \begin{bmatrix} k_1 z_2 + k_2 z_1 + k_3 z_1 z_2 + k_4 z_1^2 z_2 + k_5 z_1 z_2^2 \end{bmatrix} =$$
$$= 2z_1^3 z_2 + z_1 z_2^2 + 2z_1 z_2 \qquad /5.83/$$

Comparison of the coefficients of like $z_1^i z_2^j$ terms of both sides of /5.83/ yields /5.80/ with

$$\overline{M} = \begin{bmatrix} 1 & 0 & 0 & 0 & 0 & 0 & 0 & 0 & 0 \\ 0 & 0 & 1 & 0 & 0 & 0 & 0 & 0 & 0 \\ 0 & 1 & 0 & 0 & 1 & 0 & 0 & 0 & 0 \\ 0 & 0 & 0 & 0 & 0 & 1 & 1 & 0 & 0 \\ 0 & 0 & 0 & 0 & 0 & 0 & 0 & 1 & 0 \\ 0 & 0 & 0 & 0 & 0 & 0 & 0 & 0 & 1 \\ 0 & 0 & 0 & 0 & 0 & 0 & 1 & 0 & 0 \end{bmatrix}, \quad \overline{d} = \begin{bmatrix} 0 \\ 0 \\ 2 \\ 0 \\ 1 \\ 0 \\ 2 \end{bmatrix} \qquad /5.84/$$

It is easy to check that the condition /5.80/ is satisfied for /5.84/. The equation /5.79/ with /5.84/ has the solution

$$k_1 = \begin{bmatrix} 0 \\ 1 \end{bmatrix}, \quad k_2 = \begin{bmatrix} 0 \\ 0 \end{bmatrix}, \quad k_3 = k_5 = \begin{bmatrix} 1 \\ 0 \end{bmatrix}, \quad k_4 = \begin{bmatrix} 0 \\ 2 \end{bmatrix}$$

Hence

$$k/z_1,z_2/ = \frac{k_1}{z_1} + \frac{k_2}{z_2} + k_3 + z_1 k_4 + z_2 k_5 = \begin{bmatrix} 1 + z_2 \\ \frac{1}{z_1} + 2z_1 \end{bmatrix}$$

and

$$d_c/z_1,z_2/ = \det\left[Z - A - b k^T/z_1,z_2/\right] = \begin{vmatrix} z_1 & -1 \\ -z_2 & -z_1^{-1} - 2z_1 + z_2 + 1 \end{vmatrix} =$$

$$= -2z_1^2 + z_1 z_2 + z_1 - z_2 - 1$$

Therefore, the closed-loop characteristic polynomial has the desired form /5.82/.

Problem 2.

For a single-input system /5.73/ has the form

$$d_c/z_1,z_2/ = \det\left[Z - A - b f^T/z_1,z_2/C\right] \qquad /5.85/$$

where

$$f/z_1,z_2/ = \frac{f_1}{z_1} + \frac{f_2}{z_2} + f_3 + z_1 f_4 + z_2 f_5 \qquad /5.86/$$

$f_i \in R^{1 \times 1}$ for $i = 1, 2, \ldots, 5$.

In a similar way as for Problem 1 we obtain

$$M_c^T/z_1,z_2/ \left[\frac{f_1}{z_1} + \frac{f_2}{z_2} + f_3 + z_1 f_4 + z_2 f_5\right] = d/z_1,z_2/ \qquad /5.87/$$

or

$$M_c^T/z_1,z_2/ \left[f_1 z_2 + f_2 z_1 + f_3 z_1 z_2 + f_4 z_1^2 z_2 + f_5 z_1 z_2^2\right] = z_1 z_2 d/z_1,z_2/ \qquad /5.88/$$

where

$$M_c/z_1,z_2/ = C \cdot \text{adj}[Z - A] \cdot b = \sum_{i=0}^{n_1} \sum_{j=0}^{n_2} \hat{m}_{ij} z_1^i z_2^j \quad \text{with} \quad \hat{m}_{n_1 n_2} = 0 \qquad /5.89/$$

and $d/z_1,z_2/$ is defined by /5.77/.

Substitution of /5.89/ and /5.77/ into /5.88/ yields

$$\sum_{i=0}^{n_1} \sum_{j=0}^{n_2} \hat{m}_{ij} \left[f_1 z_1^i z_2^{j+1} + f_2 z_1^{i+1} z_2^j + f_3 z_1^{i+1} z_2^{j+1} + f_4 z_1^{i+2} z_2^{j+1} + \right.$$

$$+ f_5 z_1^{i+1} z_2^{j+2} \Big] = \sum_{i=0}^{n_1+1} \sum_{j=0}^{n_2+1} d_{ij} z_1^{i+1} z_2^{j+1} \qquad /5.90/$$

Comparison of the coefficients of like $z_1^i z_2^j$ terms of both sides of /5.90/ yields

$$\hat{M} \hat{f} = \hat{d} \qquad /5.91/$$

where

$$\hat{M} = \begin{bmatrix} \hat{m}_{00}^T & 0 & 0 & 0 & 0 \\ 0 & \hat{m}_{00}^T & 0 & 0 & 0 \\ \hat{m}_{10}^T & \hat{m}_{01}^T & \hat{m}_{00}^T & 0 & 0 \\ \hat{m}_{20}^T & \hat{m}_{11}^T & \hat{m}_{10}^T & \hat{m}_{00}^T & 0 \\ \hat{m}_{11}^T & \hat{m}_{02}^T & \hat{m}_{01}^T & 0 & \hat{m}_{00}^T \\ \cdot & \cdot & \cdot & \cdot & \cdot \\ 0 & 0 & 0 & \hat{m}_{n_1,n_2-1}^T & 0 \\ 0 & 0 & 0 & 0 & \hat{m}_{n_1-1,n_2}^T \end{bmatrix}, \quad \hat{d} = \begin{bmatrix} 0 \\ 0 \\ d_{00} \\ d_{10} \\ d_{01} \\ \cdots \\ d_{n_1+1,n_2-1} \\ d_{n_1-1,n_2+1} \end{bmatrix}$$

$$\hat{f} = \begin{bmatrix} f_1 \\ f_2 \\ \cdots \\ f_5 \end{bmatrix}$$

The equation /5.91/ is a linear system of $(n_1+2)(n_2+2) - 1$ equations with 51 unknowns.

We have thus established the following

### Theorem 5.6

Problem 5 for single-input system has a solution if and only if the condition

$$\operatorname{rank} \hat{M} = \operatorname{rank} [\hat{M} \; \hat{d}] \qquad /5.92/$$

is satisfied.

## Example 5.9

Given

$$A = \begin{bmatrix} 0 & 1 \\ -1 & -2 \end{bmatrix}, \quad b = \begin{bmatrix} 0 \\ 1 \end{bmatrix}, \quad C = \begin{bmatrix} 0 & -1 \end{bmatrix}$$

and

$$d_c/z_1,z_2/ = 2z_1^2 + 2z_1 z_2 + 3z_1 + 2 \qquad /5.93/$$

find $f/z_1,z_2/$ of the PID controller such that the closed-loop characteristic polynomial is equal to /5.93/.
In this case

$$d_o/z_1,z_2/ = \det[Z - A] = \begin{vmatrix} z_1 & -1 \\ 1 & z_2+2 \end{vmatrix} = z_1 z_2 + 2z_1 + 1$$

$$M_c/z_1,z_2/ = C \, \text{adj}[Z - A] \, b = \begin{bmatrix} 0 & -1 \end{bmatrix} \begin{bmatrix} z_2+2 & 1 \\ -1 & z_1 \end{bmatrix} \begin{bmatrix} 0 \\ 1 \end{bmatrix} = -z_1$$

and

$$d/z_1,z_2/ = d_o/z_1,z_2/ - d_c/z_1,z_2/ = -2z_1^2 - z_1 z_2 - z_1 - 1$$

Equation /5.88/ takes the form

$$f_1 z_2 + f_2 z_1 + f_3 z_1 z_2 + f_4 z_1^2 z_2 + f_5 z_1 z_2^2 = 2z_1^2 z_2 + z_1 z_2^2 + z_1 z_2 + z_2 \qquad /5.94/$$

Comparison of the coefficients of like $z_1^i z_2^j$ terms of both sides of /5.94/ yields

$$f_1 = 1, \quad f_2 = 0, \quad f_3 = 1, \quad f_4 = 2, \quad f_5 = 1$$

Hence

$$f/z_1,z_2/ = \frac{1}{z_1} + 1 + 2z_1 + z_2$$

and

$$d_c/z_1,z_2/ = \det[Z - A - b\, f^T/z_1,z_2/\, C] = \begin{vmatrix} z_1 & -1 \\ 1 & z_1^{-1} + 2z_1 + 2z_2 + 3 \end{vmatrix} =$$

$$= 2z_1^2 + 2z_1 z_2 + 3z_1 + 2$$

Therefore, the closed-loop characteristic polynomial has the desired form /5.93/.

### 5.3.3 Problem solutions for multi-input systems.

#### Problem 4.

We assume that $K_i$ /i=1,2,...,5/ have the dyadic structure

$$K_i = qk_i^T \quad /i=1,2,\ldots,5/ \qquad /5.95/$$

where $q \in R^{m \times 1}$ and $k_i \in R^{n \times 1}$.

Note that the closed-loop system matrix $A_c = A + BK/z_1,z_2/$ has the same form $A_c = A + bk^T/z_1,z_2/$ as the single-input system if $b = Bq$. Therefore the method developed for single-input system can be applied for finding $k/z_1,z_2/$.

Example 5.10

Given

$$A = \begin{bmatrix} A_{11} & A_{12} \\ A_{21} & A_{22} \end{bmatrix} = \begin{bmatrix} 0 & 1 & | & 0 \\ -1 & -2 & | & 1 \\ \hline 0 & -1 & | & -1 \end{bmatrix}, \quad B = \begin{bmatrix} 0 & 0 \\ 0 & 1 \\ \hline -1 & 0 \end{bmatrix}$$

and

$$d_c/z_1,z_2/ = -2z_1 z_2^2 - z_1^2 z_2 - 5z_1 z_2 - 2z_1^2 - 2z_1 - 2z_2 - 4 \qquad /5.96/$$

find $K/z_1,z_2/$ of the PID controller such that the closed-loop characteristic polynomial is equal to /5.96/.

We have

$$d_o/z_1,z_2/ = \det[Z - A] = \begin{vmatrix} z_1 & -1 & 0 \\ 1 & z_2+2 & -1 \\ 0 & 1 & z_2+1 \end{vmatrix} = z_1^2 z_2 + z_1^2 + 2z_1 z_2 + 3z_1 +$$

$$+ z_2 + 1$$

Let $q = \begin{bmatrix} -1 \\ 1 \end{bmatrix}$.

Then $b = Bq = \begin{bmatrix} 0 \\ 1 \\ \hline 1 \end{bmatrix}$ and

$$M/z_1,z_2/ = \text{adj}[Z - A] \cdot b = \begin{bmatrix} z_2 + 2 \\ z_1 z_2 + 2z_1 \\ z_1^2 + z_1 + 1 \end{bmatrix}$$

Taking into account that

$$d/z_1,z_2/ = d_o/z_1,z_2/ - d_c/z_1,z_2/ = 2z_1^2 z_2 + 2z_1 z_2^2 + 3z_1^2 + 7z_1 z_2 + 5z_1 +$$

$$+ 3z_2 + 5$$

we obtain equation /5.76/ in the form

$$[z_2+2 \quad z_1z_2+2z_1 \quad z_1^2+z_1+1][k_1z_2+k_2z_1+k_3z_1z_2+k_4z_1^2z_2+k_5z_1z_2^2] =$$
$$= 2z_1^3z_2^2 + 2z_1^2z_2^3 + 3z_1^3z_2 + 7z_1^2z_2^2 + 5z_1^2z_2 + 3z_1z_2^2 + 5z_1z_2 \qquad /5.97/$$

Comparison of the coefficients of like $z_1^i z_2^j$ terms of both sides of /5.97/ yields /5.79/ with

$$\bar{M} = \begin{bmatrix} 2 & 0 & 1 & 0 & 0 & 0 & 0 & 0 & 0 & 0 & 0 & 0 & 0 & 0 & 0 \\ 0 & 0 & 0 & 2 & 0 & 1 & 0 & 0 & 0 & 0 & 0 & 0 & 0 & 0 & 0 \\ 0 & 2 & 1 & 1 & 0 & 0 & 2 & 0 & 1 & 0 & 0 & 0 & 0 & 0 & 0 \\ 1 & 0 & 0 & 0 & 0 & 0 & 0 & 0 & 0 & 0 & 0 & 0 & 0 & 0 & 0 \\ 0 & 0 & 0 & 0 & 2 & 1 & 0 & 0 & 0 & 0 & 0 & 0 & 0 & 0 & 0 \\ 0 & 0 & 1 & 0 & 1 & 0 & 0 & 2 & 1 & 2 & 0 & 1 & 0 & 0 & 0 \\ 0 & 1 & 0 & 0 & 0 & 0 & 1 & 0 & 0 & 0 & 0 & 2 & 0 & 1 & 0 \\ 0 & 0 & 0 & 0 & 0 & 1 & 0 & 0 & 0 & 0 & 0 & 0 & 0 & 0 & 0 \\ 0 & 0 & 0 & 0 & 0 & 0 & 0 & 1 & 0 & 1 & 0 & 0 & 0 & 2 & 1 \\ 0 & 0 & 0 & 0 & 0 & 0 & 0 & 0 & 0 & 0 & 0 & 1 & 0 & 0 & 0 \\ 0 & 0 & 0 & 0 & 0 & 0 & 0 & 1 & 0 & 2 & 1 & 0 & 0 & 0 & 0 \\ 0 & 0 & 0 & 0 & 0 & 0 & 0 & 0 & 0 & 1 & 0 & 0 & 0 & 0 & 1 \\ 0 & 0 & 0 & 0 & 0 & 0 & 0 & 0 & 0 & 0 & 0 & 0 & 1 & 0 & 0 \\ 0 & 0 & 0 & 0 & 0 & 0 & 0 & 0 & 0 & 0 & 1 & 0 & 0 & 0 & 0 \end{bmatrix}, \quad \bar{d} = \begin{bmatrix} 0 \\ 0 \\ 5 \\ 0 \\ 0 \\ 5 \\ 3 \\ 0 \\ 7 \\ 0 \\ 3 \\ 2 \\ 2 \\ 0 \end{bmatrix} \qquad /5.98/$$

It is easy to check that condition /5.80/ is satisfied for /5.98/. The equation /5.79/ with /5.98/ has the solution

$$k_1 = \begin{bmatrix} 0 \\ 1 \\ 0 \end{bmatrix}, \quad k_2 = \begin{bmatrix} 0 \\ 0 \\ 0 \end{bmatrix}, \quad k_3 = \begin{bmatrix} 1 \\ 2 \\ 1 \end{bmatrix}, \quad k_4 = \begin{bmatrix} 0 \\ 1 \\ 0 \end{bmatrix}, \quad k_5 = \begin{bmatrix} 0 \\ 2 \\ 1 \end{bmatrix}$$

Hence

$$K/z_1,z_2/ = q\,k^T/z_1,z_2/ = \begin{bmatrix} -1 \\ 1 \end{bmatrix}\begin{bmatrix} 1 & z_1^{-1}+2+z_1+2z_2 & 1+z_2 \end{bmatrix} =$$
$$= \begin{bmatrix} -1 & -z_1^{-1}-2-z_1-2z_2 & -1-z_2 \\ 1 & z_1^{-1}+2+z_1+2z_2 & 1+z_2 \end{bmatrix}$$

Problem 5.

We assume that $F/z_1,z_2/$ has the dyadic structure

$$F_i = q\,f_i^T \quad /i=1,2,\ldots,5/ \qquad /5.99/$$

where $q \in R^{m \times 1}$ and $f_i \in R^{1 \times 1}$.

Note that the closed-loop system matrix $A_c = A + BF/z_1,z_2/C$ has the same form $A_c = A + b\,f^T/z_1,z_2/C$ as the single-input system if $b=Bq$. Therefore the method developed for single-input system can be applied for finding $f/z_1,z_2/$.

## 5.5 EIGENVALUE ASSIGNMENT

### 5.5.1 Eigenvalue assignment for 2-D systems.

**Problem formulation.**

Consider RM described by /5.1/ and /5.2/ with the state feedback law /5.3/. The closed-loop system equation is

$$x' = A_c x + Bv \qquad /5.100/$$

where

$$A_c = A + BK \qquad /5.101/$$

The 2-D characteristic polynomial $d_o/z_1,z_2/$ of A is defined by

$$d_o/z_1,z_2/ = \det\begin{bmatrix} I_{n_1}z_1 - A_{11} & -A_{12} \\ -A_{21} & I_{n_2}z_2 - A_{22} \end{bmatrix} = \sum_{i=0}^{n_1}\sum_{j=0}^{n_2} a_{ij} z_1^i z_2^j \qquad /5.102/$$

with $a_{n_1 n_2} = 1$.

**Definition 5.1**

The 2-D characteristic polynomial /5,102/ is called separable iff

$$d_o/z_1,z_2/ = d_1/z_1/\, d_2/z_2/ \qquad /5.103/$$

where

$$d_i/z_i/ = z_i^{n_i} + a_{n_i-1}^i z_i^{n_i-1} + \ldots + a_1^i z_i + a_0^i \qquad /i=1,2/$$

According Definition 1.2 the set of 2-D eigenvalues for the separable $d_o/z_1,z_2/$ is given by

$$E = \left\{ /z_{1i},z_{2j}/ \,:\, /0,0/ \leqslant /i,j/ \leqslant /n_1,n_2/ \right\} \qquad /5.104/$$

where $z_{10} = z_{20} = 0$.

The eigenvalue assignment problem for RM can be stated as follows:

Given A,B and the set /5.104/, find K such that the closed-loop characteristic polynomial

$$d_c/z_1,z_2/ = \det[Z - A - BK] \qquad /5.105/$$

is separable with the desired set of 2-D eigenvalues /5.104/.

Problem solution.

Two methods of solving the problem will be presented.

Method 1. [6]

This method is based on the following

Theorem 5.7 [6]

The eigenvalue assignment problem has a solution if

i. $[I_{n_1} - B_1 B_1^g] A_{12} = 0$ or $[I_{n_2} - B_2 B_2^g] A_{21} = 0$

ii. the pairs $/A_{11}, B_1/$, $/\bar{A}_{22}, \bar{B}_2/$ or $/A_{22}, B_2/$, $/\bar{A}_{11}, \bar{B}_1/$ are reachable where

$$\bar{A}_{22} = A_{22} - B_2 B_1^g A_{12}, \qquad \bar{B}_2 = B_2[I_m - B_1^g B_1]$$
$$\bar{A}_{11} = A_{11} - B_1 B_2^g A_{21}, \qquad \bar{B}_1 = B_1[I_m - B_2^g B_2] \qquad /5.106/$$

and $B_1^g$ /$B_2^g$/ is the generalized inverse matrix of $B_1$ /$B_2$/ satisfying the condition $B_1 B_1^g B_1 = B_1$ / $B_2 B_2^g B_2 = B_2$ /.

Proof. For

$$K = [K_1 \quad K_2], \quad K_i \in R^{m \times n_i} \quad /i=1,2/ \qquad /5.107/$$

the matrix /5.101/ takes the form

$$A_c = \begin{bmatrix} A_{11} + B_1 K_1 & A_{12} + B_1 K_2 \\ A_{21} + B_2 K_1 & A_{22} + B_2 K_2 \end{bmatrix} \qquad /5.108/$$

The closed-loop characteristic polynomial is separable if

$$B_1 K_2 = -A_{12} \quad \text{or} \quad B_2 K_1 = -A_{21} \qquad /5.109/$$

It is easy to show that /5.109/ have solutions if and only if the condition i. holds. A general solution to $B_1 K_2 = -A_{12}$ is

$$K_2 = -B_1^g A_{12} + [I_m - B_1^g B_1] K_{20} \qquad /5.110/$$

where $K_{20}$ is an arbitrary matrix of appropriate dimensions.
Note that the matrix $K_{20}$ can be chosen so that the matrix

$$A_{22} + B_2 K_2 = \bar{A}_{22} + \bar{B}_2 K_{20} \qquad /5.111/$$

has the desired eigenvalues $z_{21}, z_{22}, \ldots, z_{2n_2}$ if and only if the pair $/\bar{A}_{22}, \bar{B}_2/$ is reachable.

Similarly, the matrix $K_1$ can be chosen so that $A_{11} + B_1 K_1$ has the desired eigenvalues $z_{11}, z_{12}, \ldots, z_{1n_1}$ if and only if the pair $/A_{11}, B_1/$ is reachable.

The second part of the theorem can be proved in a similar way. □

From Kronecker-Capelli's theorem it follows that /5.109/ have solutions if and only if

$$\text{rank } B_1 = \text{rank}[B_1 \quad A_{12}]$$

and  /5.112/

$$\text{rank } B_2 = \text{rank}[B_2 \quad A_{21}],$$

respectively.

Therefore, condition i. of Theorem 5.7 is equivalent to /5.112/

Let $\text{rank } B_1 = r \leq \min/n_1, m/$. It is well known that one can find full rank matrices $B_{11} \in R^{n_1 \times r}$, $B_{12} \in R^{r \times m}$ such that $B_1 = B_{11} B_{12}$ and

$$B_1^g = B_{12}^g [B_{12} B_{12}^T]^{-1} [B_{11}^T B_{11}]^{-1} B_{11}^T \qquad /5.113/$$

If conditions i. and ii. are satisfied the matrix K can be found by the use of the following

### Algorithm 5.2

**Step 1.** Choose $K_1$ such that $A_{11} + B_1 K_1$ has the given desired eigenvalues $z_{11}, z_{12}, \ldots, z_{1n_1}$.

**Step 2.** Find $B_1^g$ and $\bar{A}_{22}, \bar{B}_2$ or $B_2^g$ and $\bar{A}_{11}, \bar{B}_1$ using /5.113/ and /5.106/.

**Step 3.** Choose $K_{20}$ such that $\bar{A}_{22} + \bar{B}_2 K_{20}$ has the desired eigenvalues $z_{21}, z_{22}, \ldots, z_{2n_2}$.

**Step 4.** Find $K_2$ and $K$ using /5.110/ and /5.107/.

### Example 5.11

Given

$$A = \begin{bmatrix} A_{11} & A_{12} \\ A_{21} & A_{22} \end{bmatrix} = \begin{bmatrix} 2 & 1 & | & 1 & 3 & -1 \\ -4 & -4 & | & 0 & 0 & 0 \\ -- & -- & | & -- & -- & -- \\ 3 & 4 & | & -2 & 1 & 2 \\ -1 & -4 & | & 3 & 3 & -2 \\ 2 & 0 & | & 0 & 0 & -4 \end{bmatrix}, \quad B = \begin{bmatrix} B_1 \\ B_2 \end{bmatrix} = \begin{bmatrix} 1 & 0 \\ 0 & 0 \\ -- & -- \\ 0 & 1 \\ 1 & -1 \\ 1 & 0 \end{bmatrix}$$

and
$$E = \{z_{11} = z_{12} = -2, \ z_{21} = z_{22} = z_{23} = -1\} \qquad /5.114/$$
find K such that the closed-loop characteristic polynomial is separable with the desired set of 2-D eigenvalues /5.114/.
It is easy to check that conditions i. and ii. of Theorem 5.7 are satisfied.

Step 1. For
$$K_1 = \begin{bmatrix} -2 & 0 \\ 0 & 0 \end{bmatrix}$$
the matrix
$$A_{11} + B_1 K_1 = \begin{bmatrix} 2 & 1 \\ -4 & -4 \end{bmatrix} + \begin{bmatrix} 1 & 0 \\ 0 & 0 \end{bmatrix} K_1 = \begin{bmatrix} 0 & 1 \\ -4 & -4 \end{bmatrix}$$
has the desired eigenvalues $z_{11} = z_{12} = -2$.

Step 2. Taking into account that
$$B_1 = B_{11} B_{12} = \begin{bmatrix} 1 \\ 0 \end{bmatrix} \begin{bmatrix} 1 & 0 \end{bmatrix}$$
from /5.112/ and /5.106/ we obtain
$$B_1^g = B_{12}^T \left[ B_{12} B_{12}^T \right]^{-1} \left[ B_{11}^T B_{11} \right]^{-1} B_{11}^T = \begin{bmatrix} 1 & 0 \\ 0 & 0 \end{bmatrix}$$
and the reachable pair
$$\bar{A}_{22} = A_{22} - B_2 B_1^g A_{12} = \begin{bmatrix} -2 & 1 & 2 \\ 2 & 0 & -1 \\ -1 & -3 & -3 \end{bmatrix}$$
$$\bar{B}_2 = B_2 \left[ I_m - B_1^g B_1 \right] = \begin{bmatrix} 0 & 1 \\ 0 & -1 \\ 0 & 0 \end{bmatrix}$$

Step 3. For
$$K_{20} = \begin{bmatrix} 0 & 0 & 0 \\ 2 & 0 & -2 \end{bmatrix}$$
the matrix
$$\bar{A}_{22} + \bar{B}_2 K_{20} = \begin{bmatrix} -2 & 1 & 2 \\ 2 & 0 & -1 \\ -1 & -3 & -3 \end{bmatrix} + \begin{bmatrix} 0 & 1 \\ 0 & -1 \\ 0 & 0 \end{bmatrix} K_{20} = \begin{bmatrix} 0 & 1 & 0 \\ 0 & 0 & 1 \\ -1 & -3 & -3 \end{bmatrix}$$
has the desired eigenvalues
$$z_{21} = z_{22} = z_{23} = -1.$$

Step 4.

$$K_2 = -B_1^g A_{12} + [I_m - B_1^g B_1] K_{20} = \begin{bmatrix} -1 & -3 & 1 \\ 2 & 0 & -2 \end{bmatrix}$$

and

$$K = [K_1 \quad K_2] = \begin{bmatrix} -2 & 0 & -1 & -3 & 1 \\ 0 & 0 & 2 & 0 & -2 \end{bmatrix}$$

The closed-loop system matrix

$$A_c = A + BK = \begin{bmatrix} 0 & 1 & | & 0 & 0 & 0 \\ -4 & -4 & | & 0 & 0 & 0 \\ \hline 3 & 4 & | & 0 & 1 & 0 \\ 1 & -4 & | & 0 & 0 & 1 \\ 0 & 0 & | & -1 & -3 & -3 \end{bmatrix}$$

has the separable characteristic polynomial with the desired set of 2-D eigenvalues /5.114/.

## Method 2. [4]

We assume that

$$K = K_t + K_p \qquad /5.115/$$

where

$$K_t = [K_{t1} \quad K_{t2}], \quad K_{ti} \in R^{m \times n_i} \quad /i=1,2/ \qquad /5.116/$$

$$K_p = [K_{p1} \quad K_{p2}], \quad K_{pi} \in R^{m \times n_i} \quad /i=1,2/ \qquad /5.116b/$$

Matrix /5.116a/ is chosen so that

$$A_t = A + BK_t = \begin{bmatrix} A_{11} + B_1 K_{t1} & A_{12} + B_1 K_{t2} \\ 0 & A_{22} + B_2 K_{t2} \end{bmatrix} \qquad /5.117a/$$

or

$$A_t = A + BK_t = \begin{bmatrix} A_{11} + B_1 K_{t1} & 0 \\ A_{21} + B_2 K_{t1} & A_{22} + B_2 K_{t2} \end{bmatrix} \qquad /5.117b/$$

The matrix $A_t$ has the form /5.117a/ (or /5.117b/) if and only if

$$B_2 K_{t1} = -A_{21} \quad / \text{ or } \quad B_1 K_{t2} = -A_{12} / \qquad /5.118/$$

From Kronecker-Capelli's theorem it follows that the equation /5.118/ has a solution if and only if

$$\text{rank } B_2 = \text{rank} [B_2 \quad A_{21}] \quad / \text{ or rank } B_1 = \text{rank} [B_1 \quad A_{12}] / \qquad /5.119/$$

Note that $K_{t2}$ /$K_{t1}$/ can be chosen arbitrary, for example $K_{t2} = 0$.
Let $B_{ij}$ be the j-th column of the matrix $B_i$ /i=1,2 ; j=1,2,...,m/.
Let us assume that

$1^o$ the pair /$A_{11}+BK_{t1}$, $B_{1i}$/ is reachable,

$2^o$ there exists $j \neq i$ such that the pair /$A_{22}+B_2K_{t2}$, $B_{2j}$/ is reachable and $B_{2i} = 0$.

Without loss of generality we can assume $j=1$ and $i=2$.
Hence

$$B_2 = \begin{bmatrix} B_{21} & 0 & B_{23} & \cdots & B_{2m} \end{bmatrix} \qquad /5.120/$$

Reachability of the pair /$A_{t11}$, $B_{12}$/ implies non-singularity of the matrix

$$R_1 = \begin{bmatrix} B_{12} & A_{t11}B_{12} & \cdots & A_{t11}^{n_1-1}B_{12} \end{bmatrix} \qquad /5.121/$$

where $A_{t11} = A_{11} + B_1 K_{t1}$

Let $t_1$ be the $n_1$-th row of $R_1^{-1}$ and

$$T_1 = \begin{bmatrix} t_1 \\ t_1 A_{t11} \\ \vdots \\ t_1 A_{t11}^{n_1-1} \end{bmatrix} . \qquad /5.122/$$

It is well known that

$$\bar{A}_{t11} = T_1 A_{t11} T_1^{-1} = \begin{bmatrix} 0 & | & I_{n_1-1} \\ -- & | & -- \\ & -a_1 & \end{bmatrix} \qquad /5.123/$$

$$\bar{B}_1 = T_1 B_1 = \begin{bmatrix} \bar{B}_{11} & e_{n_1} & \bar{B}_{13} & \cdots & \bar{B}_{1m} \end{bmatrix} \qquad /5.123b/$$

where $a_1 = \begin{bmatrix} a_{10} & a_{11} & \cdots & a_{1,n_1-1} \end{bmatrix}$ and $e_{n_1}$ is the $n_1$-th column of $I_{n_1}$.

It can be easily verified that for

$$K_{p1} T_1^{-1} = \begin{bmatrix} 0 \\ a_1 - c_1 \\ 0 \\ \vdots \\ 0 \end{bmatrix} \qquad /5.124/$$

the matrix $\bar{A}_{c11}$ has the form

$$A_{c11} = T_1\left[A_{t11} + B_1 K_{p1}\right] T_1^{-1} = \left[\begin{array}{c|c} 0 & I_{n_1-1} \\ \hline & -c_1 \end{array}\right] \qquad /5.125/$$

where

$$c_1 = \left[c_{10} \quad c_{11} \quad \cdots \quad c_{1,n_1-1}\right]$$

Similarly, the reachability of the pair $/A_{t22}, B_{21}/$ implies non-singularity of the matrix

$$R_2 = \left[B_{21} \quad A_{t22} B_{21} \quad \cdots \quad A_{t22}^{n_2-1} B_{21}\right] \qquad /5.126/$$

where $A_{t22} = A_{22} + B_2 K_{t2}$

Let $t_2$ be the $n_2$-th row of $R_2^{-1}$ and

$$T_2 = \begin{bmatrix} t_2 \\ t_2 A_{t22} \\ \vdots \\ t_2 A_{t22}^{n_2-1} \end{bmatrix}. \qquad /5.127/$$

It is well known that

$$A_{t22} = T_2 A_{t22} T_2^{-1} = \left[\begin{array}{c|c} 0 & I_{n_2-1} \\ \hline & -a_2 \end{array}\right] \qquad /5.128a/$$

and

$$\bar{B}_2 = T_2 B_2 = \left[e_{n_2} \quad 0 \quad \bar{B}_{23} \quad \cdots \quad \bar{B}_{2m}\right] \qquad /5.128b/$$

where $a_2 = \left[a_{20} \quad a_{21} \quad \cdots \quad a_{2,n_2-1}\right]$

It is easy to check that for

$$K_{p2} T_2^{-1} = \begin{bmatrix} a_2 - c_2 \\ 0 \\ \vdots \\ 0 \end{bmatrix} \qquad /5.129/$$

the matrix $\bar{A}_{c22}$ has the form

$$\bar{A}_{c22} = T_2\left[A_{t22} + B_2 K_{p2}\right] T_2^{-1} = \left[\begin{array}{c|c} 0 & I_{n_2-1} \\ \hline & -c_2 \end{array}\right] \qquad /5.130/$$

where $c_2 = \left[c_{20} \quad c_{21} \quad \cdots \quad c_{2,n_2-1}\right]$

From /5.125/ and /5.130/ it follows that

$$\overline{A}_c = \begin{bmatrix} T_1 & 0 \\ 0 & T_2 \end{bmatrix} \begin{bmatrix} A_t + BK_p \end{bmatrix} \begin{bmatrix} T_1 & 0 \\ 0 & T_2 \end{bmatrix}^{-1} = \begin{bmatrix} \overline{A}_{c11} & \overline{A}_{c12} \\ 0 & \overline{A}_{c22} \end{bmatrix} \qquad /5.131/$$

since $T_2 B_2 K_{p1} T_1^{-1} = 0$.

The matrix

$$\overline{A}_{c12} = T_1 \begin{bmatrix} A_{t12} + B_1 K_{p2} \end{bmatrix} T_2^{-1}$$

has no special form.

Note that

$$d_c/z_1, z_2/ = \det \begin{bmatrix} I_{n_1} z_1 - \overline{A}_{c11} & -\overline{A}_{c12} \\ 0 & I_{n_2} z_2 - \overline{A}_{c22} \end{bmatrix} =$$

$$= \det \begin{bmatrix} I_{n_1} z_1 - \overline{A}_{c11} & -\overline{A}_{c12} \\ 0 & I_{n_2} z_2 - \overline{A}_{c22} \end{bmatrix} = \det \begin{bmatrix} I_{n_1} z_1 - \overline{A}_{c11} \end{bmatrix} \cdot \det \begin{bmatrix} I_{n_2} z_2 - \overline{A}_{c22} \end{bmatrix} \qquad /5.132/$$

where

$$\det \begin{bmatrix} I_{n_1} z_1 - \overline{A}_{c11} \end{bmatrix} = z_1^{n_1} + c_{1,n_1-1} z_1^{n_1-1} + \ldots + c_{11} z_1 + c_{10}$$

$$\det \begin{bmatrix} I_{n_2} z_2 - \overline{A}_{c22} \end{bmatrix} = z_2^{n_2} + c_{2,n_2-1} z_2^{n_2-1} + \ldots + c_{21} z_2 + c_{20}$$

Similar conditions can be carried out for the second case.

We have thus established the following

<u>Theorem 5.8</u>

The eigenvalue assignment problem has a solution if the condition /5.119/ and the assumptions $1^o$, $2^o$ are satisfied.

If /5.119/ and $1^o$, $2^o$ are satisfied the matrix K can be found by the use of the following

<u>Algorithm 5.3</u>

<u>Step 1</u>. Find $K_{t1}$ /or $K_{t2}$/ from /5.118/.

<u>Step 2</u>. Assuming arbitrary $K_{t2}$ /or $K_{t1}$/, for example $K_{t2} = 0$, find $K_t$ and $A_t$.

<u>Step 3</u>. For given $z_{11}, z_{12}, \ldots, z_{1,n_1}$ and $z_{21}, z_{22}, \ldots, z_{2,n_2}$ find $C_1$ and $C_2$.

Step 4. Using /5.122/ and /5.127/ find $T_1$, $T_2$ and $a_1$, $a_2$.

Step 5. Using

$$K_{p1} = \begin{bmatrix} 0 \\ a_1 - c_1 \\ 0 \\ \cdot \cdot \cdot \cdot \\ 0 \end{bmatrix} T_1, \quad K_{p2} = \begin{bmatrix} a_2 - c_2 \\ 0 \\ 0 \\ \cdot \cdot \cdot \cdot \\ 0 \end{bmatrix} T_2 \qquad /5.133/$$

find $K_p = \begin{bmatrix} K_{p1} & K_{p2} \end{bmatrix}$.

Step 6. Evaluate

$$K = K_t + K_p$$

Example 5.12

Given

$$A = \begin{bmatrix} A_{11} & A_{12} \\ A_{21} & A_{22} \end{bmatrix} = \begin{bmatrix} -1 & 6 & | & 1 & 0 \\ -1 & 8 & | & 1 & -2 \\ \hline 2 & -6 & | & -3 & 3 \\ 1 & -3 & | & 1 & -1 \end{bmatrix}; \quad B = \begin{bmatrix} B_1 \\ B_2 \end{bmatrix} = \begin{bmatrix} 1 & -1 \\ 1 & 0 \\ \hline 0 & 2 \\ 0 & 1 \end{bmatrix}$$

and the set of 2-D eigenvalues

$$E = \left\{ z_{11} = 1, \ z_{12} = \tfrac{1}{2}, \ z_{21} = 1, \ z_{22} = \tfrac{1}{3} \right\} \qquad /5.134/$$

find K such that the closed-loop characteristic polynomial is separable with the desired set of 2-D eigenvalues /5.134/.

Step 1. Note that

$$K_{t1} = \begin{bmatrix} 0 & 0 \\ -1 & 3 \end{bmatrix}$$

satisfies the equation

$$\begin{bmatrix} 0 & 2 \\ 0 & 1 \end{bmatrix} K_{t1} = \begin{bmatrix} -2 & 6 \\ -1 & 3 \end{bmatrix}$$

Step 2. Assuming $K_{t2} = 0$ we obtain

$$K_t = \begin{bmatrix} K_{t1} & K_{t2} \end{bmatrix} = \begin{bmatrix} 0 & 0 & 0 & 0 \\ -1 & 3 & 0 & 0 \end{bmatrix}$$

and

$$A_t = A + BK_t = \begin{bmatrix} 0 & 3 & | & 1 & 0 \\ -1 & 8 & | & 1 & -2 \\ \hline 0 & 0 & | & -3 & 3 \\ 0 & 0 & | & 1 & -1 \end{bmatrix}$$

Step 3. For $z_{11}=1$, $z_{12}=\frac{1}{2}$, and $z_{21}=1$, $z_{22}=\frac{1}{3}$ we obtain

$$(z_1 - z_{11})(z_1 - z_{12}) = z_1^2 - \frac{3}{2}z_1 + \frac{1}{2}$$

and

$$(z_2 - z_{21})(z_2 - z_{22}) = z_2^2 - \frac{4}{3}z_2 + \frac{1}{3}$$

$$c_1 = \begin{bmatrix} \frac{1}{2} & -\frac{1}{3} \end{bmatrix}, \quad c_2 = \begin{bmatrix} \frac{1}{3} & -\frac{4}{3} \end{bmatrix}$$

Step 4. In this case $j=2$, $i=1$. Hence

$$R_1^{-1} = \begin{bmatrix} B_{11} & A_{t11}B_{11} \end{bmatrix}^{-1} = \begin{bmatrix} 1 & 3 \\ 1 & 7 \end{bmatrix}^{-1} = \frac{1}{4}\begin{bmatrix} 7 & -3 \\ -1 & 1 \end{bmatrix}$$

$$T_1 = \begin{bmatrix} t_1 \\ t_1 A_{t11} \end{bmatrix} = \frac{1}{4}\begin{bmatrix} -1 & 1 \\ -1 & 5 \end{bmatrix}$$

and

$$R_2^{-1} = \begin{bmatrix} B_{22} & A_{t22}B_{22} \end{bmatrix}^{-1} = \begin{bmatrix} 2 & -3 \\ 1 & 1 \end{bmatrix}^{-1} = \frac{1}{5}\begin{bmatrix} 1 & 3 \\ -1 & 2 \end{bmatrix}$$

$$T_2 = \begin{bmatrix} t_2 \\ t_2 A_{t22} \end{bmatrix} = \frac{1}{5}\begin{bmatrix} -1 & 2 \\ 5 & -5 \end{bmatrix}$$

To find $a_1$ and $a_2$ we calculate

$$\det[zI_2 - A_{t11}] = \begin{vmatrix} z & -3 \\ 1 & z-8 \end{vmatrix} = z^2 - 8z + 3$$

and

$$\det[zI_2 - A_{t22}] = \begin{vmatrix} z+3 & -3 \\ -1 & z+1 \end{vmatrix} = z^2 + 4z$$

Hence

$$a_1 = \begin{bmatrix} 3 & -8 \end{bmatrix}, \quad a_2 = \begin{bmatrix} 0 & 4 \end{bmatrix}$$

Step 5.

$$K_{p1} = \begin{bmatrix} a_1 - c_1 \\ 0 \end{bmatrix} T_1 = \begin{bmatrix} 1 & -7\frac{1}{2} \\ 0 & 0 \end{bmatrix}$$

$$K_{p2} = \begin{bmatrix} 0 \\ a_2 - c_2 \end{bmatrix} T_2 = \frac{1}{15}\begin{bmatrix} 0 & 0 \\ 81 & -82 \end{bmatrix}$$

and
$$K_p = \begin{bmatrix} K_{p1} & K_{p2} \end{bmatrix} = \begin{bmatrix} 1 & -7\frac{1}{2} & 0 & 0 \\ 0 & 0 & \frac{81}{15} & -\frac{82}{15} \end{bmatrix}$$

Step 6.
$$K = K_t + K_p = \begin{bmatrix} 1 & -7\frac{1}{2} & 0 & 0 \\ -1 & 3 & \frac{81}{15} & -\frac{82}{15} \end{bmatrix}$$

It is easy to check that the matrix
$$A_c = A + BK = \begin{bmatrix} 1 & -4\frac{1}{2} & -4\frac{2}{5} & 5\frac{7}{15} \\ 0 & \frac{1}{2} & 1 & -2 \\ \hline 0 & 0 & 7\frac{4}{5} & -7\frac{14}{15} \\ 0 & 0 & 6\frac{2}{5} & -6\frac{7}{15} \end{bmatrix}$$

has a separable characteristic polynomial with the desired set of 2-D eigenvalues /5.134/.

### 5.4.2 Eigenvalue assignment for 3-D systems.

#### Problem formulation.

Consider the Tzafestas-Pimenides´ model /T-PM/ described by the equations /1.32/. The closed-loop system with the state feedback law

$$u = v + Kx$$

is described by the equation

$$x' = A_c x + Bv \qquad /5.135/$$

where

$$A_c = A + BK \qquad /5.136/$$

$K \in R^{m \times n}$ /$n=n_1+n_2+n_3$/ is the feedback gain matrix and $v = v/i,j,k/ \in R^m$ is the control input vector.

The 3-D characteristic polynomial $d_o/z_1,z_2,z_3/$ of the matrix A is defined by

$$d_o/z_1,z_2,z_3/ = \det[Z - A] = \sum_{i=0}^{n_1} \sum_{j=0}^{n_2} \sum_{k=0}^{n_3} a_{ijk} z_1^i z_2^j z_3^k \qquad /5.137/$$

where $a_{n_1 n_2 n_3} = 1$ and

$$Z = \begin{bmatrix} I_{n_1} z_1 & 0 & 0 \\ 0 & I_{n_2} z_2 & 0 \\ 0 & 0 & I_{n_3} z_3 \end{bmatrix} \qquad /5.138/$$

### Definition 5.2

The 3-D characteristic polynomial /5.137/ is called separable iff

$$d_o/z_1,z_2,z_3/ = d_1/z_1/ \cdot d_2/z_2/ \cdot d_3/z_3/ \qquad /5.139/$$

where

$$d_i/z_i/ = z_i^{n_i} + a_{n_i-1}^i z_i^{n_i-1} + \ldots + a_1^i z_i + a_0^i \qquad /i=1,2,3/$$

For separable polynomial /5.139/ the 3-D eigenvalues of A are triples $/z_1,z_2,z_3/$ which satisfy $d_i/z_i/ = 0$ for $i=1,2,3$.

The eigenvalue assignment problem for T-PM can be stated as follows: Given A,B and the set of 3-D eigenvalues

$$E = \{/z_{1i},z_{2j},z_{3k}/ : i=1,2,\ldots,n_1;\ j=1,2,\ldots,n_2;\ k=1,2,\ldots,n_3\} \qquad /5.140/$$

find K such that the closed-loop characteristic polynomial

$$d_c/z_1,z_2,z_3/ = \det[Z - A - BK] \qquad /5.141/$$

is separable with the desired set of 3-D eigenvalues.

### Problem solution.

Two methods of solving the problem will be presented.

### Method 1. [7]

The method is based on the following

### Theorem 5.9

The eigenvalue assignment problem has a solution if there exists a permutation of the indeces $/i,j,k/$ on the collection $\{1,2,3\}$ such that

i. $[I_{n_1} - B_i B_i^g] A_{ij} = 0$ ,

$$\left(I_{n_i+n_j} - \begin{bmatrix}B_i\\B_j\end{bmatrix}\begin{bmatrix}B_i\\B_j\end{bmatrix}^g\right)\begin{bmatrix}A_{ik}\\A_{jk}\end{bmatrix} = 0$$

ii. the pairs $/A_{ii}, B_i/$, $/A_{jj} - B_j B_i^g A_{ij}$, $B_j(I_m - B_i^g B_i)/$ and

$$/A_{kk} - B_k\begin{bmatrix}B_i\\B_j\end{bmatrix}^g\begin{bmatrix}A_{ik}\\A_{jk}\end{bmatrix}, \; B_k\left(I_m - \begin{bmatrix}B_i\\B_j\end{bmatrix}^g\begin{bmatrix}B_i\\B_j\end{bmatrix}\right)/ \text{ are reachable}$$

where $B_i^g$ denotes any generalized inverse matrix of $B_i$ satisfying the condition $B_i B_i^g B_i = B_i$.

**Proof.** Without loss of generality we can assume $i=1$, $j=2$, $k=3$.
For $K = \begin{bmatrix} K_1 & K_2 & K_3 \end{bmatrix}$ the closed-loop system matrix /5.136/ takes the form

$$A_c = \begin{bmatrix} A_{11} + B_1 K_1 & A_{12} + B_1 K_2 & A_{13} + B_1 K_3 \\ A_{21} + B_2 K_1 & A_{22} + B_2 K_2 & A_{23} + B_2 K_3 \\ A_{31} + B_3 K_1 & A_{32} + B_3 K_2 & A_{33} + B_3 K_3 \end{bmatrix} \qquad /5.142/$$

The problem has a solution if

$$B_1 K_2 = -A_{12}, \quad \begin{bmatrix}B_1\\B_2\end{bmatrix} K_3 = -\begin{bmatrix}A_{13}\\A_{23}\end{bmatrix} \qquad /5.143/$$

and the matrices

$$A_{c11} = A_{11} + B_1 K_1, \quad A_{c22} = A_{22} + B_2 K_2, \quad A_{c33} = A_{33} + B_3 K_3 \qquad /5.144/$$

have the prespecified characteristic polynomials $d_1/z_1/$, $d_2/z_2/$ and $d_3/z_3/$, respectively.
The equations /5.143/ have solutions $K_2, K_3$ if and only if the conditions i. are satisfied. General solutions to /5.143/ have the forms

$$K_2 = -B_1^g A_{12} + (I_m - B_1^g B_1) K_{20}$$

$$K_3 = -\begin{bmatrix}B_1\\B_2\end{bmatrix}^g \begin{bmatrix}A_{13}\\A_{23}\end{bmatrix} + \left(I_m - \begin{bmatrix}B_1\\B_2\end{bmatrix}^g \begin{bmatrix}B_1\\B_2\end{bmatrix}\right) K_{30} \qquad /5.145/$$

where $K_{20}$ and $K_{30}$ are arbitrary real matrices of appropriate dimensions. The matrices $K_{20}$, $K_{30}$ can be chosen so that the matrices

$$A_{c11} = A_{11} + B_1 K_1, \quad A_{c22} = \bar{A}_{22} + \bar{B}_2 K_{20}, \quad A_{c33} = \bar{\bar{A}}_{33} + \bar{\bar{B}}_3 K_{30}$$

have the preassigned characteristic polynomials if and only if the conditions ii. are satisfied, where

$$\bar{A}_{22} = A_{22} - B_2 B_1^g A_{12} \quad , \quad \bar{B}_2 = B_2 (I_m - B_1^g B_1)$$

$$\bar{A}_{33} = A_{33} - \begin{bmatrix} B_1 \\ B_2 \end{bmatrix}^g \begin{bmatrix} A_{13} \\ A_{23} \end{bmatrix} \quad , \quad \bar{B}_3 = B_3 \left( I_m - \begin{bmatrix} B_1 \\ B_2 \end{bmatrix}^g \begin{bmatrix} B_1 \\ B_2 \end{bmatrix} \right)$$

$\square$

From /5.143/ and Kronecker-Capelli's theorem it follows that the conditions i. can be written in the equivalent form

$$\text{rank } B_i = \text{rank}\begin{bmatrix} B_i & A_{ij} \end{bmatrix}, \quad \text{rank}\begin{bmatrix} B_i \\ B_j \end{bmatrix} = \text{rank}\begin{bmatrix} B_i & A_{ik} \\ B_j & A_{jk} \end{bmatrix} \qquad /5.146/$$

Note that the conditions ii. can be satisfied only if the input matrices

$$B_j(I_m - B_i^g B_i) \quad \text{and} \quad B_k \left( I_m - \begin{bmatrix} B_i \\ B_j \end{bmatrix}^g \begin{bmatrix} B_i \\ B_j \end{bmatrix} \right)$$

are non-zero. It can be easily shown /Problem 5.1/ that

$$A(I - B^g B) = 0$$

if and only if

$$\text{rank } B = \text{rank}\begin{bmatrix} A \\ B \end{bmatrix}$$

Therefore the condition ii. can be satisfied only if

$$\text{rank } B_i < \text{rank}\begin{bmatrix} B_i \\ B_j \end{bmatrix} \quad \text{and} \quad \text{rank}\begin{bmatrix} B_i \\ B_j \end{bmatrix} < \text{rank}\begin{bmatrix} B_i \\ B_j \\ B_k \end{bmatrix} \qquad /5.147/$$

If the conditions i., ii. are satisfied then the matrix K can be found by the use of the following

Algorithm 5.4

Step 1. Choose $K_i$ so that $A_{cii} = A_{ii} + B_i K_i$ has the given eigenvalues $z_{i1}, z_{i2}, \ldots, z_{in_i}$.

Step 2. Find the matrices

$$\bar{A}_{jj} = A_{jj} - B_j B_i^g A_{ij} \quad , \quad \bar{B}_j = B_j (I_m - B_i^g B_i)$$

/5.148/

$$\bar{A}_{kk} = A_{kk} - B_k \begin{bmatrix} B_i \\ B_j \end{bmatrix}^g \begin{bmatrix} A_{ik} \\ A_{jk} \end{bmatrix} \quad , \quad \bar{B}_k = B_k \left( I_m - \begin{bmatrix} B_i \\ B_j \end{bmatrix}^g \begin{bmatrix} B_i \\ B_j \end{bmatrix} \right)$$

Step 3. Choose $K_{j0}$ and $K_{k0}$ so that
$$A_{cjj} = \bar{A}_{jj} + \bar{B}_j K_{j0} \quad \text{and} \quad A_{ckk} = \bar{A}_{kk} + \bar{B}_k K_{k0}$$
have the given eigenvalues $z_{j1}, z_{j2}, \ldots, z_{jn_j}$ and $z_{k1}, z_{k2}, \ldots, z_{kn_k}$, respectively.

Step 4. Find
$$K_j = -B_i^g A_{ij} + \left(I_m - B_i^g B_i\right) K_{j0}$$

$$K_k = -\begin{bmatrix} B_i \\ B_j \end{bmatrix}^g \begin{bmatrix} A_{ik} \\ A_{jk} \end{bmatrix} + \left(I_m - \begin{bmatrix} B_i \\ B_j \end{bmatrix}^g \begin{bmatrix} B_i \\ B_j \end{bmatrix}\right) K_{k0}$$

and
$$K = \begin{bmatrix} K_1 & K_2 & K_3 \end{bmatrix}$$

Example 5.13

Given
$$A = \begin{bmatrix} A_{11} & A_{12} & A_{13} \\ A_{21} & A_{22} & A_{23} \\ A_{31} & A_{32} & A_{33} \end{bmatrix} = \begin{bmatrix} 1 & 0 & 3 & 3 \\ 3 & 1 & -1 & 0 \\ 2 & -4 & -7 & -3 \\ 1 & -1 & 2 & 2 \end{bmatrix}, \quad B = \begin{bmatrix} B_1 \\ B_2 \\ B_3 \end{bmatrix} = \begin{bmatrix} 0 & 1 & 0 \\ 1 & 0 & 0 \\ 0 & -1 & 0 \\ 0 & 0 & 1 \end{bmatrix}$$

and the set of 3-D eigenvalues
$$E = \left\{ z_{11} = -1, \quad z_{21} = z_{22} = -2, \quad z_{31} = 3 \right\} \quad /5.149/$$
find K such that the closed-loop characteristic polynomial is separable with the desired set of 3-D eigenvalues /5.149/.

It is easy to check that conditions i. and ii. are satisfied for i=1, j=2 and k=3.

Step 1. The matrix
$$K_1 = -\begin{bmatrix} 0 \\ 2 \\ 0 \end{bmatrix}$$
satisfies the equation
$$A_{c11} = A_{11} + B_1 K_1 = 1 - \begin{bmatrix} 0 & 1 & 0 \end{bmatrix} \begin{bmatrix} 0 \\ 2 \\ 0 \end{bmatrix} = \begin{bmatrix} -1 \end{bmatrix}$$

Step 2.
$$\bar{A}_{22} = A_{22} - B_2 B_1^g A_{12} = \begin{bmatrix} 1 & -1 \\ -4 & -7 \end{bmatrix} - \begin{bmatrix} 1 & 0 & 0 \\ 0 & -1 & 0 \end{bmatrix} \begin{bmatrix} 0 \\ 1 \\ 0 \end{bmatrix} \begin{bmatrix} 0 & 3 \end{bmatrix} = \begin{bmatrix} 1 & -1 \\ -4 & -4 \end{bmatrix}$$

$$\overline{B}_2 = B_2(I_m - B_1^g B_1) = \begin{bmatrix} 1 & 0 & 0 \\ 0 & -1 & 0 \end{bmatrix} \left( \begin{bmatrix} 1 & 0 & 0 \\ 0 & 1 & 0 \\ 0 & 0 & 1 \end{bmatrix} - \begin{bmatrix} 0 \\ 1 \\ 0 \end{bmatrix} \begin{bmatrix} 0 & 1 & 0 \end{bmatrix} \right) =$$

$$= \begin{bmatrix} 1 & 0 & 0 \\ 0 & 0 & 0 \end{bmatrix}$$

$$\overline{A}_{33} = A_{33} - B_3 \begin{bmatrix} B_1 \\ B_2 \end{bmatrix}^g \begin{bmatrix} A_{13} \\ A_{23} \end{bmatrix} = [2] - [0 \ 0 \ 1] \begin{bmatrix} 0 & 1 & 0 \\ 0.5 & 0 & -0.5 \\ 0 & 0 & 0 \end{bmatrix} \begin{bmatrix} 3 \\ 0 \\ -3 \end{bmatrix} = [2]$$

$$\overline{B}_3 = B_3 \left( I_m - \begin{bmatrix} B_1 \\ B_2 \end{bmatrix}^g \begin{bmatrix} B_1 \\ B_2 \end{bmatrix} \right) = [0 \ 0 \ 1] \left( \begin{bmatrix} 1 & 0 & 0 \\ 0 & 1 & 0 \\ 0 & 0 & 1 \end{bmatrix} + \right.$$

$$\left. - \begin{bmatrix} 0 & 1 & 0 \\ 0.5 & 0 & -0.5 \\ 0 & 0 & 0 \end{bmatrix} \begin{bmatrix} 0 & 1 & 0 \\ 1 & 0 & 0 \\ 0 & -1 & 0 \end{bmatrix} \right) = [0 \ 0 \ 1]$$

Step 3. The matrices

$$K_{20} = \begin{bmatrix} -1 & 2 \\ 0 & 0 \\ 0 & 0 \end{bmatrix}, \quad K_{30} = \begin{bmatrix} 0 \\ 0 \\ 1 \end{bmatrix}$$

satisfy the equations

$$A_{c22} = \overline{A}_{22} + \overline{B}_2 K_{20} = \begin{bmatrix} 1 & -1 \\ -4 & -4 \end{bmatrix} + \begin{bmatrix} 1 & 0 & 0 \\ 0 & 0 & 0 \end{bmatrix} \begin{bmatrix} -1 & 2 \\ 0 & 0 \\ 0 & 0 \end{bmatrix} = \begin{bmatrix} 0 & 1 \\ -4 & -4 \end{bmatrix}$$

$$A_{c33} = \overline{A}_{33} + \overline{B}_3 K_{30} = [2] + [0 \ 0 \ 1] \begin{bmatrix} 0 \\ 0 \\ 1 \end{bmatrix} = [3]$$

Step 4.

$$K_2 = -B_1^g A_{12} + \left( I_m - B_1^g B_1 \right) K_{20} = - \begin{bmatrix} 0 \\ 1 \\ 0 \end{bmatrix} [0 \ 3] + \left( \begin{bmatrix} 1 & 0 & 0 \\ 0 & 1 & 0 \\ 0 & 0 & 1 \end{bmatrix} + \right.$$

$$\left. - \begin{bmatrix} 0 \\ 1 \\ 0 \end{bmatrix} [0 \ 1 \ 0] \right) \begin{bmatrix} -1 & 2 \\ 0 & 0 \\ 0 & 0 \end{bmatrix} = \begin{bmatrix} -1 & 2 \\ 0 & -3 \\ 0 & 0 \end{bmatrix}$$

$$K_3 = - \begin{bmatrix} B_1 \\ B_2 \end{bmatrix}^g \begin{bmatrix} A_{13} \\ A_{23} \end{bmatrix} + \left( I_m - \begin{bmatrix} B_1 \\ B_2 \end{bmatrix}^g \begin{bmatrix} B_1 \\ B_2 \end{bmatrix} \right) K_{20} = - \begin{bmatrix} 0 & 1 & 0 \\ 0.5 & 0 & -0.5 \\ 0 & 0 & 0 \end{bmatrix} \begin{bmatrix} 3 \\ 0 \\ -3 \end{bmatrix} +$$

$$+ \left( \begin{bmatrix} 1 & 0 & 0 \\ 0 & 1 & 0 \\ 0 & 0 & 1 \end{bmatrix} - \begin{bmatrix} 0 & 1 & 0 \\ 0.5 & 0 & -0.5 \\ 0 & 0 & 0 \end{bmatrix} \begin{bmatrix} 0 & 1 & 0 \\ 1 & 0 & 0 \\ 0 & -1 & 0 \end{bmatrix} \right) \begin{bmatrix} 0 \\ 0 \\ 1 \end{bmatrix} = \begin{bmatrix} 0 \\ -3 \\ 1 \end{bmatrix}$$

and

$$K = \begin{bmatrix} K_1 & K_2 & K_3 \end{bmatrix} = \begin{bmatrix} 0 & -1 & 2 & 0 \\ -2 & 0 & -3 & -3 \\ 0 & 0 & 0 & 1 \end{bmatrix}$$

It is easy to check that the matrix

$$A_c = A + BK = \begin{bmatrix} -1 & 0 & 0 & 0 \\ \hline 3 & 0 & 1 & 0 \\ 4 & -4 & -4 & 0 \\ \hline 1 & -1 & 2 & 3 \end{bmatrix}$$

has separable characteristic polynomial with the given set of 3-D eigenvalues /5.149/.

Method 2. [7]

We assume that

$$K = K_t + K_p \qquad /5.150/$$

where

$$K_t = \begin{bmatrix} K_{t1} & K_{t2} & K_{t3} \end{bmatrix}, \quad K_{ti} \in R^{m \times n_i} \quad /i=1,2,3/ \qquad /5.151a/$$

$$K_p = \begin{bmatrix} K_{p1} & K_{p2} & K_{p3} \end{bmatrix}, \quad K_{pi} \in R^{m \times n_i} \quad /i=1,2,3/ \qquad /5.151b/$$

Matrix /5.151a/ is chosen so that

$$A_t = A + BK_t = \begin{bmatrix} A_{11} + B_1 K_{t1} & 0 & 0 \\ A_{21} + B_2 K_{t1} & A_{22} + B_2 K_{t2} & 0 \\ A_{31} + B_3 K_{t1} & A_{32} + B_3 K_{t2} & A_{33} + B_3 K_{t3} \end{bmatrix} \qquad /5.152a/$$

or

$$A_t = A + BK_t = \begin{bmatrix} A_{11} + B_1 K_{t1} & A_{12} + B_1 K_{t2} & A_{13} + B_1 K_{t3} \\ 0 & A_{22} + B_2 K_{t2} & A_{23} + B_2 K_{t3} \\ 0 & 0 & A_{33} + B_3 K_{t3} \end{bmatrix} \qquad /5.152b/$$

or

$$A_t = A + BK_t = \begin{bmatrix} A_{11} + B_1 K_{t1} & A_{12} + B_1 K_{t2} & 0 \\ 0 & A_{22} + B_2 K_{t2} & 0 \\ A_{31} + B_3 K_{t1} & A_{32} + B_3 K_{t2} & A_{33} + B_3 K_{t3} \end{bmatrix} \quad /5.152c/$$

The matrix $A_t$ has the form /5.152a/ if and only if

$$B_1 K_{t2} = -A_{12} \quad \text{and} \quad \begin{bmatrix} B_1 \\ B_2 \end{bmatrix} K_{t3} = -\begin{bmatrix} A_{13} \\ A_{23} \end{bmatrix} \quad /5.153/$$

From Kronecker-Capelli's theorem it follows that the equations /5.153/ have solutions if and only if

$$\text{rank } B_1 = \text{rank}\begin{bmatrix} B_1 & A_{12} \end{bmatrix}, \quad \text{rank}\begin{bmatrix} B_1 \\ B_2 \end{bmatrix} = \text{rank}\begin{bmatrix} B_1 & A_{13} \\ B_2 & A_{23} \end{bmatrix} \quad /5.154/$$

Note that $K_{t1}$ can be chosen arbitrary, for example $K_{t1} = 0$.
Similar results can be obtained for /5.152b/ and /5.152c/.
Let us assume that there exist $i \neq j_1 \neq j_2$ such that
1° $B_{1k} = 0$ for $k = j_1, j_2$ and the pair $/A_{11} + B_1 K_{t1}, B_{1i}/$ is reachable,
2° $B_{2j_2} = 0$ and the pairs $/A_{22} + B_2 K_{t2}, B_{2j_1}/$, $/A_{23} + B_3 K_{t3}, B_{3j_2}/$ are reachable.

Without loss of generality we can assume $i = 1$, $j_1 = 2$, $j_2 = 3$. Hence

$$B_1 = \begin{bmatrix} B_{11} & 0 & 0 & B_{14} & \cdots & B_{1m} \end{bmatrix}$$

The reachability of the pair $/A_{t11}, B_{11}/$ implies nonsingularity of the matrix

$$R_1 = \begin{bmatrix} B_{11} & A_{t11} B_{11} & \cdots & A_{t11}^{n_1 - 1} B_{11} \end{bmatrix} \quad /5.155/$$

where $A_{t11} = A_{11} + B_1 K_{t1}$

Let $t_1$ be the $n_1$-th row of $R_1^{-1}$ and

$$T_1 = \begin{bmatrix} t_1 \\ t_1 A_{t11} \\ \vdots \\ t_1 A_{t11}^{n_1 - 1} \end{bmatrix} \quad /5.156/$$

It is well known that

$$T_1 A_{t11} T_1^{-1} = \left[\begin{array}{c|c} 0 & I_{n_1-1} \\ \hline & -a_1 \end{array}\right] \qquad /5.157a/$$

and

$$T_1 B_1 = \left[\begin{array}{cccccc} e_{n_1} & 0 & 0 & \bar{B}_{14} & \cdots & \bar{B}_{1m} \end{array}\right] \qquad /5.157b/$$

where

$$a_1 = \left[\begin{array}{cccc} a_{10} & a_{11} & \cdots & a_{1,n_1-1} \end{array}\right]$$

and $e_{n_1}$ is the $n_1$-th column of $I_{n_1}$.
It can be easily verified that for

$$K_{p1} T_1^{-1} = \left[\begin{array}{c} a_1 - c_1 \\ 0 \\ \cdots \\ 0 \end{array}\right] \qquad /5.158/$$

the matrix $\bar{A}_{c11}$ has the form

$$\bar{A}_{c11} = T_1 \left(A_{t11} + B_1 K_{p1}\right) T_1^{-1} = \left[\begin{array}{c|c} 0 & I_{n_1-1} \\ \hline & -c_1 \end{array}\right] \qquad /5.159/$$

where

$$c_1 = \left[\begin{array}{cccc} c_{10} & c_{11} & \cdots & c_{1,n_1-1} \end{array}\right]$$

Similarly, the reachability of the pair $/A_{t22}, B_{22}/$ implies non-singularity of the matrix

$$R_2 = \left[\begin{array}{cccc} B_{22} & A_{t22} B_{22} & \cdots & A_{22}^{n_2-1} B_{22} \end{array}\right] \qquad /5.160/$$

where $A_{t22} = A_{22} + B_2 K_{t2}$.
Let $t_2$ be the $n_2$-th row of $R_2^{-1}$ and

$$T_2 = \left[\begin{array}{c} t_2 \\ t_2 A_{t22} \\ \cdots \\ t_2 A_{t22}^{n_2-1} \end{array}\right] \qquad /5.161/$$

It is well known that

$$T_2 A_{t22} T_2^{-1} = \left[\begin{array}{c|c} 0 & I_{n_2-1} \\ \hline & -a_2 \end{array}\right] \qquad /5.162a/$$

and

$$T_2 B_2 = \left[\begin{array}{cccccc} \bar{B}_{21} & e_{n_2} & 0 & \bar{B}_{24} & \cdots & \bar{B}_{2m} \end{array}\right] \qquad /5.162b/$$

where
$$a_2 = \begin{bmatrix} a_{20} & a_{21} & \cdots & a_{2,n_2-1} \end{bmatrix}$$
and $e_{n_2}$ is the $n_2$-th column of $I_{n_2}$.

It can be easily verified that for
$$K_{p2}T_2^{-1} = \begin{bmatrix} 0 \\ a_2 - c_2 \\ 0 \\ \vdots \\ 0 \end{bmatrix} \qquad /5.163/$$

the matrix $\bar{A}_{c22}$ has the form
$$\bar{A}_{c22} = T_2 \left( A_{t22} + B_2 K_{p2} \right) T_2^{-1} = \begin{bmatrix} 0 & | & I_{n_2-1} \\ \hline & -c_2 & \end{bmatrix} \qquad /5.164/$$

where
$$c_2 = \begin{bmatrix} c_{20} & c_{21} & \cdots & c_{2,n_2-1} \end{bmatrix}$$
and
$$T_1 B_1 K_{p2} T_2^{-1} = 0$$

Similarly, the reachability of the pair $/A_{t33}, B_{33}/$ implies nonsingularity of the matrix
$$R_3 = \begin{bmatrix} B_{33} & A_{t33}B_{33} & \cdots & A_{t33}^{n_3-1} B_{33} \end{bmatrix} \qquad /5.165/$$

where $A_{t33} = A_{33} + B_3 K_{t3}$

Let $t_3$ be the $n_3$-th row of $R_3^{-1}$ and
$$T_3 = \begin{bmatrix} t_3 \\ t_3 A_{t33} \\ \vdots \\ t_3 A_{t33}^{n_3-1} \end{bmatrix} \qquad /5.166/$$

It is well known that
$$T_3 A_{t33} T_3^{-1} = \begin{bmatrix} 0 & | & I_{n_3-1} \\ \hline & -a_3 & \end{bmatrix} \qquad /5.167a/$$

and
$$T_3 B_3 = \begin{bmatrix} \bar{B}_{31} & \bar{B}_{32} & e_{n_3} & \bar{B}_{34} & \cdots & \bar{B}_{3m} \end{bmatrix}$$
where
$$a_3 = \begin{bmatrix} a_{30} & a_{31} & \cdots & a_{3,n_3-1} \end{bmatrix}$$
and $e_{n_3}$ is the $n_3$-th column of $I_{n_3}$.

It can be easily verified that for

$$K_{p3}T_3^{-1} = \begin{bmatrix} 0 \\ 0 \\ a_3 - c_3 \\ 0 \\ \vdots \\ 0 \end{bmatrix} \qquad /5.168/$$

the matrix $\bar{A}_{c33}$ has the form

$$\bar{A}_{c33} = T_3\left(A_{t33} + B_3 K_{p3}\right) T_3^{-1} = \left[\begin{array}{c|c} 0 & I_{n_3-1} \\ \hline & -c_3 \end{array}\right] \qquad /5.169/$$

where

$$c_3 = \begin{bmatrix} c_{30} & c_{31} & \cdots & c_{3,n_3-1} \end{bmatrix}$$

and

$$T_1 B_1 K_{p3} T_3^{-1} = 0, \quad T_2 B_1 K_{p3} T_3^{-1} = 0$$

Note that

$$\bar{A}_c = \begin{bmatrix} T_1 & 0 & 0 \\ 0 & T_2 & 0 \\ 0 & 0 & T_3 \end{bmatrix} (A_t + BK_p) \begin{bmatrix} T_1 & 0 & 0 \\ 0 & T_2 & 0 \\ 0 & 0 & T_3 \end{bmatrix}^{-1} = \begin{bmatrix} \bar{A}_{c11} & 0 & 0 \\ \bar{A}_{c21} & \bar{A}_{c22} & 0 \\ \bar{A}_{c31} & \bar{A}_{c32} & \bar{A}_{c33} \end{bmatrix}$$

where $\bar{A}_{c11}$, $\bar{A}_{c22}$, $\bar{A}_{c33}$ are given by /5.159/, /5.164/ and /5.169/, respectively and

$$\bar{A}_{c21} = T_2\left[A_{t21} + B_2 K_{p1}\right]T_1^{-1}, \quad \bar{A}_{c31} = T_3\left[A_{t31} + B_3 K_{p1}\right]T_1^{-1}$$

$$\bar{A}_{c32} = T_3\left[A_{t32} + B_3 K_{p2}\right]T_2^{-1}$$

have no special forms.
Further we have

$$\det\left[Z - A_c\right] = \det\left[Z - \bar{A}_c\right] = \bar{d}_1/z_1/\cdot \bar{d}_2/z_2/\cdot \bar{d}_3/z_3/$$

where

$$d_i/z_i/ = z_i^{n_i} + c_{i,n_i-1} z_i^{n_i-1} + \ldots + c_{i1} z_i + c_{i0} \quad /i = 1,2,3/$$

Thus, $A_c$ has the separable characteristic polynomial with the given set of 3-D eigenvalues /5.140/.
We have established the following

### Theorem 5.10

The eigenvalue assignment problem has a solution if the conditions /5.154/ and the assumptions $1^o, 2^o$ are satisfied.

If /5.154/ and assumptions $1^o, 2^o$ are satisfied the matrix K can be found by the use of the following

Algorithm 5.5

Step 1. Solving the equations /5.153/ find $K_{t2}$ and $K_{t3}$.

Step 2. Assuming arbitrary $K_{t1}$, for example $K_{t1} = 0$, find $K_t$ and $A_t = A + BK_t$.

Step 3. For the given set /5.140/ find $c_1$, $c_2$ and $c_3$.

Step 4. Using /5.155/, /5.156/, /5.161/, /5.166/ find $T_1$, $T_2$, $T_3$ and the coefficient vectors $a_1$, $a_2$, $a_3$ for characteristic polynomials of $A_{t11}$, $A_{t22}$ and $A_{t33}$, respectively.

Step 5. Using

$$K_{p1} = \begin{bmatrix} a_1 - c_1 \\ 0 \\ 0 \\ 0 \\ \cdots \\ 0 \end{bmatrix} T_1, \quad K_{p2} = \begin{bmatrix} 0 \\ a_2 - c_2 \\ 0 \\ 0 \\ \cdots \\ 0 \end{bmatrix} T_2, \quad K_{p3} = \begin{bmatrix} 0 \\ 0 \\ a_3 - c_3 \\ 0 \\ \cdots \\ 0 \end{bmatrix} T_3$$

find $K_p$ and $K = K_t + K_p$.

Example 5.14

Given

$$A = \begin{bmatrix} A_{11} & A_{12} & A_{13} \\ A_{21} & A_{22} & A_{23} \\ A_{31} & A_{32} & A_{33} \end{bmatrix} = \begin{bmatrix} 0 & 1 & 2 & 0 \\ -1 & 0 & 0 & 0 \\ 0 & 1 & 0 & 1 \\ -1 & 0 & 2 & 1 \end{bmatrix}, \quad B = \begin{bmatrix} B_1 \\ B_2 \\ B_3 \end{bmatrix} = \begin{bmatrix} 1 & 0 & 0 \\ 0 & 0 & 0 \\ 0 & 1 & 0 \\ -1 & 0 & 1 \end{bmatrix}$$

and the set of 3-D eigenvalues

$$E = \left\{ z_{11} = 1, \; z_{12} = 2, \; z_{21} = -1, \; z_{31} = -2 \right\} \qquad /5.170/$$

find K such that the closed-loop characteristic polynomial is separable with the desired set of 3-D eigenvalues /5.170/.

It is easy to check that the conditions of Theorem 5.10 are satisfied.

Step 1. The matrices

$$K_{t2} = -\begin{bmatrix} 2 \\ 0 \\ 0 \end{bmatrix}, \quad K_{t3} = -\begin{bmatrix} 0 \\ 1 \\ 0 \end{bmatrix}$$

satisfy the equations

$$\begin{bmatrix} 1 & 0 & 0 \\ 0 & 0 & 0 \end{bmatrix} K_{t2} = -\begin{bmatrix} 2 \\ 0 \end{bmatrix}, \quad \begin{bmatrix} 1 & 0 & 0 \\ 0 & 0 & 0 \\ 0 & 1 & 0 \end{bmatrix} K_{t3} = -\begin{bmatrix} 0 \\ 0 \\ 1 \end{bmatrix}$$

Step 2. Assuming $K_{t1} = 0$ we obtain

$$K_t = \begin{bmatrix} K_{t1} & K_{t2} & K_{t3} \end{bmatrix} = \begin{bmatrix} 0 & 0 & -2 & 0 \\ 0 & 0 & 0 & -1 \\ 0 & 0 & 0 & 0 \end{bmatrix}$$

and

$$A_t = A + BK_t = \begin{bmatrix} 0 & 1 & 0 & 0 \\ -1 & 0 & 0 & 0 \\ 0 & 1 & 0 & 0 \\ -1 & 0 & 4 & 1 \end{bmatrix}$$

Step 3. For /5.170/ we have

$$(z_1 - z_{11})(z_1 - z_{12}) = z_1^2 - 3z_1 + 2$$
$$z_2 - z_{21} = z_2 + 1$$
$$z_3 - z_{31} = z_3 + 2$$

and $c_1 = \begin{bmatrix} 2 & -3 \end{bmatrix}$, $c_2 = \begin{bmatrix} 1 \end{bmatrix}$, $c_3 = \begin{bmatrix} 2 \end{bmatrix}$.

Step 4.

$$R_1^{-1} = \begin{bmatrix} B_{11} & A_{t11}B_{11} \end{bmatrix}^{-1} = \begin{bmatrix} 1 & 0 \\ 0 & -1 \end{bmatrix}^{-1} = \begin{bmatrix} 1 & 0 \\ 0 & -1 \end{bmatrix}$$

and

$$T_1 = \begin{bmatrix} t_1 \\ t_1 A_{t11} \end{bmatrix} = \begin{bmatrix} 0 & -1 \\ 1 & 0 \end{bmatrix}, \quad T_2 = \begin{bmatrix} 1 \end{bmatrix}, \quad T_3 = \begin{bmatrix} 1 \end{bmatrix}.$$

To find $a_1$ we calculate

$$\det\begin{bmatrix} I_2 z - A_{t11} \end{bmatrix} = \begin{vmatrix} z & -1 \\ 1 & z \end{vmatrix} = z^2 + 1.$$

Hence $a_1 = \begin{bmatrix} 1 & 0 \end{bmatrix}$ and $a_2 = \begin{bmatrix} 0 \end{bmatrix}$, $a_3 = \begin{bmatrix} -1 \end{bmatrix}$.

Step 5.

$$K_{p1} = \begin{bmatrix} a_1 - c_1 \\ 0 \\ 0 \end{bmatrix} T_1 = \begin{bmatrix} 3 & 1 \\ 0 & 0 \\ 0 & 0 \end{bmatrix}, \quad K_{p2} = \begin{bmatrix} 0 \\ a_2 - c_2 \\ 0 \end{bmatrix} T_2 = \begin{bmatrix} 0 \\ -1 \\ 0 \end{bmatrix},$$

$$K_{p3} = \begin{bmatrix} 0 \\ 0 \\ a_3 - c_3 \end{bmatrix} T_3 = \begin{bmatrix} 0 \\ 0 \\ -3 \end{bmatrix}.$$

Therefore

$$K_p = \begin{bmatrix} K_{p1} & K_{p2} & K_{p3} \end{bmatrix} = \begin{bmatrix} 3 & 1 & 0 & 0 \\ 0 & 0 & -1 & 0 \\ 0 & 0 & 0 & -3 \end{bmatrix}$$

and

$$K = K_t + K_p = \begin{bmatrix} 3 & 1 & -2 & 0 \\ 0 & 0 & -1 & -1 \\ 0 & 0 & 0 & -3 \end{bmatrix}$$

It is easy to check that the matrix

$$A_c = A + BK = \left[ \begin{array}{cc|c|c} 3 & 2 & 0 & 0 \\ -1 & 0 & 0 & 0 \\ \hline 0 & 1 & -1 & 0 \\ \hline -4 & -1 & 4 & -2 \end{array} \right]$$

has the separable characteristic polynomial with the desired set of 3-D eigenvalues /5.170/.

The above considerations can be extended for n-D systems [11].

PROBLEMS

1. Show that $A[I - B^g B] = 0$ and $[I - B B^g]A = 0$ if and only if rank $B$ = rank $\begin{bmatrix} A \\ B \end{bmatrix}$ and rank $B$ = rank $[A \quad B]$, respectively.

   Hint: Find nonsingular matrix $T$ such that
   $$\begin{bmatrix} A \\ B \end{bmatrix} T = \begin{bmatrix} A_1 & A_2 \\ B_1 & 0 \end{bmatrix}$$
   where $B_1$ has full column rank.
   Note that $A_2 = 0$ if and only if
   $$\text{rank } B = \text{rank } \begin{bmatrix} A \\ B \end{bmatrix}.$$
   Then
   $$B^g = T \begin{bmatrix} B_1^g \\ 0 \end{bmatrix}$$
   and
   $$A[I - B^g B] = [A_1 \quad A_2]\left( I - \begin{bmatrix} I & 0 \\ 0 & 0 \end{bmatrix} \right)T^{-1} = [0 \quad A_2] T^{-1}$$

2. Given
   $$A = \begin{bmatrix} A_{11} & A_{12} \\ A_{21} & A_{22} \end{bmatrix} = \begin{bmatrix} 0 & 3 & 0 & 1 \\ -2 & 0 & 5 & -2 \\ 4 & 0 & 0 & 1 \\ 5 & 2 & -3 & -2 \end{bmatrix}, \quad B = \begin{bmatrix} B_1 \\ B_2 \end{bmatrix} = \begin{bmatrix} 1 & 0 \\ 0 & 1 \\ 0 & 0 \\ 1 & 0 \end{bmatrix}$$
   and the set of 2-D eigenvalues
   $$E = \{ z_{11} = -2, \quad z_{21} = z_{22} = z_{23} = 1 \}$$
   find $K$ such that the closed-loop characteristic polynomial is separable with the given $E$.
   Hint: Use Method 1.
   Answer:
   $$K = \begin{bmatrix} 2 & 3 & 0 & 1 \\ 0 & 0 & 4 & -2 \end{bmatrix}$$

3. Show that the eigenvalue assignment problem for 3-D systems has a solution if there exists $i, j$ /$i \neq j$/ such that
   $$\begin{bmatrix} B_i \\ B_j \end{bmatrix} \quad /i, j = 1, 2, 3/ \text{ is of full row rank}$$

and

$$B_k \begin{bmatrix} B_i \\ B_j \end{bmatrix}^g \begin{bmatrix} A_{ik} \\ A_{jk} \end{bmatrix} = A_{kk} - A_{ckk} \quad /k \neq i, k \neq j/$$

where $B^g$ is the generalized inverse matrix of $B$ satisfying the condition $B B^g B = B$.

**Hint:** Use

$$\begin{bmatrix} B_i \\ B_j \end{bmatrix} K_l = \begin{bmatrix} A_{il} - A_{cil} \\ A_{jl} - A_{cjl} \end{bmatrix} \quad \text{for } l = 1,2,3$$

and assume $A_c$ which given a separable closed-loop characteristic polynomial with desired eigenvalues.

4. Show that if

$$\det[Z - \bar{A}_c] = d_1/z_1/ \cdot d_2/z_2/ \cdot d_3/z_3/$$

then the solution to the eigenvalue assignment problem for 3-D systems can be reduced to finding a nonsingular matrix

$$T = \begin{bmatrix} T_1 & 0 & 0 \\ 0 & T_2 & 0 \\ 0 & 0 & T_3 \end{bmatrix}$$

and a feedback gain matrix $K$ such that

$$A + BF = T\bar{A}_c T^{-1}$$

**Hint:** Note that $\det[Z - \bar{A}_c] = \det[Z - T\bar{A}_c T^{-1}]$

5. For F-MM I of the form

$$x/i+1,j+1/ = \begin{bmatrix} 1 & -1 \\ 0 & 2 \end{bmatrix} x/i,j/ + \begin{bmatrix} 0 & 1 \\ -1 & 0 \end{bmatrix} x/i+1,j/ + \begin{bmatrix} -1 & 0 \\ 0 & 1 \end{bmatrix} x/i,j+1/ +$$

$$+ \begin{bmatrix} 1 & 0 \\ 0 & 1 \end{bmatrix} u/i,j/$$

find a matrix $K$ of the state feedback law $u/i,j/ = Kx/i,j/$ such that the closed-loop characteristic polynomial is

$$d_c/z_1,z_2/ = z_1^2 z_2^2 - z_1^2 + z_2^2 - 2z_1 z_2 + 2z_1$$

**Hint:** Compare the coefficients of like terms $z_1^i z_2^j$ of both sides of the relation

$$\det[Iz_1 z_2 - A_o - BK - A_1 z_2 - A_2 z_1] = z_1^2 z_2^2 - z_1^2 + z_2^2 - 2z_1 z_2 + 2z_1$$

Answer:
$$K = \begin{bmatrix} 1 & 1 \\ 0 & -2 \end{bmatrix}$$

6. Show that if for RM the conditions
$$\text{rank } B_1 = \text{rank}[B_1 \quad A_{12}]$$
and
$$\text{rank } B_2 = \text{rank}[B_2 \quad A_{21}]$$
are satisfied then a state-feedback matrix K can be chosen so that $y/i,j/ = 0$ for $i > M$ and $j > N$, where M,N are positive integers such that $v/i,j/ = 0$, $x^h/0,j/ = 0$ and $x^v/i,0/ = 0$ for $i > M$, $j > N$.

Hint: Note that
$$\begin{bmatrix} A_{c11} & 0 \\ 0 & A_{c22} \end{bmatrix}^{i,j} = 0 \quad \text{for} \quad i > 0 \text{ and } j > 0.$$

REFERENCES

[1] Eising R.: Realization and stabilization of 2-D systems. IEEE Trans. Automat. Control vol. AC-23, Oct. 1978, pp.793-799.

[2] Emre E. and Khargonekar P.P.: Regulation of split linear systems over rings; coefficient-assignment and observers. IEEE Trans. Automat. Control vol. AC-27, Febr. 1982, pp.104-113.

[3] Kaczorek T.: Das Polverschiebungsproblem in 2-D linearen Systemen. Wissenschaftliche Berichte der Technischen Hochschule Leipzig, Heft 2, 1983, pp.8-12.

[4] Kaczorek T.: Pole assignment problem in two-dimensional linear systems. Int.J.Control vol.37, no.1, 1983, pp.183-190.

[5] Kaczorek T.: Pole assignment of 3-D linear systems with separable characteristic polynomials. Foundations of Control Engineering vol.8, no.2, 1983, pp.81-91.

[6] Kaczorek T.: Eigenvalue assignment problem for 2-D systems with separable characteristic polynomials. Bull.Acad.Polon.Sci. Ser. Sci.Techn. vol.32, no.1-2, 1984 /in press/.

[7] Kaczorek T.: Eigenvalue assignment of 3-D systems. Multivariable Control: Concepts and Tools. Editor S.G. Tzafestas. Reidel Publishing Company 1984.

[8] Kaczorek T.: Polynomial assignment via output dynamic feedback of 3-D systems. Bull.Acad.Polon.Sci. Ser. Sci.Techn. vol.31, no.5-6, 1984 /in press/.

[9] Kaczorek T.: Control Theory, vol.I, PWN Warszawa 1974 /in Polish/.

[10] Kaczorek T.: Zeroing of 2-D linear system output by state feedback. Bull.Acad.Polon.Sci. Ser.Sci.Techn. vol.30, no.3-4, 1982, pp. 59-64.

[11] Kaczorek T., Kurek J.: Separability-assignment problem for q-dimensional linear discrete-time systems. Int.J.Control vol.39, no.6, 1984, pp.1375-1382.

[12] Mertzios B.G.: Pole assignment of 2-D systems for separable characteristic equations. Int.J.Control vol.39, no.5, 1984, pp. 879-889.

[13] Paraskevopoulos P.N.: Characteristic polynomial assignment and determination of the residual polynomial in 2-D systems. IEEE Trans. Automat. Control vol. AC-26, 1981, pp.541-543.

[14] Paraskevopoulos P.N.: Eigenvalue assignment of linear 2-dimensional systems. Proc. IEE vol.126, 1979, pp.1204-1208.

[15] Paraskevopoulos P.N., Kosmidou O.I.: Eigenvalue assignment of two-dimensional systems using PID controllers. Int.J.Systems Sci. vol.12, 1981, pp.407-422.

[16] Pringle R.M., Rayner A.A.: Generalized inverse matrices with applications to statistics. Griffin, London 1971.

[17] Šebek M.: On 2-D pole placement. IEEE Trans. Automat. Control vol. AC-28, 1984 /in press/.

[18] Tzafestas S.G., Pimenides T.G.: Exact model matching control of three-dimensional systems using state and output feedback. Int.J.Systems Sci. vol.13, no.11, 1982, pp.1171-1187.

# 6. OBSERVERS, EXACT MODEL MATCHING AND DECOUPLING

## 6.1. Asymptotic and deadbeat observers

### 6.1.1. Definitions and sufficient conditions

Consider the Roesser's model /RM/ described by the equations

$$x' = Ax + Bu \qquad /6.1/$$
$$y = Cx \qquad /6.2/$$

where

$$x' = \begin{bmatrix} x^h/i+1,j/ \\ x^v/i,j+1/ \end{bmatrix}, \quad x = \begin{bmatrix} x^h/i,j/ \\ x^v/i,j/ \end{bmatrix}$$

$$A = \begin{bmatrix} A_{11} & A_{12} \\ A_{21} & A_{22} \end{bmatrix}, \quad B = \begin{bmatrix} B_1 \\ B_2 \end{bmatrix}, \quad C = \begin{bmatrix} C_1 & C_2 \end{bmatrix}$$

$x^h/i,j/ \in R^{n_1}$ is the horizontal state vector,
$x^v/i,j/ \in R^{n_2}$ is the vertical state vector,
$u = u/i,j/ \in R^m$ is the input vector,
$y = y/i,j/ \in R^l$ is the output vector,
and $A_{ij}$, $B_i$, $C_i$ are real constant matrices of appropriate dimensions. The boundary condition are given by

$$x^h/0,j/, \; j=0,1,2,\ldots, \text{ and } x^v/i,0/, \; i=0,1,2,\ldots \qquad /6.3/$$

Further let us consider the following 2-D system

$$z' = Fz + Gu + Hy \qquad /6.4/$$
$$x = Lz + Ky \qquad /6.5/$$

where

$$z' = \begin{bmatrix} z^h/i+1,j/ \\ z^v/i,j+1/ \end{bmatrix}, \quad z = \begin{bmatrix} z^h/i,j/ \\ z^v/i,j/ \end{bmatrix}$$

$$F = \begin{bmatrix} F_{11} & F_{12} \\ F_{21} & F_{22} \end{bmatrix}, \quad G = \begin{bmatrix} G_1 \\ G_2 \end{bmatrix}, \quad H = \begin{bmatrix} H_1 \\ H_2 \end{bmatrix}$$

$z^h/i,j/ \in R^{p_1}$ is the horizontal state vector.

$z^v/i,j/ \in R^{p_2}$ is the vertical state vector,

$\hat{x} \in R^n$ /$n=n_1+n_2$/ is an estimate /approximation/ of x, and

$F_{ij}$, $G_i$, $H_i$, L, K are real constant matrices of appropriate dimensions.

## Definition 6.1

The 2-D system described by /6.4/ and /6.5/ is called an asymptotic observer of the vector x of RM if

$$\lim_{i,j \to \infty} \hat{x}/i,j/ = \lim_{i,j \to \infty} x/i,j/ \qquad /6.6/$$

independently of the known /measurable/ input vector u and of the /known or unknown/ boundary conditions /6.3/.

## Definition 6.2

The 2-D system described by /6.4/ and /6.5/ is called a deadbeat observer of the vector x of RM if there exist some finite positive integers M, N such that

$$\hat{x}/i,j/ = x/i,j/ \text{ for } i > M \text{ and } j > N \qquad /6.7/$$

independently of the known /measurable/ input vector u and of the /known or unknown/ boundary conditions /6.3/.

Let us define the observer error as

$$e = z - Tx \qquad /6.8/$$

where

$$e = \begin{bmatrix} e^h/i,j/ \\ e^v/i,j/ \end{bmatrix}, \quad T \in R^{p \times n} \quad /p = p_1+p_2 < n/$$

$e^h/i,j/ \in R^{p_1}$ is the horizontal error vector,

$e^v/i,j/ \in R^{p_2}$ is the vertical error vector.

## Theorem 6.1

The 2-D system described by /6.4/ and /6.5/ is an asymptotic observer for RM if the following conditions hold

$$TA = FT + HC \qquad /6.9a/$$

$$G = TB \qquad /6.9b/$$

$$LT + KC = I_n \qquad /6.9c/$$

If in addition the boundary conditions /6.3/ are unknown then

$$\lim_{i \to \infty} F^{i,j} = 0 \qquad \text{for } j = 0,1,2,\ldots$$
$$\lim_{j \to \infty} F^{i,j} = 0 \qquad \text{for } i = 0,1,2,\ldots \qquad /6.10/$$

**Proof**

Substitution of /6.4/, /6.8/ and /6.1/, /6.2/ into

$$e' = z' - Tx'$$

yields

$$e' = F/e+Tx/ + Gu + HCx - T/Ax+Bu/ = Fe + /FT+HC-TA/x + /G-TB/u \qquad /6.11/$$

If /6.9a/ and /6.9b/ hold, then

$$e' = Fe \qquad /6.12/$$

Substituting /6.8/ and /6.2/ into /6.5/ we obtain

$$\hat{x} = L/e+Tx/ + KCx = Le + /LT+KC/x \qquad /6.13/$$

If /6.9c/ holds, then

$$\hat{x} = Le + x \qquad /6.14/$$

The solution to /6.12/ is given by

$$\begin{bmatrix} e^h/i,j/ \\ e^v/i,j/ \end{bmatrix} = \sum_{l=0}^{j} F^{i,j-1} \begin{bmatrix} e^h/0,1/ \\ 0 \end{bmatrix} + \sum_{k=0}^{i} F^{i-k,j} \begin{bmatrix} 0 \\ e^v/k,0/ \end{bmatrix} \qquad /6.15/$$

If the conditions /6.10/ are satisfied, then

$$\lim_{i,j \to \infty} e/i,j/ = 0$$

and it is seen from /6.14/ that /6.6/ holds for any bounded $e^h/0,1/$ and $e^v/k,0/$, $k, l = 0,1,2,\ldots$

If the boundary conditions /6.3/ are known then from /6.8/ the boundary conditions of the observer $z^h/0,1/$ and $z^v/k,0/$, $k,1 = 0,1,2,...$ can be determined so that $e^h/0,1/ = 0$ and $e^v/k,0/ = 0$ for $k,1 = 0,1,2,...$, i.e.

$$\begin{bmatrix} z^h/0,1/ \\ z^v/k,0/ \end{bmatrix} = T \begin{bmatrix} x^h/0,1/ \\ x^v/k,0/ \end{bmatrix}$$

In this case $e/i,j/ = 0$ for $i,j \geqslant 0$. □

### 6.1.2. Design of observers

The design problem of the observers may be formulated as follows: Given the matrices A, B, C of RM, find matrices F, G, H, L, K of the observer. To solve the problem let us assume that the full rank matrix C has the form

$$C = \begin{bmatrix} 0 & I_1 \end{bmatrix} \qquad /6.16/$$

and

$$T = \begin{bmatrix} I_p & \bar{K} \end{bmatrix} \qquad /6.17/$$

where $\bar{K} \in R^{p \times l}$.

Note that /6.9c/ is satisfied if

$$K = \begin{bmatrix} -\bar{K} \\ I_1 \end{bmatrix}, \qquad L = \begin{bmatrix} I_p \\ 0 \end{bmatrix} \qquad /6.18/$$

Let

$$A = \begin{bmatrix} A_1 & A_2 \\ A_3 & A_4 \end{bmatrix}, \qquad B = \begin{bmatrix} \bar{B}_1 \\ \bar{B}_2 \end{bmatrix} \qquad /6.19/$$

where

$A_1 \in R^{p \times p}$, $A_4 \in R^{l \times l}$, $\bar{B}_1 \in R^{p \times m}$, $\bar{B}_2 \in R^{l \times m}$.

Substitution of /6.16/, /6.16/, /6.19/ into /6.9a/ and /6.9b/ yields

$$\begin{bmatrix} I_p & \bar{K} \end{bmatrix} \begin{bmatrix} A_1 & A_2 \\ A_3 & A_4 \end{bmatrix} = F \begin{bmatrix} I_p & \bar{K} \end{bmatrix} + H \begin{bmatrix} 0 & I_1 \end{bmatrix} \qquad /6.20/$$

and
$$G = \begin{bmatrix} I_p & \bar{K} \end{bmatrix} \begin{bmatrix} \bar{B}_1 \\ \bar{B}_2 \end{bmatrix} = \bar{B}_1 + \bar{K}\bar{B}_2 \qquad /6.21/$$

Comparing the suitable submatrices of /6.20/ we obtain

$$F = A_1 + \bar{K} A_3 \qquad /6.22/$$

$$H = A_2 + \bar{K}A_4 - F\bar{K} = A_2 + \bar{K}A_4 - A_1\bar{K} - \bar{K}A_3\bar{K} \qquad /6.23/$$

It can be easily shown that if

$$F = \begin{bmatrix} F_{11} & 0 \\ F_{21} & F_{22} \end{bmatrix} \qquad /6.24/$$

then

$$F^{i,0} = \begin{bmatrix} F_{11}^i & 0 \\ 0 & 0 \end{bmatrix}, \quad F^{0,j} = \begin{bmatrix} 0 & 0 \\ F_{22}^{j-1} F_3 & F_{22}^j \end{bmatrix} \qquad /6.25a/$$

and

$$F^{i,j} = \begin{bmatrix} 0 & 0 \\ F_{22}^{j-1} F_{21} F_{11}^i & 0 \end{bmatrix} \quad \text{for } i,j \geqslant 1 \qquad /6.25b/$$

Substitution of /6.25/ into /6.15/ yields

$$e^h/i,j/ = F_{11}^i \, e^h/0,j/ \qquad /6.26a/$$

and

$$e^v/i,j/ = \sum_{l=0}^{j-1} F_{22}^{j-l-1} F_{21} F_{11}^i \, e^h/0,l/ \qquad /6.26b/$$

From /6.26/ it follows that if the eigenvalues of $F_{11}$ and $F_{22}$ are all inside the unit disk, then $\lim_{i \to \infty} e^h/i,j/ = 0$, $\lim_{i \to \infty} e^v/i,j/ = 0$ and /6.6/ holds.

Therefore the design problem of an asymptotic observer for RM is reduced to finding $\bar{K}$ from /6.22/ such that the eigenvalues of $F_{11}$ and $F_{22}$ are all inside the unit disk. If $\bar{K}$ is known, then the matrices F, G, H, K and L may be determined from /6.22/, /6.23/, /6.21/ and /6.18/, respectively. For finding $\bar{K}$ all methods presented in Chapter 5 may be used.

If rank $A_3 = p \leq 1$, then solving /6.22/ we obtain

$$K = /F-A_1/ \left[ A_3^T A_3 \right]^{-1} A_3^T \qquad /6.27/$$

In this case $\bar{K}$ can be determined from /6.27/ for the desired matrix F. Note that if the matrices $F_{11}$ and $F_{22}$ are nilpotent/all eigenvalues at the origin/ then from /6.25/ we have $F^{i,j} = 0$ for $i > M$ and $j > N$, where M, N are some finite positive integers. In this case /6.7/ is satisfied and we have a deadbeat observer.

From above considerations the following algorithm for design of the asymptotic /or deadbeat/ observer follows

### Algorithm 6.1

Step 1  Find $p = n-1$ and $A_1$, $A_2$, $A_3$, $A_4$, $\bar{B}_1$, $\bar{B}_2$.

Step 2  Using one of the methods presented in Chapter 5 or /6.27/ find $\bar{K}$ from /6.22/ such that the eigenvalues of $F_{11}$ and $F_{22}$ are all inside the unit disk /asymptotic observer/ or $F_{11}$ and $F_{22}$ are the nilpotent matrices /deadbeat observer/

Step 3  Find F, H, G, K and L using /6.22/, /6.23/, /6.21/ and /6.18/.

Step 4  Write the equations /6.4/ and /6.5/ of the observer

### Example 6.1

Given RM with

$$A = \begin{bmatrix} A_{11} & A_{12} \\ A_{21} & A_{22} \end{bmatrix} = \begin{bmatrix} 0 & 1 & | & 0 & 1 \\ -1 & 0 & | & 0 & 0 \\ \hline 1 & 0 & | & 0 & 1 \\ 0 & 1 & | & -2 & -1 \end{bmatrix}, \quad B = \begin{bmatrix} B_1 \\ B_2 \end{bmatrix} = \begin{bmatrix} 1 \\ 0 \\ \hline 0 \\ -1 \end{bmatrix}$$

$$C = \begin{bmatrix} C_1 & C_2 \end{bmatrix} = \begin{bmatrix} 0 & 0 & | & 1 & 0 \\ 0 & 0 & | & 0 & 1 \end{bmatrix}$$

find the equations /6.4/ and /6.5/ of

a/ an asymptotic observer

b/ a deadbeat observer.

In this case $n_1 = n_2 = 2$, $n = n_1 + n_2 = 4$, $m = 1$ and $l = 2$

Step 1  We have p = n-1 = 2 and

$$A_1 = A_{11} = \begin{bmatrix} 0 & 1 \\ -1 & 0 \end{bmatrix}, \quad A_2 = A_{12} = \begin{bmatrix} 0 & 1 \\ 0 & 0 \end{bmatrix}, \quad A_3 = A_{21} = \begin{bmatrix} 1 & 0 \\ 0 & 1 \end{bmatrix}$$

$$A_4 = A_{22} = \begin{bmatrix} 0 & 1 \\ -2 & -1 \end{bmatrix}, \quad \bar{B}_1 = B_1 = \begin{bmatrix} 1 \\ 0 \end{bmatrix}, \quad \bar{B}_2 = B_2 = \begin{bmatrix} 0 \\ -1 \end{bmatrix}$$

Step 2  a/ For asymptotic observer we assume $F_{11} = F_{22} = 0,1$ and $F_{12} = F_{21} = 0$. Hence

$$\bar{K} = /F-A_1/\begin{bmatrix} A_3^T & A^T \end{bmatrix}^{-1} A_3^T = \begin{bmatrix} 0,1 & -1 \\ 1 & 0,1 \end{bmatrix}$$

b/ For deadbeat observer we assume $F_{11} = F_{12} = F_{22} = 0$ and $F_{21} = 1$. Hence

$$\bar{K} = /F-A_1/\begin{bmatrix} A_3^T & A^T \end{bmatrix}^{-1} A_3^T = \begin{bmatrix} 0 & -1 \\ 2 & 0 \end{bmatrix}$$

Step 3  a/ $F = A_1 + \bar{K}A_3 = \begin{bmatrix} 0,1 & 0 \\ 0 & 0,1 \end{bmatrix}$

$$H = A_2 + \bar{K}A_4 - F\bar{K} = \begin{bmatrix} 1,99 & 2,2 \\ -0,3 & 0,89 \end{bmatrix}$$

$$G = \bar{B}_1 + \bar{K}\bar{B}_2 = \begin{bmatrix} 2 \\ -0,1 \end{bmatrix}, \quad K = \begin{bmatrix} -\bar{K} \\ I_1 \end{bmatrix} = \begin{bmatrix} -0,1 & 1 \\ -1 & -0,1 \\ 1 & 0 \\ 0 & 1 \end{bmatrix},$$

$$L = \begin{bmatrix} I_p \\ 0 \end{bmatrix} = \begin{bmatrix} 1 & 0 \\ 0 & 1 \\ 0 & 0 \\ 0 & 0 \end{bmatrix}$$

b/

$$F = A_1 + \bar{K}A_3 = \begin{bmatrix} 0 & 0 \\ 1 & 0 \end{bmatrix}, \quad H = A_2 + \bar{K}A_4 - F\bar{K} = \begin{bmatrix} 2 & 2 \\ 0 & 3 \end{bmatrix}$$

$$G = \bar{B}_1 + \bar{K}\bar{B}_2 = \begin{bmatrix} 2 \\ 0 \end{bmatrix}, \quad K = \begin{bmatrix} -\bar{K} \\ I_1 \end{bmatrix} = \begin{bmatrix} 0 & 1 \\ -2 & 0 \\ 1 & 0 \\ 0 & 1 \end{bmatrix}, \quad L = \begin{bmatrix} I_p \\ 0 \end{bmatrix} = \begin{bmatrix} 1 & 0 \\ 0 & 1 \\ 0 & 0 \\ 0 & 0 \end{bmatrix}$$

**Step 4** The desired equations /6.4/ and /6.5/ are the following

a/ for asymptotic observer

$$\begin{bmatrix} z^h/i+1,j/ \\ z^v/i,j+1/ \end{bmatrix} = \begin{bmatrix} 0,1 & 0 \\ 0 & 0,1 \end{bmatrix} \begin{bmatrix} z^h/i,j/ \\ z^v/i,j/ \end{bmatrix} +$$

$$+ \begin{bmatrix} 2 \\ -0,1 \end{bmatrix} u/i,j/ + \begin{bmatrix} 1,99 & 2,2 \\ -0,3 & 0,89 \end{bmatrix} \begin{bmatrix} y_1/i,j/ \\ y_2/i,j/ \end{bmatrix}$$

$$\begin{bmatrix} \hat{x}^h/i,j/ \\ \hat{x}^v/i,j/ \end{bmatrix} = \begin{bmatrix} 1 & 0 \\ 0 & 1 \\ \hline 0 & 0 \\ 0 & 0 \end{bmatrix} \begin{bmatrix} z^h/i,j/ \\ z^v/i,j/ \end{bmatrix} + \begin{bmatrix} -0,1 & 1 \\ -1 & -0,1 \\ \hline 1 & 0 \\ 0 & 1 \end{bmatrix} \begin{bmatrix} y_1/i,j/ \\ y_2/i,j/ \end{bmatrix}$$

b/ for deadbeat observer

$$\begin{bmatrix} z^h/i+1,j/ \\ z^v/i,j+1/ \end{bmatrix} = \begin{bmatrix} 0 & 0 \\ 1 & 0 \end{bmatrix} \begin{bmatrix} z^h/i,j/ \\ z^v/i,j/ \end{bmatrix} + \begin{bmatrix} 2 \\ 0 \end{bmatrix} u/i,j/ + \begin{bmatrix} 2 & 2 \\ 0 & 3 \end{bmatrix} \begin{bmatrix} y_1/i,j/ \\ y_2/i,j/ \end{bmatrix}$$

$$\begin{bmatrix} \hat{x}^h/i,j/ \\ \hat{x}^v/i,j/ \end{bmatrix} = \begin{bmatrix} 1 & 0 \\ 0 & 1 \\ \hline 0 & 0 \\ 0 & 0 \end{bmatrix} \begin{bmatrix} z^h/i,j/ \\ z^v/i,j/ \end{bmatrix} + \begin{bmatrix} 0 & 1 \\ -2 & 0 \\ \hline 1 & 0 \\ 0 & 1 \end{bmatrix} \begin{bmatrix} y_1/i,j/ \\ y_2/i,j/ \end{bmatrix}$$

The stabilization of 2-D systems using 2-D asymptotic observers is considered in [1] and the state observer design problem for 3-D systems is considered in [3]. A design procedure of a minimum order observer for implementation of a state feedback law for 2-D systems in developed in [2, 4].

### 6.1.3. Kawaji's method for design of minimal order observer

Following [2] without loss of generality it may be assumed that

$$\text{rank } C = 1 \qquad\qquad /6.28/$$

and

$$\text{rank } C_1 = l_1, \quad \text{rank } C_2 = l_2$$

We shall show that there exist nonsingular matrices $P$, $Q_1$ and $Q_2$ such that

$$PCQ = \begin{bmatrix} I_{1-l_2} & 0 & 0 & 0 \\ C_3 & C_4 & 0 & I_{l_2} \end{bmatrix} \quad Q = \begin{bmatrix} Q_1 & 0 \\ 0 & Q_2 \end{bmatrix} \qquad /6.29a/$$

or

$$PCQ = \begin{bmatrix} I_{l_1} & 0 & \bar{C}_3 & \bar{C}_4 \\ 0 & 0 & 0 & I_{1-l_1} \end{bmatrix} \qquad /6.29b/$$

Since rank $C_2 = l_2$, it is always possible to find nonsingular matrices $P$ and $Q_2$ such that

$$PC_2 Q_2 = \begin{bmatrix} 0 & 0 \\ 0 & I_{l_2} \end{bmatrix}$$

Let

$$PC_1 = \begin{bmatrix} C_{11} & C_{12} \\ \hline C_{21} & C_{22} \end{bmatrix} \begin{matrix} \}1-l_2 \\ \}l_2 \end{matrix}$$

$$\underbrace{\phantom{C_{11}}}_{1-l_2} \underbrace{\phantom{C_{12}}}_{n_1-l+l_2}$$

Then rank $\begin{bmatrix} C_{11} & C_{12} \end{bmatrix} = 1-l_2$, since rank $\begin{bmatrix} C_1 & C_2 \end{bmatrix} = 1$.
Therefore, if we choose $\hat{C}$ so that $\begin{bmatrix} C_{11} & C_{12} \\ \hat{C} \end{bmatrix}$ is nonsingular, then it is easy to verify that

$$PC_1 Q_1 = \begin{bmatrix} I_{1-l_2} & 0 \\ C_3 & C_4 \end{bmatrix}$$

where

$$Q_1 = \begin{bmatrix} C_{11} & C_{12} \\ \hat{C} & \end{bmatrix}^{-1}$$

In a similar way we can prove /6.29b/.

Let us define

$$\bar{A} = QAQ = \begin{bmatrix} \bar{A}_{11} & \bar{A}_{12} & \bar{A}_{13} & \bar{A}_{14} \\ \bar{A}_{21} & \bar{A}_{22} & \bar{A}_{23} & \bar{A}_{24} \\ \bar{A}_{31} & \bar{A}_{32} & \bar{A}_{33} & \bar{A}_{34} \\ \bar{A}_{41} & \bar{A}_{42} & \bar{A}_{43} & \bar{A}_{44} \end{bmatrix} \begin{matrix} \}l-l_2 \\ \}n_1-l+l_2 \\ \}n_2-l_2 \\ \}l_2 \end{matrix} \qquad /6.30a/$$

$$\bar{B} = Q^{-1}B = \begin{bmatrix} \bar{B}_{11} \\ \bar{B}_{12} \\ \bar{B}_{21} \\ \bar{B}_{22} \end{bmatrix} \begin{matrix} \}l-l_2 \\ \}n_1-l+l_2 \\ \}n_2-l_2 \\ \}l_2 \end{matrix} \qquad \bar{C} = PCQ$$

and

$$\bar{T} = TQ \qquad /6.30b/$$

Taking into account /6.9/ and /6.30/ it can be easily checked that

$$\bar{T}\bar{A} = F\bar{T} + HP^{-1}\bar{C} \qquad /6.31a/$$

$$G = \bar{T}\bar{B} \qquad /6.31b/$$

$$Q^{-1}L\bar{T} + Q^{-1}KP^{-1}\bar{C} = I_n \qquad /6.31c/$$

From /6.31c/ we have

$$Q^{-1}\begin{bmatrix} L & KP^{-1} \end{bmatrix}\begin{bmatrix} \bar{T} \\ \bar{C} \end{bmatrix} = I_n$$

and

$$\text{rank}\begin{bmatrix} \bar{T} \\ \bar{C} \end{bmatrix} = n$$

Note that the matrix $\bar{T}$ may be assumed in the form

$$\bar{T} = \begin{bmatrix} M_1 & I_{n_1-l+l_2} & 0 & 0 \\ 0 & 0 & I_{n_2-l_2} & M_2 \end{bmatrix}$$

where $M_1$ and $M_2$ are arbitrary matrices of appropriate sizes.

From /6.31a/ it follows that

$$\bar{T}\bar{A} = \begin{bmatrix} F & HP^{-1} \end{bmatrix} \begin{bmatrix} \bar{T} \\ \bar{C} \end{bmatrix}$$

and

$$F = \bar{T}\bar{A} \begin{bmatrix} \bar{T} \\ \bar{C} \end{bmatrix}^{-1} \begin{bmatrix} I_{n-1} \\ 0 \end{bmatrix}$$

Taking into account that

$$\begin{bmatrix} \bar{T} \\ \bar{C} \end{bmatrix} = \begin{bmatrix} M_1 & I_{n_1-1+l_2} & 0 & 0 \\ 0 & 0 & I_{n_2-l_2} & M_2 \\ I_{1-l_2} & 0 & 0 & 0 \\ C_3 & C_4 & 0 & I_{l_2} \end{bmatrix}$$

and

$$\begin{bmatrix} \bar{T} \\ \bar{C} \end{bmatrix}^{-1} = \begin{bmatrix} 0 & 0 & I_{1-l_2} & 0 \\ I_{n_1-l_2+l_2} & 0 & -M_1 & 0 \\ M_2C_4 & I_{n_2-l_2} & M_2/C_3-C_4M_1/ & -M_2 \\ -C_4 & 0 & C_4M_1-C_3 & I_{l_2} \end{bmatrix}$$

we obtain

$$F = \begin{bmatrix} M_1\bar{A}_{12}+\bar{A}_{22}+/M_1\bar{A}_{13}+\bar{A}_{23}/M_2C_4-/M_1\bar{A}_{14}+\bar{A}_{24}/C_4, & M_1\bar{A}_{13}+\bar{A}_{23} \\ \bar{A}_{32}+M_2\bar{A}_{42}+/\bar{A}_{33}+M_2\bar{A}_{43}/M_2C_4-/\bar{A}_{34}+M_2\bar{A}_{44}/C_4, & \bar{A}_{33}+M_2\bar{A}_{43} \end{bmatrix} = \begin{bmatrix} F_{11} & F_{12} \\ F_{21} & F_{22} \end{bmatrix}$$

/6.32/

Theorem 6.2

A minimal order observer for RM exists if

/i/ $\bar{A}_{23}\bar{A}_{13}^g\bar{A}_{13} = \bar{A}_{23}$

/ii/ the pair $/\bar{A}_{23}\bar{A}_{13}^g/\bar{A}_{14}C_4-\bar{A}_{12}/+\bar{A}_{22}-\bar{A}_{24}C_4$, $/I_{1-l_2}-\bar{A}_{13}\bar{A}_{13}^g//\bar{A}_{12}-\bar{A}_{14}C_4//$
is detectable,

/iii/ the pair $/\bar{A}_{33}, \bar{A}_{43}/$ is detectable,

where $\bar{A}_{13}^g$ is the generalized inverse matrix of $\bar{A}_{13}$.

## Proof

We shall show that if the conditions /i/-/iii/ are satisfied then there exist matrices $M_1$ and $M_2$ such that the matrix /6.32/ is asymptotically stable. If /i/ is satisied, then the equation

$$M_1 \bar{A}_{13} + \bar{A}_{23} = 0$$

has a solution in the form

$$M_1 = -\bar{A}_{23} \bar{A}_{13}^g + N / I_{1-l_2} - \bar{A}_{13} \bar{A}_{13}^g / \qquad /6.33/$$

where N is an arbitrary matrix of appropriate size.

Substitution of /6.33/ into $F_{11}$ yields

$$F_{11} = M_1 / \bar{A}_{12} - \bar{A}_{14} C_4 / + \bar{A}_{22} - \bar{A}_{24} C_4 =$$

$$= \bar{A}_{22} \bar{A}_{13}^g / \bar{A}_{14} C_4 - \bar{A}_{12} / + \bar{A}_{22} - \bar{A}_{24} C_4 + N / I_{1-l_2} - \bar{A}_{13} \bar{A}_{13}^g / / \bar{A}_{12} - \bar{A}_{14} C_4 / \qquad /6.34/$$

If /ii/ is satisfied then we can find a matrix N such that the eigenvalues of $F_{11}$ we inside the unit disk.

Also /iii/ implies the existence of a matrix $M_2$ such that $F_{22}$ is asymptotically stable. Therefore, F is asymptotically stable and $\lim_{i,j \to \infty} e/i,j/ = 0$. □

For the case of /6.29b/ we can prove in a similar way the following

## Theorem 6.2´

A minimal order observer for RM exists if

/i/   $\hat{A}_{32} \hat{A}_{42}^g \hat{A}_{42} = \hat{A}_{32}$

/ii/  the pair $/\hat{A}_{33} - \hat{A}_{31} \bar{C}_3 - \hat{A}_{32} \hat{A}_{42}^g / \hat{A}_{43} - \hat{A}_{41} \bar{C}_3 /, / I_{1-l_1} - \hat{A}_{42} \hat{A}_{42}^g / / \hat{A}_{43} - \hat{A}_{41} \bar{C}_3 //$
      is detectable,

/iii/ the pair $/\hat{A}_{12}, \hat{A}_{22}/$ is detectable,

where

$$Q^{-1} A Q = \begin{bmatrix} \hat{A}_{11} & \hat{A}_{12} & \hat{A}_{13} & \hat{A}_{14} \\ \hat{A}_{21} & \hat{A}_{22} & \hat{A}_{23} & \hat{A}_{24} \\ \hat{A}_{31} & \hat{A}_{32} & \hat{A}_{33} & \hat{A}_{34} \\ \hat{A}_{41} & \hat{A}_{42} & \hat{A}_{43} & \hat{A}_{44} \end{bmatrix} \begin{matrix} \} \, l_1 \\ \} \, n_1 - l_1 \\ \} \, n_2 - l + l_1 \\ \} \, l - l_1 \end{matrix}$$

From /6.31/ we have

$$\begin{bmatrix} F & HP^{-1} \\ Q^{-1}L & Q^{-1}KP^{-1} \end{bmatrix} \begin{bmatrix} \bar{T} \\ \bar{C} \end{bmatrix} = \begin{bmatrix} \bar{T}\bar{A} \\ I_n \end{bmatrix}$$

and

$$\begin{bmatrix} F & H \\ L & K \end{bmatrix} = \begin{bmatrix} I_{n-1} & 0 \\ 0 & Q \end{bmatrix} \begin{bmatrix} \bar{T}\bar{A} \\ I_n \end{bmatrix} \begin{bmatrix} \bar{T} \\ \bar{C} \end{bmatrix}^{-1} \begin{bmatrix} I_{n-1} & 0 \\ 0 & P \end{bmatrix} \qquad /6.35/$$

$$G = \bar{T}\bar{B} \qquad /6.36/$$

The minimal order of the observer is n-1.

If the conditions of theorem 6.2 are satisfied the following algorithm may be used for finding F, G, H, L and K of the minimal order observer for RM.

### Algorithm 6.2

**Step 1**  Choose nonsingular matrices P, $Q_1$, $Q_2$ and find $\bar{A}$, $\bar{B}$, $\bar{C}$ given by /6.30a/.

**Step 2**  Using the algorithm 5.2 find matrices $M_1$ and $M_2$ such that $F_{12} = 0$ and $F_{11}$, $F_{22}$ are asymptotically stable.

**Step 3**  Using /6.35/ and /6.36/ find F, H, L, K and G of the observer.

### Example 6.2

Given RM with [2]

$$A = \begin{bmatrix} A_{11} & A_{12} \\ A_{21} & A_{22} \end{bmatrix} = \begin{bmatrix} 0 & 0 & 1 & | & 1 & 1 & 0 \\ 0 & 1 & -1 & | & 0 & 0 & 1 \\ 0 & 1 & 1 & | & 0 & 0 & 0 \\ \hline 0 & 1 & 0 & | & 2 & 0 & -1 \\ -1 & 0 & 1 & | & 1 & 1 & 0 \\ 2 & -1 & 1 & | & 0 & 1 & 0 \end{bmatrix}, \quad B = \begin{bmatrix} B_1 \\ B_2 \end{bmatrix} = \begin{bmatrix} 1 \\ 0 \\ 1 \\ \hline 2 \\ 0 \\ -1 \end{bmatrix},$$

$$C = \begin{bmatrix} C_1 & C_2 \end{bmatrix} = \begin{bmatrix} 1 & 0 & 0 & | & 0 & 0 & 0 \\ 0 & 0 & 0 & | & 0 & 1 & 0 \\ 1 & 0 & 0 & | & 0 & 0 & 1 \\ 0 & 0 & 1 & | & 0 & 1 & 0 \end{bmatrix}$$

In this case $n_1 = n_2 = 3$, $l_1 = l_2 = 2$ and $l = 4$.

Step 1  We choose

$$P = \begin{bmatrix} 0 & -1 & 0 & 1 \\ 1 & 0 & 0 & 0 \\ 0 & 1 & 0 & 0 \\ 0 & 0 & 1 & 0 \end{bmatrix}, \quad Q_1 = \begin{bmatrix} 0 & 1 & 0 \\ 0 & 0 & 1 \\ 1 & 0 & 0 \end{bmatrix}, \quad Q_2 = \begin{bmatrix} 1 & 0 & 0 \\ 0 & 1 & 0 \\ 0 & 0 & 1 \end{bmatrix}$$

and we calculate

$$\bar{A} = Q^{-1}AQ = \left[\begin{array}{cc|c|cc|c} 1 & 0 & 1 & 0 & 0 & 0 \\ 1 & 0 & 0 & 1 & 1 & 0 \\ \hline -1 & 0 & 1 & 0 & 0 & 1 \\ \hline 0 & 0 & 1 & 1 & 2 & 0 & -1 \\ \hline 1 & -1 & 0 & 1 & 1 & 0 \\ 1 & 2 & -1 & 0 & 1 & 0 \end{array}\right], \quad \bar{B} = Q^{-1}B = \begin{bmatrix} 1 \\ 1 \\ \hline 0 \\ \hline 2 \\ \hline 0 \\ -1 \end{bmatrix},$$

$$C = PCQ = \left[\begin{array}{cc|cc|cc} 1 & 0 & 0 & 0 & 0 & 0 \\ 0 & 1 & 0 & 0 & 0 & 0 \\ \hline 0 & 0 & 0 & 0 & 1 & 0 \\ 0 & 1 & 0 & 0 & 0 & 1 \end{array}\right]$$

In this case

$$A_{13} = \begin{bmatrix} 0 \\ 1 \end{bmatrix}, \quad \bar{A}_{13}^g = \begin{bmatrix} 0 & 1 \end{bmatrix}, \quad \bar{A}_{23} = 0, \quad C_3 = \begin{bmatrix} 0 & 0 \\ 0 & 1 \end{bmatrix}, \quad C_4 = \begin{bmatrix} 0 \\ 0 \end{bmatrix}$$

$$\bar{A}_{23}\bar{A}_{13}^g / \bar{A}_{14}C_4 - \bar{A}_{12}/ + \bar{A}_{22} - \bar{A}_{24}C_4 = 1,$$

$$/I_{1-1_2} - \bar{A}_{13}\bar{A}_{13}^g / /\bar{A}_{12} - \bar{A}_{14}C_4/ = \begin{bmatrix} 1 \\ 0 \end{bmatrix} \quad \text{and } \bar{A}_{33} = 2, \quad \bar{A}_{43} = \begin{bmatrix} 1 \\ 0 \end{bmatrix}.$$

It is easy to check that the conditions /i/-/iii/ of theorem 6.2 are satisfied.

Step 2  We assume $F_{11} = F_{22} = 0.5$. Let $M_1 = \begin{bmatrix} m_{11} & m_{12} \end{bmatrix}$, $M_2 = \begin{bmatrix} m_{21} & m_{22} \end{bmatrix}$ and $N = \begin{bmatrix} n_{11} & n_{12} \end{bmatrix}$. Then from /6.34/ we have

$$F_{11} = 1 + \begin{bmatrix} n_{11} & n_{12} \end{bmatrix} \begin{bmatrix} 1 \\ 0 \end{bmatrix} = 1 + n_{11} = 0.5$$

and

$$F_{22} = 2 + \begin{bmatrix} m_{21} & m_{22} \end{bmatrix} \begin{bmatrix} 1 \\ 0 \end{bmatrix} = 2 + m_{21} = 0.5$$

Hence, assuming $n_{12} = m_{22} = 0$ we obtain

$$N = \begin{bmatrix} -0.5 & 0 \end{bmatrix}, \qquad M_2 = \begin{bmatrix} -1.5 & 0 \end{bmatrix}$$

and from /6.33/

$$M_1 = -\bar{A}_{23}\bar{A}^g_{13} + N/I_{1-1_2} - \bar{A}_{13}\bar{A}^g_{13}/ = \begin{bmatrix} -0.5 & 0 \end{bmatrix}$$

<u>Step 3</u>  Taking into account that

$$\bar{T} = \begin{bmatrix} M_1 & I_{n_1-1+l_2} & 0 & 0 \\ 0 & 0 & I_{n_2-l_2} & M_2 \end{bmatrix} = \begin{bmatrix} -0.5 & 0 & 1 & 0 & 0 & 0 \\ 0 & 0 & 0 & 1 & -1.5 & 0 \end{bmatrix}$$

$$\begin{bmatrix} \bar{T} \\ \bar{C} \end{bmatrix} = \begin{bmatrix} -0.5 & 0 & 1 & 0 & 0 & 0 \\ 0 & 0 & 0 & 1 & -1.5 & 0 \\ 1 & 0 & 0 & 0 & 0 & 0 \\ 0 & 1 & 0 & 0 & 0 & 0 \\ 0 & 0 & 0 & 0 & 1 & 0 \\ 0 & 1 & 0 & 0 & 0 & 1 \end{bmatrix}$$

and

$$\begin{bmatrix} \bar{T} \\ \bar{C} \end{bmatrix}^{-1} = \begin{bmatrix} 0 & 0 & 1 & 0 & 0 & 0 \\ 0 & 0 & 0 & 1 & 0 & 0 \\ 1 & 0 & 0.5 & 0 & 0 & 0 \\ 0 & 1 & 0 & 0 & 1.5 & 0 \\ 0 & 0 & 0 & 0 & 1 & 0 \\ 0 & 0 & 0 & -1 & 0 & 1 \end{bmatrix}$$

from /6.35/ and /6.36/ we obtain

$$\begin{bmatrix} F & H \\ L & K \end{bmatrix} = \begin{bmatrix} I_{n-1} & 0 \\ 0 & Q \end{bmatrix} \begin{bmatrix} \bar{T}\bar{A} \\ I_n \end{bmatrix} \begin{bmatrix} \bar{T} \\ \bar{C} \end{bmatrix}^{-1} \begin{bmatrix} I_{n-1} & 0 \\ 0 & P \end{bmatrix} =$$

$$= \begin{bmatrix} 0.5 & 0 & | & -1 & 1.25 & 1 & -1.25 \\ 1 & 0.5 & | & 2.5 & 0.25 & -1 & -1 \\ \hline 0 & 0 & | & 0 & 1 & 0 & 0 \\ 1 & 0 & | & 0.5 & 0 & 0 & 0 \\ 0 & 0 & | & 1 & 0 & 0 & 0 \\ 0 & 1 & | & 0 & 0 & 1.5 & 0 \\ 0 & 0 & | & 0 & 0 & 1 & 0 \\ 0 & 0 & | & 0 & -1 & 0 & 1 \end{bmatrix}$$

and
$$G = \bar{T}\bar{B} = \begin{bmatrix} -0.5 \\ 2 \end{bmatrix}$$

## Theorem 6.3

If a full order observer can be designed for RM then a minimal order observer can be designed too.

## Proof

It can be shown /see Problem 6.1/ that a full order observer for RM exists if:

/i/ $A_{12} C_2^g C_2 = A_{12}$ /6.37a/

/ii/ the pair $//I-C_2 C_2^g/C_1, A_{11}-A_{12} C_2^g C_1/$ is detectable /6.37b/

/iii/ the pair $/C_2, A_{22}/$ is detectable. /6.37c/

We shall show that the conditions /6.37/ imply the conditions /i/-/iii/ of theorem 6.2.

From /6.30a/ we have

$$A_{12} = Q_1 \begin{bmatrix} \bar{A}_{13} & \bar{A}_{14} \\ \bar{A}_{23} & \bar{A}_{24} \end{bmatrix} Q_2^{-1} \;,\quad A_{22} = Q_2 \begin{bmatrix} \bar{A}_{33} & \bar{A}_{34} \\ \bar{A}_{43} & \bar{A}_{44} \end{bmatrix} Q_2^{-1} \qquad /6.38a/$$

and

$$C_1 = P^{-1} \begin{bmatrix} I_{1-l_2} & 0 \\ C_3 & C_4 \end{bmatrix} Q_1^{-1} \;,\quad C_2 = P^{-1} \begin{bmatrix} 0 & 0 \\ 0 & I_{l_2} \end{bmatrix} Q_2^{-1} \qquad /6.38b/$$

The generalized inverse matrix of $C_2$ is

$$C_2^g = Q_2 \begin{bmatrix} 0 & 0 \\ 0 & I_{l_2} \end{bmatrix} P \qquad /6.39/$$

Substitution of /6.38/ and /6.39/ into /6.37a/ yields

$$A_{12} C_2^g C_2 = Q_1 \begin{bmatrix} 0 & \bar{A}_{14} \\ 0 & \bar{A}_{24} \end{bmatrix} Q_2^{-1} = Q_1 \begin{bmatrix} \bar{A}_{13} & \bar{A}_{14} \\ \bar{A}_{23} & \bar{A}_{24} \end{bmatrix} Q_2^{-1} \qquad /6.40/$$

From /6.40/ it follows that $\bar{A}_{13} = 0$, $\bar{A}_{23} = 0$, which implies that /i/ of theorem 6.2 holds.

Taking into account that

$$/I_1-C_2C_2^g/C_1 = P^{-1}\begin{bmatrix} I_{1-l_2} & 0 \\ 0 & 0 \end{bmatrix} Q_1^{-1}$$

$$A_{11}-A_{12}C_2^g C_1 = Q_1 \begin{bmatrix} \bar{A}_{11}-\bar{A}_{14}C_3, & \bar{A}_{12}-\bar{A}_{14}C_4 \\ \bar{A}_{21}-\bar{A}_{24}C_3, & \bar{A}_{22}-\bar{A}_{24}C_4 \end{bmatrix} Q_1^{-1}$$

we obtain

$$\bar{A}_{23}\bar{A}_{13}^g/\bar{A}_{14}C_4-\bar{A}_{12}/ + \bar{A}_{22}-\bar{A}_{24}C_4 = \bar{A}_{22}-\bar{A}_{24}C_4$$

$$/I_{1-l_2}-\bar{A}_{13}\bar{A}_{13}^g//\bar{A}_{12}-\bar{A}_{14}C_4/ = \bar{A}_{12}-\bar{A}_{14}C_4 \qquad /6.41/$$

We shall show by contradiction that the pair /6.41/ is detectable. Suppose that /6.41/ is not detectable. Then

$$\operatorname{rank} \begin{bmatrix} \lambda I-\bar{A}_{22}+\bar{A}_{24}C_4 \\ -\bar{A}_{12}+\bar{A}_{14}C_4 \end{bmatrix} < n_1-l+l_2 \qquad /6.42/$$

holds for any $\lambda \in C^+$, where $C^+$ is the exterior of the unit circle of the complex plane.

Using /6.42/ we obtain

$$\operatorname{rank} \begin{bmatrix} \lambda I-\bar{A}_{11}+\bar{A}_{14}C_3 & -\bar{A}_{12}+\bar{A}_{14}C_4 \\ -\bar{A}_{21}+\bar{A}_{24}C_3 & \lambda I-\bar{A}_{22}+\bar{A}_{24}C_4 \\ I_{1-l_2} & 0 \\ 0 & 0 \end{bmatrix} < n_1 \qquad /6.43/$$

which implies that the pair $/P/I_1-C_2C_2^g/C_1Q_1, Q_1^{-1}/A_{11}-A_{12}C_2^g C_1/Q_1/$ is not detectable. But this contradicts with /6.37b/.

Similarly it can be shown that /6.37c/ implies /iii/ of theorem 6.2. By theorem 6.2 the proof is thus completed. □

## 6.2. EXACT MODEL MATCHING VIA STATIC STATE FEEDBACK

### 6.2.1. Problem formulation

Consider RM described by the equations /6.1/ and /6.2/. The transfer function matrix of RM is given by

$$G/z_1,z_2/ = C\left[Z-A\right]^{-1}B = \frac{N/z_1,z_2/}{d/z_1,z_2/} \qquad /6.44/$$

where

$$N/z_1,z_2/ = C\,\text{adj}\left[Z-A\right]B = \sum_{i=0}^{n_1}\sum_{j=0}^{n_2} N_{ij}z_1^i z_2^j \;,\quad N_{ij}\in R^{1\times m} \qquad /6.45a/$$

$$d/z_1,z_2/ = \det\left[Z-A\right] = \sum_{i=0}^{n_1}\sum_{j=0}^{n_2} d_{ij}z_1^i z_2^j \qquad /6.45b/$$

and

$$Z = \begin{bmatrix} I_{n_1}z_1 & 0 \\ 0 & I_{n_2}z_2 \end{bmatrix} \qquad /6.45c/$$

The control law applied to RM /Fig.6.1/ is of the form

$$u = Kx + Hv \qquad /6.46/$$

where $K\in R^{m\times n}$, $H\in R^{m\times m}$ is a nonsingular matrix and $v = v/i,j/\in R^m$ is a new command input vector.

The closed-loop transfer function matrix is

$$G_c/z_1,z_2/ = G/z_1,z_2/\left[I_m - K\bar{G}/z_1,z_2/\right]^{-1}H \qquad /6.47/$$

where

$$\bar{G}/z_1,z_2/ = \left[Z-A\right]^{-1}B = \frac{\bar{N}/z_1,z_2/}{d/z_1,z_2/} \qquad /6.48/$$

$$\bar{N}/z_1,z_2/ = \text{adj}\left[Z-A\right]B = \sum_{i=0}^{n_1}\sum_{j=0}^{n_2} \bar{N}_{ij}z_1^i z_2^j \;,\quad \bar{N}_{ij}\in R^{n\times m} \qquad /6.49/$$

The exact model-matching problem may be stated as follows [5,6]. Given RM with /6.44/ and the model transfer function matrix

$$G_m/z_1,z_2/ = \frac{N_m/z_1,z_2/}{d_m/z_1,z_2/} \in R^{1\times m}/z_1,z_2/ \qquad /6.50/$$

find the matrices K and H such that

$$G_c/z_1,z_2/ = G_m/z_1,z_2/ \qquad /6.51/$$

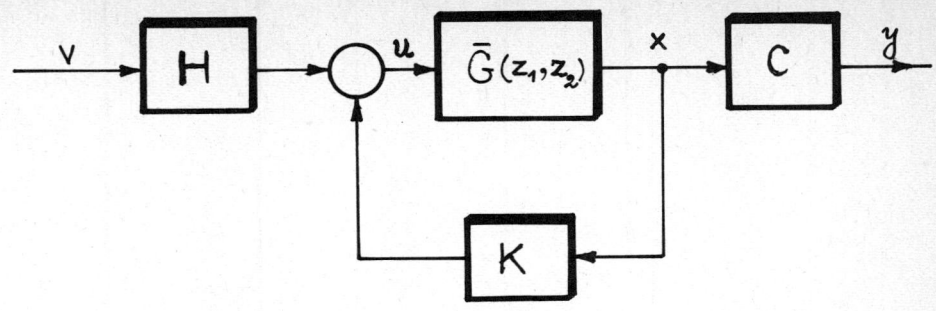

Fig. 6.1.

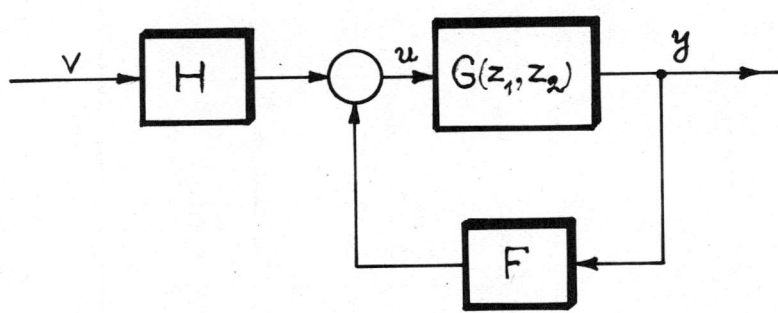

Fig. 6.2.

## 6.2.2. Problem solution

Postmultiplying /6.47/ by $H^{-1}\left[I_m - K\bar{G}/z_1,z_2/\right]$ and using /6.51/ we obtain

$$G_m/z_1,z_2/\bar{H} + G_m/z_1,z_2/\bar{K}\,\bar{G}/z_1,z_2/ = G/z_1,z_2/ \qquad /6.52/$$

where

$$\bar{H} = H^{-1} \qquad /6.53a/$$

$$\bar{K} = -H^{-1}K \qquad /6.53b/$$

Substitution of /6.44/, /6.48/ and /6.50/ into /6.52/ yields

$$d/z_1,z_2/N_m/z_1,z_2/\bar{H} + N_m/z_1,z_2/\bar{K}\,\bar{N}/z_1,z_2/ = d_m/z_1,z_2/N/z_1,z_2/ \qquad /6.54/$$

It is easy to show that using the Kronecker product $A \otimes B$ of matrices
$A = \left[a_{ij}\right]_{\substack{i=1,\ldots,m \\ j=1,\ldots,n}}$, $B$ defined by

$$A \otimes B = \begin{bmatrix} a_{11}B & \cdots & a_{1n}B \\ \cdots & \cdots & \cdots \\ a_{m1}B & \cdots & a_{mn}B \end{bmatrix}$$

we may write /6.52/ in the form

$$M/z_1,z_2/\bar{h} + L/z_1,z_2/\bar{k} = p/z_1,z_2/ \qquad /6.55/$$

where

$$M/z_1,z_2/ = d/z_1,z_2/N_m/z_1,z_2/ \otimes I_m = \sum_{i=0}^{\bar{n}_1} \sum_{j=0}^{\bar{n}_2} M_{ij} z_1^i z_2^j \qquad /6.56a/$$

$$L/z_1,z_2/ = N_m/z_1,z_2/ \otimes \bar{N}^T/z_1,z_2/ = \sum_{i=0}^{\bar{n}_1} \sum_{j=0}^{\bar{n}_2} L_{ij} z_1^i z_2^j \qquad /6.56b/$$

$$P/z_1,z_2/ = d_m/z_1,z_2/\,N/z_1,z_2/ = \sum_{i=0}^{\bar{n}_1} \sum_{j=0}^{\bar{n}_2} P_{ij} z_1^i z_2^j$$

$M_{ij} \in R^{lmxm^2}$, $L_{ij} \in R^{lmxnm}$, $P_{ij} \in R^{lxm}$

$\bar{n}_i = n_i + \hat{n}_i$, $\hat{n}_i = \deg_{z_i} d_m/z_1,z_2/$ for $i = 1,2$

$\bar{h} = \left[\bar{H}_1\ \bar{H}_2\ \cdots\ \bar{H}_m\right]^T \in R^{m^2}$, $\bar{k} = \left[\bar{K}_1\ \bar{K}_2\ \cdots\ \bar{K}_m\right] \in R^{nm}$

$$p/z_1,z_2/ = \left[P_1/z_1,z_2/, P_2/z_1,z_2/, \ldots, P_l/z_1,z_2/\right]^T =$$

$$= \sum_{i=0}^{\bar{n}_1} \sum_{j=0}^{\bar{n}_2} p_{ij} z_1^i z_2^j \qquad /6.56c/$$

$\bar{H}_i/\bar{K}_i, P_i/z_1,z_2//$ is the ith row of $H/\bar{K}, P/z_1,z_2//$

Substitution of /6.56/ into /6.55/ yields

$$\sum_{i=0}^{\bar{n}_1} \sum_{j=0}^{\bar{n}_2} M_{ij} \bar{h} z_1^i z_2^j + \sum_{i=0}^{\bar{n}_1} \sum_{j=0}^{\bar{n}_2} L_{ij} \bar{k} z_1^i z_2^j = \sum_{i=0}^{\bar{n}_1} \sum_{j=0}^{\bar{n}_2} p_{ij} z_1^i z_2^j \qquad /6.57/$$

Equating coefficients of like $z_1^i z_2^j$ terms of both sides of /6.57/ we obtain

$$M_{ij} \bar{h} + L_{ij} \bar{k} = p_{ij} \qquad /6.58/$$

for $i = 0,1,\ldots,\bar{n}_1$; $j = 0,1,\ldots,\bar{n}_2$.

The equations /6.58/ may be written in the form

$$Qx = p \qquad /6.59/$$

where

$$Q = \begin{bmatrix} M_{00} & L_{00} \\ M_{10} & L_{10} \\ M_{01} & L_{01} \\ M_{11} & L_{11} \\ \cdot & \cdot \\ M_{\bar{n}_1\bar{n}_2} & L_{\bar{n}_1\bar{n}_2} \end{bmatrix} \in R^{lm\bar{n} \times m/n+m/}, \quad x = \begin{bmatrix} \bar{h} \\ \bar{k} \end{bmatrix} \in R^{m/n+m/}$$

$$p = \begin{bmatrix} p_{00} \\ p_{10} \\ p_{01} \\ p_{11} \\ \cdot \cdot \cdot \\ p_{\bar{n}_1\bar{n}_2} \end{bmatrix} \in R^{lm\bar{n}}, \quad \bar{n} = /\bar{n}_1+1//\bar{n}_2+1/$$

The equation /6.59/ is the linear system of lnm equations with /n+m/ unknowns.

Let rank $Q = r$ and $U$ be a $l\bar{n}m \times lm\bar{n}$ invertible matrix of elementary row operations such that

$$UQ = \begin{bmatrix} Q_1 \\ 0 \end{bmatrix} \qquad /6.60/$$

where $Q_1$ is a $rxm/n+m/$ matrix involving $r$ linearly independent rows of $Q$. Premultiplying /6.60/ by $U$ and using /6.61/ we obtain

$$\begin{bmatrix} Q_1 \\ 0 \end{bmatrix} x = \begin{bmatrix} P_1 \\ P_2 \end{bmatrix}$$

and

$$Q_1 = P_1 \qquad /6.61/$$

$$P_2 = 0 \qquad /6.62/$$

where

$$\begin{bmatrix} P_1 \\ P_2 \end{bmatrix} = Up, \qquad P_1 \in R^r, \qquad P_2 \in R^{lmn-r}$$

We have thus established the following [5, 6]

Theorem 6.4

The exact model-matching problem has a solution if the condition /6.62/ is satisfied.

If /6.62/ holds, then from /6.61/ we obtain

$$x = Q_1^T \left[ Q_1 Q_1^T \right]^{-1} P_1 \qquad /6.63/$$

The solution /6.63/ yields the matrices $\bar{H}$, $\bar{K}$ and the desired matrices $H$ and $K$ can be determined from /6.53/, i.e.

$$H = \bar{H}^{-1} \qquad /6.64a/$$

and

$$K = -\bar{H}^{-1} \bar{K} \qquad /6.64b/$$

If /6.46/ is satisfied the following algorithm may be used for finding the matrices $H$ and $K$.

## Algorithm 6.3

**Step 1** Using /6.44/ and /6.48/ find $G/z_1,z_2/$ and $\bar{G}/z_1,z_2/$ for given A, B, C.

**Step 2** Find $M/z_1,z_2/$, $L/z_1,z_2/$, $p/z_1,z_2/$ defined by /6.56/ and $M_{ij}$, $L_{ij}$, $p_{ij}$.

**Step 3** Find Q, p and r = rank Q.

**Step 4** Find an elementary row operations matrix U and $Q_1$, $P_1$.

**Step 5** Using /6.63/ find x and $\bar{H}$, $\bar{K}$.

**Step 6** Find H and K from /6.64/.

## Example 6.3

Given

$$A = \begin{bmatrix} A_{11} & A_{12} \\ A_{21} & A_{22} \end{bmatrix} = \left[\begin{array}{cc|c} 0 & 1 & 1 \\ -1 & 0 & 1 \\ \hline 1 & 0 & -1 \end{array}\right], \quad B = \begin{bmatrix} B_1 \\ B_2 \end{bmatrix} = \left[\begin{array}{cc} 1 & 0 \\ 0 & 1 \\ \hline 1 & 0 \end{array}\right],$$

$$C = \begin{bmatrix} C_1 & C_2 \end{bmatrix} = \begin{bmatrix} 1 & 0 & | & 1 \end{bmatrix}$$

/6.65/

and

$$G_m/z_1,z_2/ = \frac{1}{z_1^2 z_2 + 2z_1^2 + 2z_1 - 4} \begin{bmatrix} z_1 z_2 + z_1^2 + 3z_1 + 2, & z_2 + 4 \end{bmatrix}$$

find

$$H = \begin{bmatrix} h_1 & h_{12} \\ h_{21} & h_{22} \end{bmatrix}, \quad K = \begin{bmatrix} k_{11} & k_{12} & k_{13} \\ k_{21} & k_{22} & k_{23} \end{bmatrix}$$

such that /6.51/ holds.

**Step 1**

$$G/z_1,z_2/ = \begin{bmatrix} Z-A \end{bmatrix}^{-1} B = \begin{bmatrix} z_1 & -1 & 0 \\ 1 & z_1 & -1 \\ -1 & 0 & z_2+1 \end{bmatrix}^{-1} \begin{bmatrix} 1 & 0 \\ 0 & 1 \\ 1 & 0 \end{bmatrix} =$$

$$= \frac{1}{z_1^2 z_2 + z_1^2 + z_2} \begin{bmatrix} z_1 z_2 + z_1 + 1, & z_2 + 1 \\ z_1, & -z_2, & z_1 z_2 + z_1 \\ z_1^2 + z_1 + 1, & 1 \end{bmatrix}$$

/6.66/

and

$$G/z_1,z_2/ = C[Z-A]^{-1}B = \begin{bmatrix} 1 & 0 & 1 \end{bmatrix} \frac{1}{z_1^2 z_2 + z_1^2 + z_2} \begin{bmatrix} z_1 z_2 + z_1 + 1, & z_2 + 1 \\ z_1 - z_2, & z_1 z_2 + z_1 \\ z_1^2 + z_1 + 1, & 1 \end{bmatrix} =$$

$$= \frac{1}{z_1^2 z_2 + z_1^2 + z_2} \begin{bmatrix} z_1 z_2 + z_1^2 + 2z_1 + 2, & z_2 + 2 \end{bmatrix} \qquad /6.67/$$

Step 2

$$M/z_1,z_2/ = d/z_1,z_2/\; N_m/z_1,z_2/ \otimes I_m =$$
$$= /z_1^2 z_2 + z_1^2 + z_2/ \begin{bmatrix} z_1 z_2 + z_1^2 + 3z_1 + 2, & 0 & , & z_2 + 4, & 0 \\ 0 & , & z_1 z_2 + z_1^2 + 3z_1 + 2, & 0 & , & z_2 + 4 \end{bmatrix} \qquad /6.68/$$

$$L/z_1,z_2/ = N_m/z_1,z_2/ \otimes \bar{N}^{-T}/z_1,z_2/ = \begin{bmatrix} l_{11} & l_{12} & l_{13} & l_{14} & l_{15} & l_{16} \\ l_{21} & l_{22} & l_{23} & l_{24} & l_{25} & l_{26} \end{bmatrix} \qquad /6.69/$$

$l_{11} = /z_1 z_2 + z_1^2 + 3z_1 + 2//z_1 z_2 + z_1 + 1/, \quad l_{12} = /z_1 z_2 + z_1^2 + 3z_1 + 2//z_1 - z_2/,$

$l_{13} = /z_1 z_2 + z_1^2 + 3z_1 + 2//z_1^2 + z_1 + 1/, \quad l_{14} = /z_2 + 4//z_1 z_2 + z_1 + 1/,$

$l_{15} = /z_2 + 4//z_1 - z_2/, \quad l_{16} = /z_2 + 4//z_1^2 + z_1 + 1/,$

$l_{21} = /z_1 z_2 + z_1^2 + 3z_1 + 2//z_2 + 1/, \quad l_{22} = /z_1 z_2 + z_1^2 + 3z_1 + 2//z_1 z_2 + z_1/,$

$l_{23} = z_1 z_2 + z_1^2 + 3z_1 + 2, \quad l_{24} = /z_2 + 4//z_2 + 1/, \quad l_{25} = /z_2 + 4//z_1 z_2 + z_1/,$

$l_{26} = z_2 + 4$

$$P/z_1,z_2/ = p^T/z_1,z_2/ = d_m/z_1,z_2/\; N/z_1,z_2/ =$$
$$= /z_1^2 z_2 + 2z_1^2 + 2z_2 - 4/ \begin{bmatrix} z_1 z_2 + z_1^2 + 2z_1 + 2, & z_2 + 2 \end{bmatrix} \qquad /6.70/$$

and

$M_{00} = 0$, $\quad M_{12} = \begin{bmatrix} 1 & 0 & 0 & 0 \\ 0 & 1 & 0 & 0 \end{bmatrix}$, $\quad M_{20} = \begin{bmatrix} 2 & 0 & 4 & 0 \\ 0 & 2 & 0 & 4 \end{bmatrix}$,

$M_{01} = \begin{bmatrix} 2 & 0 & 4 & 0 \\ 0 & 2 & 0 & 4 \end{bmatrix}$, $\quad M_{21} = \begin{bmatrix} 3 & 0 & 5 & 0 \\ 0 & 3 & 0 & 5 \end{bmatrix}$, $\quad M_{22} = \begin{bmatrix} 0 & 0 & 1 & 0 \\ 0 & 0 & 0 & 1 \end{bmatrix}$,

$M_{10} = 0$, $\quad M_{30} = \begin{bmatrix} 3 & 0 & 0 & 0 \\ 0 & 3 & 0 & 0 \end{bmatrix}$, $\quad M_{31} = \begin{bmatrix} 4 & 0 & 0 & 0 \\ 0 & 4 & 0 & 0 \end{bmatrix}$,

$$M_{11} = \begin{bmatrix} 3 & 0 & 0 & 0 \\ 0 & 3 & 0 & 0 \end{bmatrix}, \quad M_{32} = \begin{bmatrix} 1 & 0 & 0 & 0 \\ 0 & 1 & 0 & 0 \end{bmatrix}, \quad M_{40} = \begin{bmatrix} 1 & 0 & 0 & 0 \\ 0 & 1 & 0 & 0 \end{bmatrix},$$

$$M_{02} = \begin{bmatrix} 0 & 0 & 1 & 0 \\ 0 & 0 & 0 & 1 \end{bmatrix}, \quad M_{41} = \begin{bmatrix} 1 & 0 & 0 & 0 \\ 0 & 1 & 0 & 0 \end{bmatrix}, \quad M_{42} = 0$$

$$L_{00} = \begin{bmatrix} 2 & 0 & 2 & 4 & 0 & 4 \\ 2 & 0 & 2 & 4 & 0 & 4 \end{bmatrix}, \quad L_{30} = \begin{bmatrix} 1 & 1 & 4 & 0 & 0 & 0 \\ 0 & 1 & 0 & 0 & 0 & 0 \end{bmatrix}, \quad P_{20} = \begin{bmatrix} 4 & 4 \end{bmatrix}$$

$$L_{01} = \begin{bmatrix} 0 & -2 & 0 & 1 & -4 & 1 \\ 2 & 0 & 0 & 4 & 0 & 1 \end{bmatrix}, \quad L_{31} = \begin{bmatrix} 1 & 0 & 1 & 0 & 0 & 0 \\ 0 & 1 & 0 & 0 & 0 & 0 \end{bmatrix}, \quad P_{21} = \begin{bmatrix} 4 & 4 \end{bmatrix}$$

$$L_{10} = \begin{bmatrix} 5 & 2 & 5 & 4 & 4 & 4 \\ 3 & 2 & 3 & 0 & 4 & 0 \end{bmatrix}, \quad L_{32} = 0, \quad P_{22} = \begin{bmatrix} 0 & 1 \end{bmatrix}$$

$$L_{11} = \begin{bmatrix} 3 & -3 & 1 & 5 & 1 & 1 \\ 4 & 2 & 1 & 0 & 5 & 0 \end{bmatrix}, \quad L_{40} = \begin{bmatrix} 0 & 0 & 1 & 0 & 0 & 0 \\ 0 & 0 & 0 & 0 & 0 & 0 \end{bmatrix}, \quad P_{30} = \begin{bmatrix} 6 & 6 \end{bmatrix}$$

$$L_{02} = \begin{bmatrix} 0 & 0 & 0 & 0 & -1 & 0 \\ 0 & 0 & 0 & 1 & 0 & 0 \end{bmatrix}, \quad L_{41} = 0, \quad P_{31} = \begin{bmatrix} 4 & 0 \end{bmatrix}$$

$$L_{12} = \begin{bmatrix} 0 & -1 & 0 & 1 & 0 & 0 \\ 1 & 0 & 0 & 0 & 1 & 0 \end{bmatrix}, \quad L_{42} = 0, \quad P_{32} = \begin{bmatrix} 1 & 0 \end{bmatrix}$$

$$L_{20} = \begin{bmatrix} 4 & 3 & 6 & 0 & 0 & 4 \\ 1 & 3 & 1 & 0 & 0 & 0 \end{bmatrix}, \quad P_{00} = \begin{bmatrix} -8 & -8 \end{bmatrix}, \quad P_{40} = \begin{bmatrix} 2 & 0 \end{bmatrix}$$

$$L_{21} = \begin{bmatrix} 4 & 0 & 1 & 0 & 0 & 1 \\ 1 & 4 & 0 & 0 & 0 & 0 \end{bmatrix}, \quad P_{10} = \begin{bmatrix} -4 & 4 \end{bmatrix}, \quad P_{41} = \begin{bmatrix} 1 & 0 \end{bmatrix}$$

$$P_{01} = \begin{bmatrix} 0 & -4 \end{bmatrix}$$

$$L_{22} = \begin{bmatrix} 1 & 0 & 0 & 0 & 0 & 0 \\ 0 & 1 & 0 & 0 & 0 & 0 \end{bmatrix}, \quad P_{11} = \begin{bmatrix} -4 & 2 \end{bmatrix}, \quad P_{42} = 0$$

$$P_{02} = 0, \quad P_{12} = 0,$$

Step 3

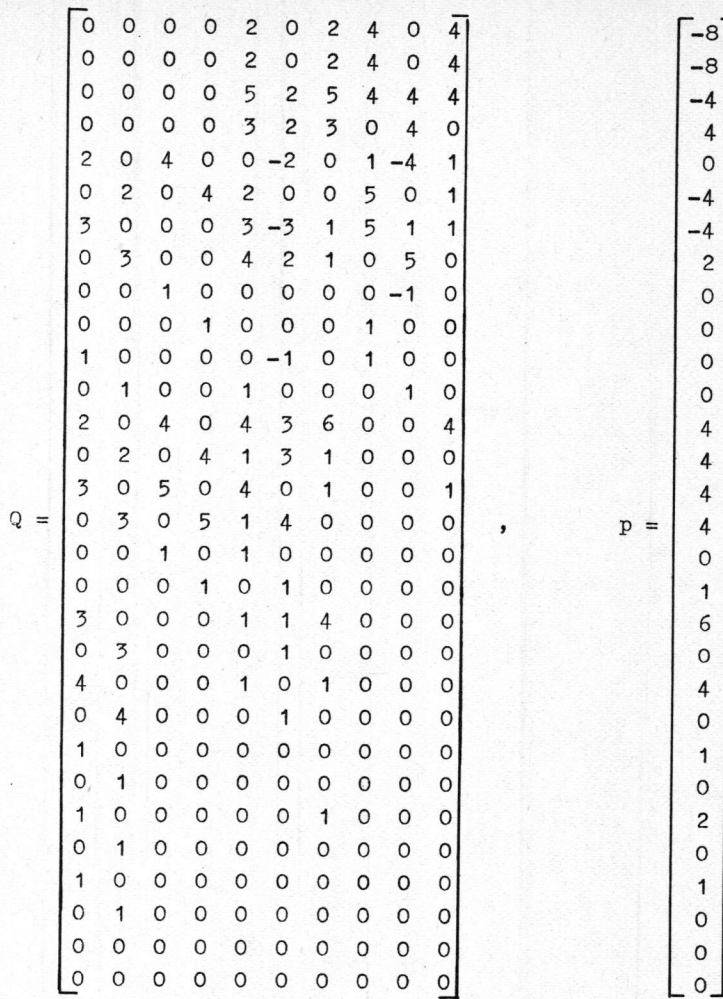

r = rank Q = 10.

Step 4  Carrying out elementary pow operations we can check that the condition /6.62/ is satisfied and

$$\begin{bmatrix} 0 & 0 & 0 & 0 & 0 & 0 & 0 & 1 & 0 & 0 \\ 0 & 0 & 0 & 0 & 0 & 0 & 0 & 0 & 1 & 0 \\ 0 & 0 & 0 & 0 & 0 & 0 & 0 & 0 & 0 & 1 \\ 0 & 0 & 1 & 0 & 0 & 0 & 0 & 0 & 0 & 0 \\ 0 & 0 & 0 & 1 & 0 & 0 & 0 & 0 & 0 & 0 \\ 0 & 0 & 0 & 0 & 1 & 0 & 0 & 0 & 0 & 0 \\ 0 & 0 & 0 & 0 & 0 & 1 & 0 & 0 & 0 & 0 \\ 0 & 0 & 0 & 0 & 0 & 0 & 1 & 0 & 0 & 0 \\ 0 & 1 & 0 & 0 & 0 & 0 & 0 & 0 & 0 & 0 \\ 1 & 0 & 0 & 0 & 0 & 0 & 0 & 0 & 0 & 0 \end{bmatrix} \begin{bmatrix} \bar{h}_{11} \\ \bar{h}_{12} \\ \bar{h}_{21} \\ \bar{h}_{22} \\ \bar{k}_{11} \\ \bar{k}_{12} \\ \bar{k}_{13} \\ \bar{k}_{21} \\ \bar{k}_{22} \\ \bar{k}_{23} \end{bmatrix} = \begin{bmatrix} -1 \\ 1 \\ -1 \\ 1 \\ 1 \\ -1 \\ 0 \\ 1 \\ 0 \\ 1 \end{bmatrix} \quad /6.71/$$

**Step 5** Solving /6.71/ we obtain

$$\bar{H} = \begin{bmatrix} \bar{h}_{11} & \bar{h}_{12} \\ \bar{h}_{21} & \bar{h}_{22} \end{bmatrix} = \begin{bmatrix} 1 & 0 \\ 1 & 1 \end{bmatrix}, \quad \bar{K} = \begin{bmatrix} \bar{k}_{11} & \bar{k}_{12} & \bar{k}_{13} \\ \bar{k}_{21} & \bar{k}_{22} & \bar{k}_{23} \end{bmatrix} = \begin{bmatrix} -1 & 0 & 1 \\ -1 & 1 & -1 \end{bmatrix}$$

**Step 6** The desired matrices are

$$H = \bar{H}^{-1} = \begin{bmatrix} 1 & 0 \\ -1 & 1 \end{bmatrix}$$

$$K = -\bar{H}^{-1}\bar{K} = \begin{bmatrix} 1 & 0 & -1 \\ 0 & -1 & 2 \end{bmatrix}$$

## 6.3. Exact model matching via static output feedback

Consider RM with transfer function matrix /6.44/. The control law applied to RM /Fig. 6.2/ is of the form

$$u = Fy + Hv \quad /6.72/$$

where $F \in R^{m \times l}$, $H \in R^{m \times m}$ is a non-singular matrix and $v = v/i,j/ \in R^m$ is a new command input vector.

The closed-loop transfer function matrix is

$$G_c/z_1,z_2/ = G/z_1,z_2/\left[I_m - FG/z_1,z_2/\right]^{-1} H \quad /6.73/$$

where

$$G/z_1,z_2/ = C\left[Z-A\right]^{-1} B \quad /6.74/$$

The exact model-matching problem may be stated as follows [5, 6].
Given RM with /6.74/ and the model transfer function matrix /6.50/ find

the matrices F and H such that /6.51/ holds.
To solve the problem we postmultiply /6.73/ by $H^{-1}\left[I_m - FG/z_1, z_2/\right]$ and we use /6.51/

$$G_m/z_1,z_2/ \: \bar{H} + G_m/z_1,z_2/ \: \bar{F}G/z_1,z_2/ = G/z_1,z_2/ \qquad /6.75/$$

where

$$\bar{H} = H^{-1} \quad \text{and} \quad \bar{F} = -H^{-1}F \qquad /6.76/$$

Substitution of /6.44/ and /6.50/ into /6.75/ yields

$$d/z_1,z_2/N_m/z_1,z_2/\bar{H} + N_m/z_1,z_2/\bar{F}N/z_1,z_2/ = d/z_1,z_2/N/z_1,z_2/ \qquad /6.77/$$

The equation /6.77/ has the same form as /6.54/. Therefore the matrices $\bar{H}$, $\bar{F}$ and H, F can be found in an analogous manner as H and K in state feedback case.

From comparison of /6.47/ and /6.73/ it follows that

$$FC = K \qquad /6.78/$$

An another approach to the problem is to find K by the use of the algorithm 6.4 and next solve /6.78/ for F.

## 6.4. Exact model-matching via dynamic output feedback

Consider RM with transfer function matrix /6.44/. The control law applied to RM is of the form

$$U/z_1,z_2/ = F/z_1,z_2/ \: Y/z_1,z_2/ + H \: V/z_1,z_2/ \qquad /6.79/$$

where

$$F/z_1,z_2/ = \frac{F_1}{z_1} + \frac{F_2}{z_2} + F_3 + z_1 F_4 + z_2 F_5 \qquad /6.80/$$

$F_i \in R^{m \times l}$, $i = 1, \ldots, 5$, $H \in R^{m \times m}$ is a non-singular matrix, $U/z_1,z_2/$, $Y/z_1,z_2/$ and $V/z_1,z_2/$ are 2-D Z-transforms of $u/i,j/$, $y/i,j/$ and $v/i,j/$ respectively.

The closed-loop transfer function matrix is

$$G_c/z_1,z_2/ = \left[I_1 - G/z_1,z_2/F/z_1,z_2/\right]^{-1} G/z_1,z_2/H \qquad /6.81/$$

where $G/z_1,z_2/$ is given by /6.74/.

The exact model-matching problem via dynamic output feedback may be stated as follows [7].
Given RM with /6.44/ and the model transfer function matrix /6.50/, find the matrices $F_1$, $F_2$, $F_3$, $F_4$, $F_5$ and H such that /6.51/ holds.
Premultiplying /6.81/ by $I_1 - G/z_1,z_2/F/z_1,z_2/$ and using /6.51/ we obtain

$$G/z_1,z_2/H + G/z_1,z_2/F/z_1,z_2/G_m/z_1,z_2/ = G_m/z_1,z_2/ \qquad /6.82/$$

Substitution of /6.44/, /6.50/ and /6.80/ into /6.82/ yields

$$z_1 z_2 d_m/z_1,z_2/N/z_1,z_2/H + N/z_1,z_2/\left[z_2 F_1 + z_1 F_2 + z_1 z_2 F_3 + z_1^2 z_2 F_4 + z_1 z_2^2 F_5\right] N_m/z_1,z_2/ = z_1 z_2 d/z_1,z_2/N_m/z_1,z_2/ \qquad /6.83/$$

Using the Kronecker product we may write /6.88/ in the form

$$\bar{M}/z_1,z_2/h + L_1/z_1,z_2/f_1 + L_2/z_1,z_2/f_2 + L_3/z_1,z_2/f_3 + L_4/z_1,z_2/f_4 + L_5/z_1,z_2/f_5 = \bar{p}/z_1,z_2/ \qquad /6.84/$$

where

$$\bar{M}/z_1,z_2/ = z_1 z_2 d_m/z_1,z_2/N/z_1,z_2/\otimes I_m = \sum_{i=0}^{\bar{n}_1+1}\sum_{j=0}^{\bar{n}_2+1} \bar{M}_{ij} z_1^i z_2^j \qquad /6.85a/$$

$$L_1/z_1,z_2/ = z_2 N/z_1,z_2/\otimes N_m^T/z_1,z_2/ = \sum_{i=0}^{\bar{n}_1}\sum_{j=0}^{\bar{n}_2+1} L_{ij}^{/1/} z_1^i z_2^j \qquad /6.85b/$$

$$L_2/z_1,z_2/ = z_1 N/z_1,z_2/\otimes N_m^T/z_1,z_2/ = \sum_{i=0}^{\bar{n}_1+1}\sum_{j=0}^{\bar{n}_2} L_{ij}^{/2/} z_1^i z_2^j \qquad /6.85c/$$

$$L_3/z_1,z_2/ = z_1 z_2 N/z_1,z_2/\otimes N_m^T/z_1,z_2/ = \sum_{i=0}^{\bar{n}_1+1}\sum_{j=0}^{\bar{n}_2+1} L_{ij}^{/3/} z_1^i z_2^j \qquad /6.85d/$$

$$L_4/z_1,z_2/ = z_1^2 z_2 N/z_1,z_2/\otimes N_m^T/z_1,z_2/ = \sum_{i=0}^{\bar{n}_1+1}\sum_{j=0}^{\bar{n}_2+1} L_{ij}^{/4/} z_1^i z_2^j \qquad /6.85e/$$

$$L_5/z_1,z_2/ = z_1 z_2^2 N/z_1,z_2/\otimes N_m^T/z_1,z_2/ = \sum_{i=0}^{\bar{n}_1+1}\sum_{j=0}^{\bar{n}_2+2} L_{ij}^{/5/} z_1^i z_2^j \qquad /6.85f/$$

$$h = [H_1 H_2 \ldots H_m]$$
$$f_i = [F_{i1} F_{i2} \ldots F_{im}]^T \qquad \qquad /6.85g/$$

$H_i / F_{ji}/$ is the ith row of $H$ $/F_j/$

$$\bar{p}/z_1,z_2/ = [\bar{p}_1/z_1,z_2/, \bar{p}_2/z_1,z_2/, \ldots, \bar{p}_l/z_1,z_2/]^T =$$
$$= \sum_{i=0}^{\bar{n}_1} \sum_{j=0}^{\bar{n}_2} \bar{p}_{ij} z_1^i z_2^j \qquad \qquad /6.85h/$$

$\bar{p}_i/z_1,z_2$ is the ith row of

$$\bar{P}/z_1,z_2/ = z_1 z_2 d/z_1,z_2/N_m/z_1,z_2/$$

$\bar{n}_i = n_i + \hat{n}_i$, $\hat{n}_i = \deg_{z_i} d_m/z_1,z_2/$ for $i = 1,2$.

Substitution of /6.85/ into /6.84/ yields

$$\sum_{i=0}^{\bar{n}_1+1} \sum_{j=0}^{\bar{n}_2+1} \bar{M}_{ij} h z_1^i z_2^j + \sum_{i=0}^{\bar{n}_1} \sum_{j=0}^{\bar{n}_2+1} L_{ij}^{/1/} f_1 z_1^i z_2^j + \sum_{i=0}^{\bar{n}_1+1} \sum_{j=0}^{\bar{n}_2} L_{ij}^{/2/} f_2 z_1^i z_2^j +$$
$$+ \sum_{i=0}^{\bar{n}_1+1} \sum_{j=0}^{\bar{n}_2+1} L_{ij}^{/3/} f_3 z_1^i z_2^j + \sum_{i=0}^{\bar{n}_1+2} \sum_{j=0}^{\bar{n}_2+1} L_{ij}^{/4/} f_4 z_1^i z_2^j + \sum_{i=0}^{\bar{n}_1+1} \sum_{j=0}^{\bar{n}_2+2} L_{ij}^{/5/} f_5 z_1^i z_2^j =$$
$$= \sum_{i=0}^{\bar{n}_1} \sum_{j=0}^{\bar{n}_2} \bar{p}_{ij} z_1^i z_2^j \qquad \qquad /6.86/$$

Equating coefficients of like $z_1^i z_2^j$ terms of both sides of /6.86/ we obtain

$$\bar{M}_{ij} h + L_{ij}^{/1/} f_1 + L_{ij}^{/2/} f_2 + L_{ij}^{/3/} f_3 + L_{ij}^{/4/} f_4 + L_{ij}^{/5/} f_5 = \bar{p}_{ij} \qquad /6.87/$$

for $i = 0,1,\ldots,\tilde{n}_1 = \bar{n}_1+2$, $j = 0,1,\ldots,\tilde{n}_2 = \bar{n}_2+2$.

The equations /6.87/ may be written in the form

$$\bar{Q}\bar{x} = \bar{p} \qquad \qquad /6.88/$$

where

$$Q = \begin{bmatrix} \bar{M}_{00} & L_{00}^{/1/} & \cdots & L_{00}^{/5/} \\ \bar{M}_{10} & L_{10}^{/1/} & \cdots & L_{10}^{/5/} \\ \bar{M}_{01} & L_{01}^{/1/} & \cdots & L_{01}^{/5/} \\ \bar{M}_{11} & L_{11}^{/1/} & \cdots & L_{11}^{/5/} \\ \cdots & \cdots & \cdots & \cdots \\ \bar{M}_{\tilde{n}_1 \tilde{n}_2} & L_{\tilde{n}_1 \tilde{n}_2}^{/1/} & \cdots & L_{\tilde{n}_1 \tilde{n}_2}^{/5/} \end{bmatrix} \in R^{lm\tilde{n} \times m/m+51/}$$

$$\bar{x} = \begin{bmatrix} h \\ f_1 \\ f_2 \\ f_3 \\ f_4 \\ f_5 \end{bmatrix} \in R^{m/m+51/}, \qquad \bar{p} = \begin{bmatrix} \bar{p}_{00} \\ \bar{p}_{10} \\ \bar{p}_{01} \\ \bar{p}_{11} \\ \cdots \\ \bar{p}_{n_1 n_2} \end{bmatrix} \in R^{lm\tilde{n}}$$

$\tilde{n} = /\tilde{n}_1 + 1 / / \tilde{n}_2 + 1/$.

The equation /6.88/ is the linear system of $lm\tilde{n}$ equations with $m/m+51/$ unknowns. Let rank $\bar{Q} = \bar{r}$ and $\bar{U}$ be a $lm\tilde{n} \times lm\tilde{n}$ invertible matrix of elementary row operations such that

$$\bar{U}\bar{Q} = \begin{bmatrix} \bar{Q}_1 \\ 0 \end{bmatrix} \qquad /6.89/$$

where $\bar{Q}_1$ is a $\bar{r} \times m/m+51/$ matrix involving $\bar{r}$ linearly independent rows of $\bar{Q}$.

Premultiplying /6.88/ by $\bar{U}$ and using /6.89/ we obtain

$$\begin{bmatrix} \bar{Q}_1 \\ 0 \end{bmatrix} \bar{x} = \begin{bmatrix} \bar{p}_1 \\ \bar{p}_2 \end{bmatrix}$$

and

$$\bar{Q}_1 \bar{x} = \bar{p}_1 \qquad /6.90/$$

$$\bar{p}_2 = 0 \qquad /6.91/$$

where

$$\begin{bmatrix} \bar{p}_1 \\ \bar{p}_2 \end{bmatrix} = \bar{U}\bar{p}, \quad \bar{p}_1 \in R^{\bar{r}}, \quad \bar{p}_2 \in R^{lm\tilde{n} - \bar{r}}$$

We have thus established the following

Theorem 6.5

The exact model-matching problem has a solution if the condition /6.91/ is satisfied. If /6.91/ holds, then from /6.90/ we obtain

$$\bar{x} = \bar{Q}_1^T \left[ \bar{Q}_1 \bar{Q}_1^T \right]^{-1} \bar{p}_1 \qquad /6.92/$$

The solution /6.92/ yields the desired matrices H, $F_1, F_2, \ldots, F_5$. Note that in particular case for $F_1 = F_2 = F_4 = F_5 = 0$ and $F_3 = F$ we obtain a solution to the exact model-matching problem via static output feedback. If /6.91/ is satisfied the following algorithm may be used for finding the matrices H and $F_1, F_2, \ldots, F_5$.

Algorithm 6.4

Step 1  Find $\bar{M}/z_1,z_2/$, $L_1/z_1,z_2/,\ldots,L_5/z_1,z_2/$, $\bar{p}/z_1,z_2/$ defined by /6.85/ and $\bar{M}_{ij}$, $L_{ij}^{/k/}$ /k = 1,...,5/, $\bar{p}_{ij}$.

Step 2  Find $\bar{Q}$, $\bar{p}$ and $\bar{r} = \text{rank } \bar{Q}$.

Step 3  Find an elementary row operations matrix $\bar{U}$ and $\bar{Q}_1$, $\bar{p}_1$.

Step 4  Using /6.92/ find $\bar{x}$ and H, $F_1,\ldots,F_5$.

## 6.5. Šebek's method of exact model matching

### 6.5.1. Problem formulation

Following [10] let us consider a 2-D linear system /plant/ described by the equations

$$u = Ax_p, \quad z = Bx_p, \quad y = Cx_p \qquad /6.93/$$

where $u = u/z_1,z_2/ \in R^m/z_1,z_2/$ is the input vector,

$z = z/z_1,z_2/ \in R^n/z_1,z_2/$ is the measured output vector,

$y = y/z_1,z_2/ \in R^l/z_1,z_2/$ is the output vector

$x_p = x_p/z_1,z_2/ \in R^q/z_1,z_2/$ is the partial state vector

$A \in R^{m \times n}[z_1,z_2]$, $B \in R^{n \times m}[z_1,z_2]$ and $C \in R^{l \times m}[z_1,z_2]$.

It is assumed that A, B are right factor coprime /having only unimodular right common divisor/ and there are no hidden modes in the plant / i.e.

its characteristic polynomial is equal to det A/.

Let a linear 2-D compensator be described by the equation

$$Pu = -Qz + Ru_N \qquad /6.94/$$

where $P \in R^{mxm}[z_1,z_2]$ is invertible, $Q \in R^{mxn}[z_1,z_2]$ and $R \in R^{mxp}[z_1,z_2]$ and $u_N = u_N/z_1,z_2/ \in R^p/z_1,z_2/$ is a new input vector.

This compensator can thought of as a combination of a feedback and a feedforward /Fig.6.3/.

It is desired to find the compensator such that the resultant system is governed by the model equations

$$u_N = Fx_M, \qquad y = Gx_M \qquad /6.95/$$

where $x_M = x_M/z_1,z_2/ \in R^p/z_1,z_2/$ is the partial state vector and $F \in R^{pxp}[z_1,z_2]$ is invertible, $G \in R^{lxp}[z_1,z_2]$. F and G are assumed right factor coprime.

From /6.93/ and /6.94/ we have

$$/PA + QB/x_p = R\,u_N, \qquad y = Cx_p \qquad /6.96/$$

The transfer function matrix of the overall compensated system is given by

$$G_c/z_1,z_2/ = C/PA + QB/^{-1}R \qquad /6.97/$$

and of the model

$$G_M/z_1,z_2/ = GF^{-1} \qquad /6.98/$$

The exact model matching problem can be stated as follows. Given A, B, C and F, G, find P, Q and R such that $G_c/z_1,z_2/ = G_m/z_1,z_2/$, i.e.

$$C/PA + QB/^{-1}R = GF^{-1} \qquad /6.99/$$

6.5.2. <u>Problem solution</u>

Note that the relation /6.99/ can be satisfied if and only if

$$\text{rank } C = \text{rank} \begin{bmatrix} C \mid G \end{bmatrix} \qquad /6.100/$$

Let us denote rank $C = r \leq \min/l,m/$. Let D be a greatest common left

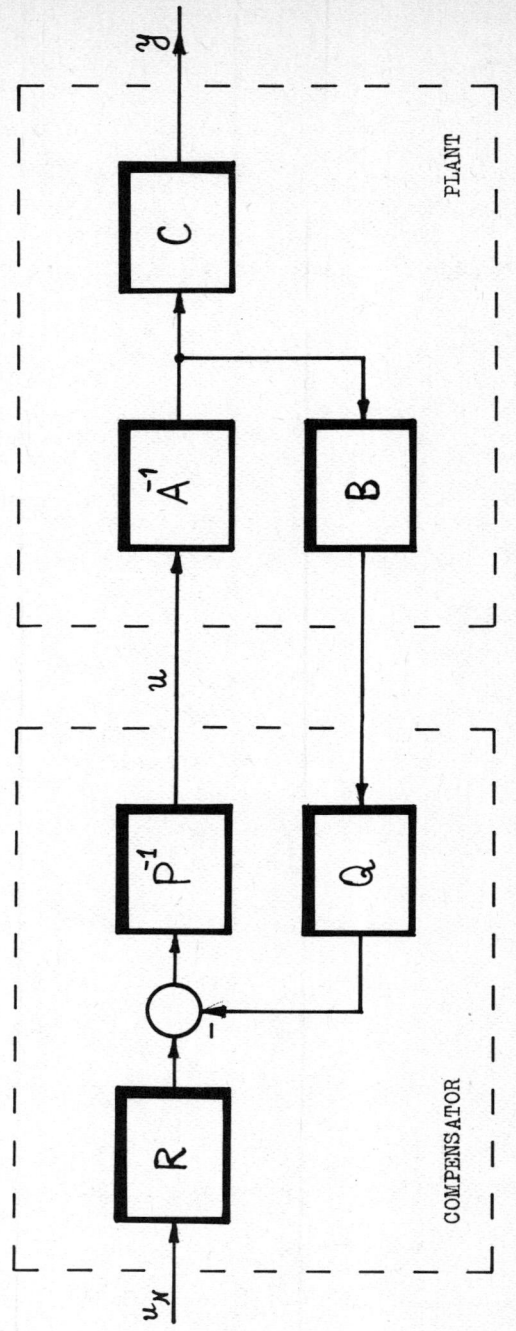

Fig. 6.3.

divisor of C and G, i.e.

$$C = D\bar{C}, \qquad G = D\hat{G} \qquad /6.101/$$

where $D \in R^{l \times r}$, $\bar{C} \in R^{r \times m}$, $\hat{G} \in R^{r \times p}$. To avoid trivia it is assumed that D is of full column rank. Further let $\hat{F} \in R^{r \times r}[z_1, z_2]$, $\hat{G} \in R^{r \times p}[z_1, z_2]$ be two left coprime matrices such that

$$\hat{F}^{-1}\hat{G} = \bar{G}F^{-1} \qquad /6.102/$$

Using /6.101/ and /6.102/, the equation /6.99/ may be written in the form

$$\hat{F}\bar{C}/PA + QB/^{-1}R = \hat{G} \qquad /6.103/$$

Let $T \in R^{m \times m}[z_1, z_2]$ denote a greatest common left divisor of PA+QB and R, i.e. there exist $X \in R^{m \times m}[z_1, z_2]$, $Y \in R^{m \times p}[z_1, z_2]$ such that

$$PA + QB = TX \qquad /6.104/$$

$$R = TY \qquad /6.105/$$

Substitution of /6.104/ and /6.105/ into /6.103/ yields

$$\hat{F}\bar{C}X^{-1}Y = \hat{G} \qquad /6.106/$$

Since $\hat{F}\bar{C}X^{-1}Y$ is a polynomial matrix and X,Y are left coprime, X must be a right divisor of $\hat{F}\bar{C}$, i.e. there exists $U \in R^{r \times m}[z_1, z_2]$ such that

$$\hat{F}\bar{C} = UX \qquad /6.107/$$

Substitution of /6.107/ into /6.106/ yields

$$\hat{G} = UY \qquad /6.108/$$

Note that $\hat{F}$ and $\hat{G}$ are left coprime by definition and $\bar{C}$ and $\hat{G}$ are so. Therefore, $\hat{F}\bar{C}$ and $\hat{G}$ are left coprime and U must be right unimodular /in $R[z_1, z_2]$/. Taking $U = [I_r \ 0]$ from /6.104/, /6.105/, /6.108/ and /6.108/ we obtain

$$PA + QB = T \begin{bmatrix} \hat{F}\bar{C} \\ T \end{bmatrix} \qquad /6.109/$$

and

$$R = T \begin{bmatrix} \hat{G} \\ \hat{T} \end{bmatrix} \qquad /6.110/$$

where $T \in R^{m \times m}[z_1, z_2]$ is an arbitrary invertible matrix,
$\bar{T} \in R^{m-r/xm}[z_1, z_2]$ complements $\hat{FC}$ to an invertible matrix and
$\hat{T} \in R^{m-r/xp}[z_1, z_2]$ is arbitrary.

### 6.5.3. Causal solution

**Definition 6.3**

A polynomial matrix $P/z_1, z_2/$ having full rank is said to be causal in one indeterminate /in $z_1$/ if and only if rank $P/0, z_2/$ = rank $P/z_1, z_2/$ and it is said to be causal in the both indeterminates if and only if rank $P/0, 0/$ = rank $P/z_1, z_2/$.

**Definition 6.4**

A 2-D system with the transfer function matrix $A^{-1}B$ /or $BA^{-1}$/ is said to be strictly causal in one indeterminate /in $z_1$/ if and only if rank $A/0, z_2/$ = rank $A/z_1, z_2/$ and $B/0, z_2/$ = 0 and it is said to be strictly causal in the both indeterminates if and only if rank $A/0, 0/$ = rank $A/z_1, z_2/$ and $B/0, 0/$ = 0.

**Definition 6.5**

A polynomial matrix is said to be unimodular if and only if it has a polynomial inverse and it is said to be right unimodular if and only if it has a polynomial right inverse.

From the practical point of view, it is often desirable to find a causal solution to the problem. This means that both the compensator and the overal system must be causal. We assume that the model /the matrix F/ is causal and the plant is strictly causal.

**Theorem 6.6**

The exact model matching problem has a causal solution if and only if $\bar{C}$ is causal.

**Proof**

From /6.99/ is follows that the overall system is causal if and only if the matrix PA+QB is causal. From /6.104/ this is equivalent to the causality of X since T can always be chosen causal. From /6.107/ this is further equaivalent to the causality of $\hat{F}$ and $\bar{C}$ because U is right unimodular and hence causal. $\hat{F}$ is causal if and only if F is causal. Therefore, the causality of PA+QB is equivalent to the causality of $\bar{C}$. We shall show that for strictly causal plant every solution to /6.104/ yields a causal compensator. For example in the case of causality in $z_1$, from /6.104/ for $z_1 = 0$ we obtain an invertible matrix of the form

$$P/0,z_2/A/0,z_2/ = T/0,z_2/X/0,z_2/$$

and

$$P/0,z_2/ = T/0,z_2/X/0,z_2/A^{-1}/0,z_2/$$

is an invertible polynomial matrix in $z_2$. Therefore, the compensator is causal /in $z_1$/. In the case of causality in the both indeterminates analogously from /6.104/ for $z_1 = z_2 = 0$ we obtain

$$F/0,0/ = T/0,0/X/0,0/A^{-1}/0,0/$$

and hence the causal compensator. □

### 6.5.4. Stable solution

#### Definition 6.6

A polynomial matrix $P/z_1,z_2/$ having full rank is said to be stable if and only if all its zeros /i.e. all $/z_1^o,z_2^o/ \in C \times C$ for which rank $P/z_1^o,z_2^o/$ < rank $P/z_1,z_2//$ fall inside the "stable region", for example bidisk.

#### Definition 6.7

Polynomial matrices $P/z_1,z_2/$, $Q/z_1,z_2/$ with the same number of columns have a common right zero $/z_1^o,z_2^o/ \in C \times C$ if and only if

$$\text{rank} \begin{bmatrix} P/z_1^o,z_2^o/ \\ Q/z_1^o,z_2^o/ \end{bmatrix} < \text{rank} \begin{bmatrix} P/z_1,z_2/ \\ Q/z_1,z_1/ \end{bmatrix}$$

## Theorem 6.7

The exact model matching problem has a stable solution if and only if

/i/     F is stable,

/ii/    $\bar{C}$ is stable,

/iii/   A and B have no unstable right zero in common.

## Proof

From /6.99/ it follows that the overall system is stable if and only if the matrix PA+QB is stable. From /6.104/ this is equivalent to the stability of X, since T can always be chosen stable. From /6.107/ this is further equivalent to the stability of $\hat{F}$ and $\bar{C}$ because U is right unimodular and hence stable. $\hat{F}$ is stable if and only if F is stable /condition /i//. The stability of $\bar{C}$ is ensured by the condition /ii/.
If the conditions /i/, /ii/ are satisfied, then X and PA+QB is stable. Note that every unstable common right zero of A and B must be a right zero of T. Therefore, the common zeros of A and B must not be unstable. On the other hand, if A and B have no unstable right zero in common, a stable T assuring solvability of /6.104/ can always be found. This can be proved using the method presented in Appendix.    □

### 6.5.5. Design procedure and algorithms

If the conditions of theorems 6.6 and 6.7 are satisfied then a stable and causal solution to the exact model matching problem can be found using the following design procedure and algorithms.

<u>Step 1</u>    Calculate any greatest common left divisor D, of C and G which has full column rank and factor it out

$$C = D\bar{C}, \qquad G = D\bar{G}$$

A greatest common left divisor D of C and G can be found by the method of Morf, Lévy and Kung [16].

**Remark:** If $\bar{C}$ is not causal or not stable, then there is no causal or no stable solution to the problem.

**Step 2** Determine left coprime polynomial matrices $\hat{F}$ and $\hat{G}$ such that /6.102/ holds. This can be done as follows. Using elementary row operations in $R/z_1/[z_2]$ perform the reduction

$$\begin{bmatrix} F & I & 0 \\ G & 0 & I \end{bmatrix} \rightarrow \begin{bmatrix} I & \hat{X}_1 & \hat{X}_2 \\ 0 & \hat{X}_3 & \hat{X}_4 \end{bmatrix}$$

where
$\hat{X}_1, \hat{X}_2, \hat{X}_3, \hat{X}_4 \in R/z_1/[z_2]$.

Denoting by $T \in R[z_1]$ a least common left denominator of the matrix $\begin{bmatrix} \hat{X}_3 & \hat{X}_4 \end{bmatrix}$ we have

$$\begin{bmatrix} \bar{X}_3 & \bar{X}_4 \end{bmatrix} = T \begin{bmatrix} \hat{X}_3 & \hat{X}_4 \end{bmatrix}$$

for polynomial matrices $\bar{X}_3, \bar{X}_4 \in R[z_1, z_2]$. Unfortunately, they may still have a nonunimodular common left factor /with determinant from $R[z_1]$/. If this is the case, we must employ the primitive factorization algorithm /A.4/ to get $[\hat{G} \; \hat{F}]$ as the primitive part of $\begin{bmatrix} \bar{X}_3 & \bar{X}_4 \end{bmatrix}$.

**Step 3** Choose a stable polynomial matrix $\bar{T} \in R^{/m-r/xm}[z_1, z_2]$ which complements $\hat{FC}$ to an invertible matrix and an arbitrary matrix $\hat{T} \in R^{/m-r/xp}[z_1, z_2]$. The choice $\bar{T}$ = a constant matrix, $\hat{T} = 0$ is recommended.

**Step 4** Choose a causal and stable polynomial matrix $T \in R^{mxm}[z_1, z_2]$ such that the equation /6.109/ is solvable and solve it for unknown matrices P and Q.

**Remark:** If A and B have some unstable right zero in common, then there is no such causal stable T and does not exist a stable solution to the problem.

To solve the equation /6.109/ one of the methods presented in Appendix can be used.

**Step 5** Find R using /6.110/.

## Example 6.4

Given
$$A = 1, \quad B = \begin{bmatrix} z_1 \\ z_2 \end{bmatrix}, \quad C = 1$$

and
$$F = \begin{bmatrix} 1 & z_2 \\ 0 & -1 \end{bmatrix}, \quad G = \begin{bmatrix} z_1, & -z_2 \end{bmatrix}$$

find P, Q and R of the linear compensator.

In this case $m = l = 1$, $n = p = 2$ and the conditions of theorems 6.6 and 6.7 are satisfied. Therefore, there exists a causal and stable solution to the problem.

**Step 1** The greatest common left divisor of

$$C = 1 \quad \text{and} \quad G = \begin{bmatrix} z_1, -z_2 \end{bmatrix}$$

is $\quad D = 1 \quad$ and $\quad \bar{C} = C = 1, \quad \bar{G} = G = \begin{bmatrix} z_1, & -z_2 \end{bmatrix}$

**Step 2** It is easy to check that

$$F = 1 \quad \text{and} \quad \hat{G} = \begin{bmatrix} z_1, z_1 z_2 + z_2 \end{bmatrix}$$

are left comprime and satisfied the equation /6.102/.

**Step 3** We choose $\tilde{T} = 0$ and $\hat{T} = 0$.

**Step 4** For $T = 1$ the equation /6.109/ is solvable and has the form

$$P \begin{bmatrix} 1 \end{bmatrix} + Q \begin{bmatrix} z_1 \\ z_2 \end{bmatrix} = 1 \qquad /6.111/$$

The general solution to /6.111/ is given by

$$P = 1 - t_1 z_1 - t_2 z_2, \quad Q = \begin{bmatrix} t_1, t_2 \end{bmatrix} \qquad /6.112/$$

where $t_1, t_2$ are arbitrary polynomials in $z_1$ and $z_2$.

**Step 5** Using /6.110/ we obtain

$$R = T \begin{bmatrix} \hat{G} \\ \hat{T} \end{bmatrix} = \hat{G} = \begin{bmatrix} z_1, & z_1 z_2 + z_2 \end{bmatrix} \qquad /6.113/$$

The desired solution is given by /6.112/ and /6.113/.

## 6.6. Decoupling by state feedback

### 6.6.1. Problem formulation

Consider RM described by the equations /6.1/ and /6.2/ for $l = m$, with control law of the form /6.46/. Substitution of /6.46/ into /6.1/ yields

$$x' = /A+BK/x + BHv \qquad /6.114/$$

The transfer function matrix of the closed-loop system described by /6.114/ and /6.2/ is

$$G_c/z_1,z_2/ = C\left[Z-A-BK\right]^{-1} BH \qquad /6.115/$$

where Z is defined by /6.45/ and $K \in R^{mxn}$, $H \in R^{mxm}$.

The decoupling problem by state feedback may be stated as follows. Given A, B, C, find the matrices K and H with $\det H \neq 0$ such that the closed-loop transfer function matrix /6.115/ is diagonal and nonsingular.

### 6.6.2. Problem solution

Following [4] we introduce the following definitions

**Definition 6.8**

For each output k /$k = 1,\ldots,l$/, let $\Gamma_k$ denote the set of all points /$p_k,q_k$/ for which the following relationship holds:

$$C_k \begin{bmatrix} A_{11}^{s-1,t} & A_{12}^{s,t-1} \\ A_{21}^{s-1,t} & A_{22}^{s,t-1} \end{bmatrix} B \begin{cases} = 0 \text{ for } \forall \ /s,t/ < /p_k,q_k/ \leqslant /n_1,n_2/ \\ \neq 0 \text{ for } \forall \ /s,t/ = /p_k,q_k/ \leqslant /n_1,n_2/ \end{cases} \qquad /6.116/$$

where

$C_k$ is the kth row of C and

$$A^{i,j} = \begin{bmatrix} A_{11}^{i,j} & A_{12}^{i,j} \\ A_{21}^{i,j} & A_{22}^{i,j} \end{bmatrix}, \quad \begin{array}{l} A_{11}^{i,j} \in R^{n_1 \times n_2} \\ A_{22}^{i,j} \in R^{n_2 \times n_2} \end{array}$$

Let /$\bar{p}_k,\bar{q}_k$/ denote the point of $\Gamma_k$ which is the nearest /i.e. the point that has the least distance/ to the i-axis and /$\hat{p}_k,\hat{q}_k$/ denote the nearest point to the j-axis /Fig. 6.4/.

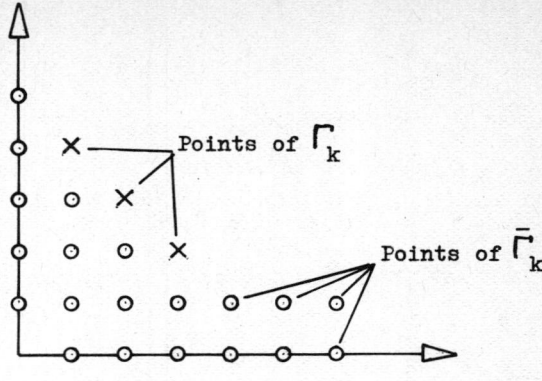

Fig. 6.4.

The set of points denoted by "x" constitutes the set of $\Gamma_k$ and the set of points denoted by "O" constitutes the set of $\bar{\Gamma}_k$

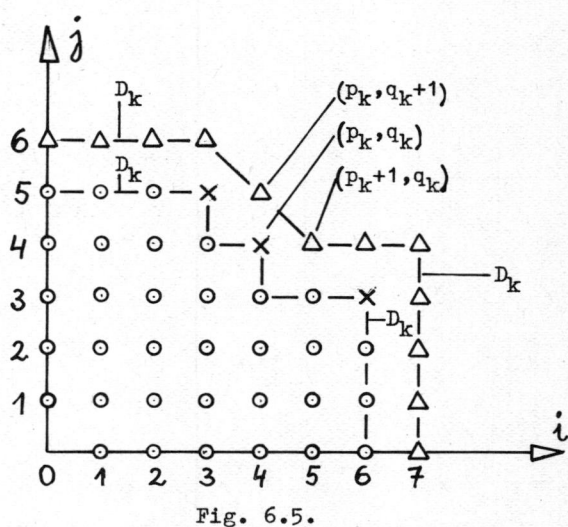

Fig. 6.5.

### Definition 6.9

For each output k let $\bar{\Gamma}_k$ denote the set of all points /s,t/ for which the following relationship holds:

$$C_k \begin{bmatrix} A_{11}^{s-1,t} & A_{12}^{s,t-1} \\ A_{21}^{s-1,t} & A_{22}^{s,t-1} \end{bmatrix} B = 0, \quad /k = 1,\ldots,l/ \qquad /6.117/$$

for $\forall /s,t/ < /p_k, q_k/$ and $/\bar{p}_k+1,0/ \leq /s,t/ \leq /n_1, \bar{q}_k-1/$, $/0, \hat{q}_k+1/ \leq /s,t/ \leq /\hat{p}_k-1, n_2/$.

### Definition 6.10

For each output k, let $\Delta_k$ denote the set of vectors

$$B_k^* = C_k \begin{bmatrix} A_{11}^{p_k-1,q_k} & A_{12}^{p_k,q_k-1} \\ A_{21}^{p_k-1,q_k} & A_{22}^{p_k,q_k-1} \end{bmatrix} B \qquad /6.118/$$

for all points of $\Gamma_k$.

### Theorem 6.8

The decoupling problem has a solution if and only if

/i/ all rows $B_k^*$ in $\Delta_k$ are proportional,

/ii/ $\det B^* \neq 0$

where

$$B^* = \begin{bmatrix} B_1^* \\ B_2^* \\ \vdots \\ B_m^* \end{bmatrix} = \begin{bmatrix} C_1 \begin{bmatrix} A_{11}^{p_1-1,q_1} & A_{12}^{p_1,q_1-1} \\ A_{21}^{p_1-1,q_1} & A_{22}^{p_1,q_1-1} \end{bmatrix} B \\ C_2 \begin{bmatrix} A_{11}^{p_2-1,q_2} & A_{12}^{p_2,q_2-1} \\ A_{21}^{p_2-1,q_2} & A_{22}^{p_2,q_2-1} \end{bmatrix} B \\ \vdots \\ C_m \begin{bmatrix} A_{11}^{p_m-1,q_m} & A_{12}^{p_m,q_m-1} \\ A_{21}^{p_m-1,q_m} & A_{22}^{p_m,q_m-1} \end{bmatrix} B \end{bmatrix} \qquad /6.119/$$

/iii/ there is a K such that

$$C_k \begin{bmatrix} /A+BK/_{11}^{s-1,t} & /A+BK/_{12}^{s,t-1} \\ /A+BK/_{21}^{s-1,t} & /A+BK/_{22}^{s,t-1} \end{bmatrix} B = c_k B_k^* \qquad /6.120/$$

for $k = 1,2,\ldots,m$ and for all $/s,t/ \leqslant /n_1,n_2/$ which do not belong to $\overline{\Gamma}_k$ and $\Gamma_k$, $c_k = c_k/s,t/$ is a proportionality constant.

<u>Proof</u>

To prove the necessity let us assume that there is a pair of matrices K and H which decouple RM. In 2.1 it was shown that the transfer function matrix of RM is given by

$$G/z_1,z_2/ = \frac{1}{d/z_1,z_2/} \left( \sum_{k=0}^{n_1} \sum_{l=0}^{n_2} \sum_{i=0}^{n_1-k} \sum_{j=0}^{n_2-1} d_{i+k,j+l} z_1^i z_2^j C \begin{bmatrix} A_{11}^{k-1,1} & A_{12}^{k,1-1} \\ A_{21}^{k-1,1} & A_{22}^{k,1-1} \end{bmatrix} B \right) =$$

$$= \frac{1}{d/z_1,z_2/} \left[ \sum_{k=0}^{n_1} \sum_{l=0}^{n_2} \sum_{i=0}^{n_1-k} \sum_{j=0}^{n_2-1} d_{i+k,j+l} z_1^i z_2^j \, C/A^{k-1,1} B^{1,0} + A^{k,1-1} B^{0,1}/ \right] \qquad /6.121/$$

where

$$d/z_1,z_2/ = \begin{vmatrix} I_{n_1} z_1 - A_{11}, & -A_{12} \\ -A_{21}, & I_{n_2} z_2 - A_{22} \end{vmatrix} = \sum_{i=0}^{n_1} \sum_{j=0}^{n_2} d_{ij} z_1^i z_2^j \,/ d_{n_1 n_2} = 1/ \qquad /6.122a/$$

$$\begin{bmatrix} A_{11}^{k-1,1} & A_{12}^{k,1-1} \\ A_{21}^{k-1,1} & A_{22}^{k,1-1} \end{bmatrix} B = A^{k-1,1} B^{1,0} + A^{k,1-1} B^{0,1} \qquad /6.122b/$$

and

$$B^{1,0} = \begin{bmatrix} B_1 \\ 0 \end{bmatrix}, \quad B^{0,1} = \begin{bmatrix} 0 \\ B_2 \end{bmatrix}$$

It can be easily shown /see Problem 6.4/ that for any square matrices P and Q the following relationship holds

$$/P+Q/^{i,j} = P^{i,j} + P^{i-1,j} Q^{1,0} + P^{i,j-1} Q^{0,1} + \ldots + P^{0,1} Q^{i,j-1} +$$
$$+ P^{1,0} Q^{i-1,j} + Q^{i,j} \quad \text{for } i,j \geqslant 0 \qquad /6.123/$$

Taking into account /6.117/, /6.122/ and /6.121/ we can write

$$C_k \begin{bmatrix} /A+BK/_{11}^{p_k-1,q_k} & /A+BK/_{12}^{p_k,q_k-1} \\ /A+BK/_{21}^{p_k-1,q_k} & /A+BK/_{22}^{p_k,q_k-1} \end{bmatrix} B =$$

$$= C_k \left[ /A+BK/^{p_k-1,q_k} B^{1,0} + /A+BK/^{p_k,q_k-1} B^{0,1} \right] =$$

$$= C_k \left[ A^{p_k-1,q_k} + A^{p_k-2,q_k} B^{1,0}_K + A^{p_k-1,q_k-1} B^{0,1}_K + \ldots \right.$$

$$+ A^{0,1}/BK/^{p_k-1,q_k-1} + A^{1,0}/BK/^{p_k-2,q_k} + /BK/^{p_k-1,q_k}/B^{1,0} +$$

$$+ /A^{p_k,q_k-1} + A^{p_k-1,q_k-1} B^{1,0}_K + A^{p_k,q_k-2} B^{0,1} K + \ldots$$

$$\left. + A^{0,1}/BK/^{p_k,q_k-2} + A^{1,0}/BK/^{p_k-1,q_k-1} + /BK/^{p_k,q_k-1}/ B^{0,1} \right] =$$

$$= C_k \left[ A^{p_k-1,q_k} B^{1,0} + A^{p_k,q_k-1} B^{0,1} \right] \quad \text{for } k = 1,\ldots,l \qquad /6.124/$$

If K and H decouple RM, then

$$G_c/z_1,z_2/ = C\left[Z-A-BK\right]^{-1} BH = \text{diag}\left[G_{c1}/z_1,z_2/,\ldots,G_{cm}/z_1,z_2/\right] \qquad /6.125/$$

with $G_{ci}/z_1,z_2/ \neq 0$ for $i = 1,\ldots,m$.

Note that /6.125/ holds only if all vectors $B_k^*$ of the set $\Delta_k$ are proportional to each other, i.e.

$$B_k^* = \lambda_k \bar{B}_k^* \qquad /6.126/$$

where

$$\bar{B}_k^* = C_k \begin{bmatrix} A_{11}^{\tilde{p}_k-1,\tilde{q}_k} & A_{12}^{\tilde{p}_k,\tilde{q}_k-1} \\ A_{21}^{\tilde{p}_k-1,\tilde{q}_k} & A_{22}^{\tilde{p}_k,\tilde{q}_k-1} \end{bmatrix} B$$

$/\tilde{p}_k,\tilde{q}_k/$ is a certain point of the set $\Gamma_k$ and $\lambda_k = \lambda_k/p_k,q_k/$ is proportionality constant.

From /6.119/, /6.121/ and /6.125/ it follows that $B^*H$ is a non-singular diagonal matrix. Since H is assumed nonsingular, thus $B^*$ is also nonsingular.

From /6.121/ and /6.125/ it follows that $G_c/z_1,z_2/$ is diagonal and nonsingular only if there is a K such that /6.120/ is satisfied. Thus, the necessity of /i/, /ii/ and /iii/ is shown. To prove the sufficiency we assume that /i/, /ii/ and /iii/ are satisfied and $H = /B^*/^{-1}$. Then using /6.126/ and /6.120/ it is easy to check that $G_c/z_1,z_2/$ is diagonal and nonsingular. □

The condition /6.120/ is a set of nonlinear algebraic equations in the elements of K. It is quite difficult to use /6.120/ to derive the feedback matrix K. In special cases an explicit expression for K, analogous to that derived for 1-D systems, may be obtained. To this end let $D_k$ be the contour /Fig. 6.5/, consisting of lines parallel to the i and j axis, such that all points of $\Gamma_k$ are to be found in its outer corner points. Let $\bar{D}_k$ be another contour /Fig. 6.5/ such that there is no point /s,t/ between $D_k$ and $\bar{D}_k$ and involves the minimum number of points /s,t/. Consider a certain point $/\bar{s},\bar{t}/ \in \bar{D}_k$. This certain point may be either the point $/p_k+1,q_k/$ or the point $/p_k,q_k+1/$. Let us define for the point $/p_k+1,q_k/$, $p_k<n_1$, the following matrices

$$A^*_{1k} = C_k \begin{bmatrix} A_{11}^{p_k-1,q_k} & A_{12}^{p_k,q_k-1} \\ A_{21}^{p_k-1,q_k} & A_{22}^{p_k,q_k-1} \end{bmatrix} \begin{bmatrix} A_{11} \\ A_{21} \end{bmatrix} \qquad /6.127a/$$

$$/k=1,\ldots,m/$$

$$A^*_{2k} = C_k \begin{bmatrix} A_{11}^{p_k,q_k-1} & A_{12}^{p_k+1,q_k-2} \\ A_{21}^{p_k,q_k-1} & A_{22}^{p_k+1,q_k-2} \end{bmatrix} \begin{bmatrix} A_{12} \\ A_{22} \end{bmatrix} \qquad /6.127b/$$

$$\Lambda_1 = \mathrm{diag}[\lambda_1 \ldots \lambda_m], \quad \Lambda_2 = \mathrm{diag}[\gamma_1 \ldots \gamma_m] \qquad /6.127c/$$

where

$$\gamma_k \begin{cases} = 0 & \text{for } /p_k+1,q_k-1/ \in \bar{\Gamma}_k \text{ or } \bar{D}_k \\ = \lambda_k/p_k+1,q_k-1/ & \text{for } /p_k+1,q_k-1/ \in \Gamma_k \end{cases}$$

If $p_k = n_1$ and $q_k < n_2$ we define

$$A^*_{1k} = C_k \begin{bmatrix} A_{11}^{p_k-2,q_k+1} & A_{12}^{p_k-1,q_k} \\ A_{21}^{p_k-2,q_k+1} & A_{22}^{p_k-1,q_k} \end{bmatrix} \begin{bmatrix} A_{11} \\ A_{21} \end{bmatrix} \quad /6.128a/$$

$$/k=1,\ldots,m/$$

$$A^*_{2k} = C_k \begin{bmatrix} A_{11}^{p_k-1,q_k} & A_{12}^{p_k,q_k-1} \\ A_{21}^{p_k-1,q_k} & A_{22}^{p_k,q_k-1} \end{bmatrix} \begin{bmatrix} A_{12} \\ A_{22} \end{bmatrix} \quad /6.128b/$$

$$\Lambda_1 = \text{diag}[\beta_1 \ldots \beta_m], \quad \Lambda_2 = \text{diag}[\lambda_1,\ldots,\lambda_m] \quad /6.128c/$$

where

$$\beta_k \begin{cases} = 0 & \text{for } /p_k-1,q_k+1/ \in \bar{\Gamma}_k \text{ or } \bar{D}_k \\ = \lambda_k /p_k-1,q_k+1/ & \text{for } /p_k-1,q_k+1/ \in \Gamma_k \end{cases}$$

For the particular point $/\bar{s},\bar{t}/ \in \bar{D}_k$ we define K and H as

$$K = -/B^*/^{-1}A^* \quad \text{and} \quad H = /B^*/^{-1} \quad /6.129/$$

where $B^*$ is defined by /6.119/ and

$$A^* = [\Lambda_1 \, A_1^* \,|\, \Lambda_2 \, A_2^*] \quad /6.130/$$

$$A_1^* = \begin{bmatrix} A_{11}^* \\ A_{12}^* \\ \vdots \\ A_{1m}^* \end{bmatrix}, \quad A_2^* = \begin{bmatrix} A_{21}^* \\ A_{22}^* \\ \vdots \\ A_{2m}^* \end{bmatrix}$$

### Theorem 6.9

The decoupling problem has a solution if the conditions /i/, /ii/ are satisfied and $K = -/B^*/^{-1}A^*$ satisfies the equation

$$C_k \begin{bmatrix} /A+BK/_{11}^{s-1,t} & /A+BK/_{12}^{s,t-1} \\ /A+BK/_{21}^{s-1,t} & /A+BK/_{22}^{s,t-1} \end{bmatrix} B = 0 \quad /6.131/$$

for all $/s,t/ \in \bar{D}_k$ and $k = 1,\ldots,m$.

### Proof

To prove the theorem it is sufficient to show that /6.131/ implies /6.120/.

Note that if /6.131/ holds, then $c_k = c_k/s,t/ = 0$ for all points $/s',t'/ \geqslant /s,t/$, where $/s,t/ \in \bar{D}_k$. Therefore the condition /6.131/ implies the condition /6.120/ and by theorem 6.8 the matrices /6.129/ decouple RM. □

If the conditions of theorem 6.8 /or theorem 6.9/ are satisfied, then the matrices K and H may be found by the use of the following

## 6.3. Algorithm 6.5

**Step 1** Using /6.116/, /6.117/ and /6.118/ find the sets $\Gamma_k$, $\bar{\Gamma}_k$ and $B_k^*$ for $k = 1,\ldots,m$.

**Step 2** Using /6.119/ find $B^*$ and $H = /B^*/^{-1}$.

**Step 3** Find K such that /6.120/ is satisfied.

In particular case when /6.131/ is satisfied, find $A^*$ using /6.130/ and $K = -/B^*/^{-1}A^*$.

## Example 6.5

Given RM with

$$A = \begin{bmatrix} A_{11} & A_{12} \\ A_{21} & A_{22} \end{bmatrix} = \begin{bmatrix} 0 & -3 & | & -3 \\ 1 & 3 & | & 1 \\ \hline 1 & 4 & | & 1 \end{bmatrix}, \quad B = \begin{bmatrix} B_1 \\ B_2 \end{bmatrix} = \begin{bmatrix} 1 & 0 \\ 2 & 1 \\ \hline -4 & -1 \end{bmatrix} \qquad /6.132/$$

$$C = \begin{bmatrix} C_1 & C_2 \end{bmatrix} = \begin{bmatrix} 2 & 1 & | & -1 \\ 1 & 0 & | & 0 \end{bmatrix}$$

find the matrices K and H which decouple the system.

In this $n_1 = n_2$, $n_2 = 1$ and $m = l = 2$.

**Step 1** First output $k = 1$

$$C_1 \begin{bmatrix} 1 & 0 & | & 0 \\ 0 & 1 & | & 0 \\ \hline 0 & 0 & | & 0 \end{bmatrix} B = \begin{bmatrix} 4 & 1 \end{bmatrix}, \quad C_1 \begin{bmatrix} 0 & 0 & | & 0 \\ 0 & 0 & | & 0 \\ \hline 0 & 0 & | & 1 \end{bmatrix} B = \begin{bmatrix} 4 & 1 \end{bmatrix}$$

Hence $\bar{\Gamma}_1 = \emptyset$ /the null set/, $\Gamma_1 = \{(1, 0), (0, 1)\}$

and

$$B_1^* = \begin{bmatrix} 4, & 1 \end{bmatrix}$$

Second output k = 2

$$C_2 = \left[\begin{array}{ccc} 1 & 0 & 0 \\ 0 & 1 & 0 \\ \hline 0 & 0 & 0 \end{array}\right] \quad B = \left[\begin{array}{c} 1 \\ 0 \end{array}\right]$$

$$C_2 = \left[\begin{array}{ccc} 0 & 0 & 0 \\ 0 & 0 & 0 \\ \hline \end{array}\right] \quad B = \left[\begin{array}{cc} 0 & 0 \end{array}\right]$$

Hence $\bar{\Gamma}_2 = \{(0, 1)\}$, $\Gamma_2 = \{(1, 0)\}$ and $B_2^* = \begin{bmatrix} 1 & 0 \end{bmatrix}$

**Step 2** From /6.119/ we have

$$B^* = \begin{bmatrix} B_1^* \\ B_2^* \end{bmatrix} = \begin{bmatrix} 4 & 1 \\ 1 & 0 \end{bmatrix}$$

and

$$H = /B^*/^{-1} = \begin{bmatrix} 0 & 1 \\ 1 & -4 \end{bmatrix}$$

**Step 3** It is easy to check that the matrix

$$K = \begin{bmatrix} 4 & 3 & 3 \\ -1 & -1 & 0 \end{bmatrix}$$

satisfies the condition /6.120/

Checking, we have

$$A+BK = \begin{bmatrix} 4 & 0 & 0 \\ 8 & 8 & 7 \\ 14 & -7 & -11 \end{bmatrix}$$

$$BH = \begin{bmatrix} 0 & 1 \\ 1 & -2 \\ -1 & 0 \end{bmatrix}$$

and

$$G_c/z_1,z_2/ = C\left[Z-A-BK\right]^{-1}BH = \frac{1}{d/z_1,z_2/}\begin{bmatrix} n_1/z_1,z_2/ & 0 \\ 0 & n_2/z_1,z_2/ \end{bmatrix}$$

where

$$d/z_1,z_2/ = \det[Z-A-BK] = z_1^2 z_2 + 11z_1^2 - 12z_1 z_2 - 83z_1 + 32z_2 + 156$$
$$n_1/z_1,z_2/ = z_1 z_2 + z_1^2 - z_1 - 4z_2 - 12$$
$$n_2/z_1,z_2/ = z_1 z_2 + 11z_1 - 8z_2 - 39$$

### 6.6.4. Mertzios method of decoupling by dynamic state feedback

**Lemma 6.1**

The closed-loop transfer function matrix /6.115/ may be written as

$$G_c/z_1,z_2/ = \frac{1}{d/z_1,z_2/} \sum_{i=0}^{n-1} CR_i BH \qquad /6.133/$$

where

$$d/z_1,z_2/ = 1 + \sum_{i=1}^{n} a_i \qquad /6.134/$$

$$R_i = Q^i - a_1 Q^{i-1} a_2 Q^{i-2} - \ldots - a_i I_n \qquad /6.135/$$

$$Q = I_n - Z + A + BK \qquad /6.136a/$$

$$a_1 = -\text{tr}Q, \quad a_2 = -\tfrac{1}{2}\text{tr}\big[Q\,R_1\big], \ldots a_n = -\tfrac{1}{n}\text{tr}\big[Q\,R_n\big] \qquad /6.136b/$$

**Proof**

Note that the closed-loop transfer function matrix /6.115/ may be written as

$$G_c/z_1,z_2/ = C\big[I_n\lambda - /I_n\lambda - Z + A + BK/\big]^{-1} BH = C\big[I_n\lambda - Q_\lambda\big]^{-1} BH \qquad /6.137/$$

where

$$Q_\lambda = I_n\lambda - Z + A + BK \qquad /6.138/$$

and $\lambda$ is a pseudo-variable.

Using the well-known Leverrier algorithm we can /6.137/ expand as

$$G_c/z_1,z_2/ = \frac{1}{d_\lambda/z_1,z_2/} C\big[\lambda^{n-1}R_0 + \lambda^{n-2}R_1 + \ldots + \lambda R_{n-2} + R_{n-1}\big]BH \qquad /6.139/$$

where

$$d_\lambda/z_1,z_2/ = \det\big[I_n\lambda - Q\big] = \lambda^n + a_1 \lambda^{n-1} + \ldots + a_n \qquad /6.140/$$

$R_i$ and $a_1$ is defined by /6.135/ and /6.136b/, respectively.

Substitution of $\lambda = 1$ into /6.139/ and /6.140/ yields the desired formulae /6.133/ and /6.134/.

## Lemma 6.2

For any square matrix

$$P = \begin{bmatrix} P_{11} & P_{12} \\ P_{21} & P_{22} \end{bmatrix} \qquad /6.141/$$

the following relationship holds

$$P^k = \sum_{i=0}^{k} P^{k-1,i} \quad \text{for } k = 1,2,\ldots,n \qquad /6.142/$$

### Proof

The lemma will be proved by induction. Note that the hypotesis is true for $k = 1$ and $k = 2$, since

$$P = P^{1,0} + P^{0,1} = \begin{bmatrix} P_{11} & P_{12} \\ 0 & 0 \end{bmatrix} + \begin{bmatrix} 0 & 0 \\ P_{21} & P_{22} \end{bmatrix}$$

and

$$P_2 = /P^{1,0}+P^{0,1}/^2 = /P^{1,0}/^2 + P^{1,0}P^{0,1}+P^{0,1}P^{1,0} + /P^{0,1}/^2 =$$

$$= P^{2,0} + P^{1,1} + P^{0,2}$$

Assuming that the hypotesis is true for $m$ $/m \geqslant 1/$ we shall show that it is also valid for $m+1$.

Using the definition of $P^{i,j}$ we can write

$$P^{m+1} = P^m/P^{1,0}+P^{0,1}/ = \sum_{i=0}^{m} P^{m-i,i}/P^{1,0}+P^{0,1}/ = \sum_{i=0}^{m+1} P^{m-i+1,i}$$

□

### Definition 6.11

For each output $k$ $/k = 1,2,\ldots,l/$, let $\Gamma_k$ denote the set of all pairs $/p_k,q_k/$ for which the following relationship holds

$$C_k \left[\overline{A}+BK\right]^{s,t} B \begin{cases} = 0 \text{ for } \forall /s,t/ < /p_k,q_k/ \\ \neq 0 \text{ for } \forall /s,t/ = /p_k,q_k/ \end{cases} \qquad /6.143/$$

where $\overline{A} = I_n-Z+A$ and $C_k$ is the kth row of $C$.

Let $/\bar{p}_k,\bar{q}_k/$ denote the point of $\Gamma_k$ which is the nearest /i.e. the point that has the least distance/ to the i-axis and $/\hat{p}_k,\hat{q}_k/$ denote the nearest point to the j-axis.

#### Definition 6.12

For each output k /k = 1,2,...,l/, let $\overline{\Gamma}_k$ denote the set of all points /s,t/ for which the following relationship holds

$$C_k[\bar{A}+BK]^{s,t}B = 0 \quad \text{for } \forall /s,t/ < /p_k,q_k/ \quad \text{and}$$

$$/\bar{p}_k+1,0/ \leq /s,t/ \leq /n_1,q_k-1/, \quad /0,\hat{q}_k+1/ \leq /s,t/ \leq /\hat{p}_k-1,n_2/$$

/6.144/

#### Definition 6.13

For each output k /k = 1,2,...,l/, let the integer $d_k$ be given by

$$d_k = \min/p_k+q_k/ \quad \text{for } /p_k,q_k/ \in \overline{\Gamma}_k \qquad /6.145/$$

Using /6.142/ we can write

$$C_k[A+BK]^m B = C_k\Big[/\bar{A}+BK/^{m,0} + /\bar{A}+BK/^{m-1,1} + \ldots$$

$$+ /\bar{A}+BK/^{p_k,q_k} + \ldots + /\bar{A}+BK/^{1,m-1} + \qquad /6.146/$$

$$+ /\bar{A}+BK/^{0,m}\Big] \qquad /k = 1,2,\ldots,l; \; m = 1,2,\ldots,n/$$

From /6.146/ and definition 6.13 it follows that

$$d_k = \min\{i: C_k \bar{A}^{-i} B \neq 0, \quad i = 0,1,\ldots,n-1\}$$

/6.147a/

or

$$d_k = n-1 \quad \text{if } C_k \bar{A}^i B = 0 \quad \text{for all } i \qquad /6.147b/$$

Using /6.147/ it can be easily shown that

$$C_k/\bar{A}+BK/^m = \begin{cases} C_k \bar{A}^m & \text{for } m = 0,1,\ldots,d_k \\ C_k \bar{A}^{d_k}/\bar{A}+BK/^{n-d_k} & \text{for } m = d_k+1,\ldots,n-1 \end{cases} \qquad /6.148/$$

#### Theorem 6.10

There exists a pair of 2-D polynomial matrices $K/z_1,z_2/$ and $H/z_1,z_2/$ which decouple RM, if and only if

$$\det B^* \neq 0 \qquad /6.149/$$

where

$$B^* = B^*/z_1,z_2/ = \begin{bmatrix} C_1 \bar{A}^{d_1} B \\ C_2 \bar{A}^{d_2} B \\ \cdots \cdots \\ C_m \bar{A}^{d_m} B \end{bmatrix} \qquad /6.150/$$

Proof

To prove the necessity let us assume that there exists a pair of polynomial matrices $K/z_1,z_2/$ and $H/z_1,z_2/$ which decouples RM, i.e. $G/z_1,z_2/$ is a nonsingular and diagonal matrix. From /6.143/, /6.145/ and /6.135/ it follows that

$$R_i = 0 \quad \text{for } i = 0,1,\ldots,d_k \qquad /6.151/$$

Taking into account /6.151/ and /6.148/ we obtain for kth row of the closed-loop transfer function matrix /6.133/

$$G_{ck}/z_1,z_2/ = \frac{1}{d/z_1,z_2/} \sum_{i=d_k}^{n-1} C_k R_i BH = \qquad /6.152/$$

$$= \frac{1}{d/z_1,z_2/} C_k \left[ \bar{A}^{d_k} + \bar{A}^{d_k}/\bar{A}+BK/ - a_1 \bar{A}^{d_k} + \right.$$

$$+ \bar{A}^{d_k}/\bar{A}+BK/^2 + \ldots + \bar{A}^{d_k}/\bar{A}+BK/^{n-d_k-1} +$$

$$\left. -a_1 \bar{A}^{d_k}/\bar{A}+BK/^{n-d_k-2} - \ldots - a_{n-d_k-1} \bar{A}^{d_k} \right] BH$$

Hence using /6.150/ we get

$$G_c/z_1,z_2/ = \frac{1}{d/z_1,z_2/} \left[ B^* + S_1 \left[ A^*/\bar{A}+BK/B - a_1 B^* \right] + \ldots \right.$$

$$+ S_{n-d_r-1} \left[ A^*/\bar{A}+BK/^{n-d_r-1} B + \right.$$

$$\left. -a_1 A^*/\bar{A}+BK/^{n-d_r-2} - \ldots - a_{n-d_r-1} B^* \right] H$$

where

$$S_j = \begin{bmatrix} S_{ij} & 0 & \cdots & 0 \\ 0 & S_{zj} & \cdots & 0 \\ 0 & 0 & \cdots & S_{nj} \end{bmatrix}, \quad j = 1, 2, \ldots, n-d_r-1$$

$$d_r = \min d_k, \quad k = 1, 2, \ldots, m$$

$$S_{kj} \begin{cases} = 1 & \text{if } n-d_k-j+1 \geqslant 0 \\ = 0 & \text{if } n-d_k-j+1 < 0 \end{cases}$$

$$A^* = \begin{bmatrix} C_1 \bar{A}^{d_1} \\ C_2 \bar{A}^{d_2} \\ \cdots \cdots \\ C_m \bar{A}^{d_m} \end{bmatrix} \qquad \qquad /6.153/$$

Note that the first term $B^*H$ in /6.152a/ is not affected by K and must be a nonsingular and diagonal matrix. This implies /6.149/, since H is nonsingular.

To prove sufficiency we assume that /6.149/ is satisfied and we choose

$$K = -/B^*/^{-1} A^* \bar{A} \qquad /6.154/$$

$$H = /B^*/^{-1} \qquad /6.155/$$

Taking into account /6.148/ we obtain

$$C_k /\bar{A}+BK/^{d_k+1} = C_k \bar{A}^{d_k}/A+BK/ = A_k^* \bar{A} + B_k^* K \qquad /6.156/$$

where $A_k^*$ and $B_k^*$ are kth row of $A^*$ and $B^*$, respectively.

Substitution of /6.154/ into /6.156/ yields

$$C_k /\bar{A}+BK/^{d_k+1} = A_k^* \bar{A} + B_k^* \left[ -/B^*/^{-1} A^* \bar{A} \right] = 0$$

Therefore

$$A^* /\bar{A}+BK/^k = 0 \quad \text{for} \quad k = 1, 2, \ldots \qquad /6.157/$$

Taking into account /6.157/ and substituting /6.155/ into /6.152/ we obtain a diagonal matrix of the form

$$G_c/z_1,z_2/ \; \frac{1}{d/z_1,z_2/} \left[ I - a_1 S_1 - \cdots - a_{n-d_r-1} S_{n-d_r-1} \right] \qquad /6.158/$$

Since each polynomial $a_{i+1}$ contains at least a term of higher power of these of $a_i$, it is clear that no one of the diagonal elements of $G_c/z_1,z_2/$ can be zero. □

If the condition /6.149/ is satisfied polynomial matrices $K = K/z_1,z_2/$, $H = H/z_1,z_2/$ which decouple RM may be found the use of the following

### Algorithm 6.6

**Step 1** Find the matrix
$$\bar{A} = I_n - Z + A$$

**Step 2** Using /6.147/ find $d_k$ for $k = 1,2,\ldots,l$.

**Step 3** Find $B^*$ and $A^*$ defined by /6.150/ and /6.153/.

**Step 4** Using /6.154/ and /6.155/ find the desired matrices K and H.

### Example 6.6

For RM with /6.132/ find the decoupling matrices K and H.

**Step 1**

$$\bar{A} = I_n - Z + A = \begin{bmatrix} 1-z_1 & -3 & -3 \\ 1 & 4-z_1 & 1 \\ \hline 1 & 4 & 2-z_2 \end{bmatrix}$$

**Step 2** First output, $k = 1$

$$C_1 B = \begin{bmatrix} 2 & 1 & -1 \end{bmatrix} \begin{bmatrix} 1 & 0 \\ 2 & 1 \\ -4 & -1 \end{bmatrix} = \begin{bmatrix} 8 & 2 \end{bmatrix}$$

and $d_1 = 0$

Second output, $k = 2$

$$C_2 B = \begin{bmatrix} 1 & 0 & 0 \end{bmatrix} \begin{bmatrix} 1 & 0 \\ 2 & 1 \\ -4 & -1 \end{bmatrix} = \begin{bmatrix} 1 & 0 \end{bmatrix}$$

and $d_2 = 0$.

Step 3

$$B^* = \begin{bmatrix} C_1\bar{A}^{d_1}B \\ C_2\bar{A}^{d_2}B \end{bmatrix} = \begin{bmatrix} C_1B \\ C_2B \end{bmatrix} = \begin{bmatrix} 8 & 2 \\ 1 & 0 \end{bmatrix}$$

and

$$A^* = \begin{bmatrix} C_1\bar{A}^{d_1} \\ C_2\bar{A}^{d_2} \end{bmatrix} = \begin{bmatrix} C_1 \\ C_2 \end{bmatrix} = \begin{bmatrix} 2 & 1 & -1 \\ 1 & 0 & 0 \end{bmatrix}$$

Step 4  The desired matrices

$$K = -/B^*/^{-1}A^*\bar{A} = -\begin{bmatrix} 0 & 1 \\ \tfrac{1}{2} & -4 \end{bmatrix}\begin{bmatrix} 2 & 1 & -1 \\ 1 & 0 & 0 \end{bmatrix}\begin{bmatrix} 1-z_1 & -3 & -3 \\ 1 & 4-z_1 & 1 \\ 1 & 4 & 2-z_2 \end{bmatrix} =$$

$$= \begin{bmatrix} z_1-1, & 3, & 3 \\ 3-3z_1, & \tfrac{1}{2}z_1-9, & -\tfrac{1}{2}z_2-8\tfrac{1}{2} \end{bmatrix}$$

and

$$H = /B^*/^{-1} = \begin{bmatrix} 0 & 1 \\ \tfrac{1}{2} & -4 \end{bmatrix}$$

Checking, we have

$$A+BK = \begin{bmatrix} z_1-1, & 0, & 0 \\ 2-z_1, & \tfrac{1}{2}z_1, & -\tfrac{1}{2}z_2-\tfrac{3}{2} \\ 2-z_1, & 1-\tfrac{1}{2}z_1, & \tfrac{1}{2}z_2-\tfrac{5}{2} \end{bmatrix}$$

and

$$G_c/z_1,z_2/ = \frac{1}{\tfrac{1}{2}z_1+\tfrac{1}{2}z_2+\tfrac{3}{2}}\begin{bmatrix} \tfrac{1}{4}z_1+\tfrac{1}{4}z_2+\tfrac{5}{4}, & 0 \\ 0, & 1 \end{bmatrix}$$

### 6.6.5. Class of decoupling matrices

We shall establish the necessary and sufficient conditions for K and H to be a decoupling pair, under the assumption that the condition /6.149/ holds.

The class of H is given by

$$H = /B^*/^{-1} D \qquad /6.159/$$

where D is a diagonal nonsingular polynomial matrix.

Let

$$T_k = T_k/K/ = \begin{bmatrix} C_k/\bar{A}+BK/^{n-1}B \\ C_k/\bar{A}+BK/^{n-2}B \\ \cdots \cdots d_k \cdots \\ C_k/\bar{A}+BK/^{d_k}B \\ 0 \end{bmatrix} \in R^{m \times n}[z_1, z_2] \qquad /6.160/$$

### Theorem 6.11

The matrices K and /6.159/ decouples RM if and only if

$$\text{rank } T_k = 1 \quad \text{for} \quad k = 1, 2, \ldots, m \qquad /6.161/$$

### Proof

To prove the necessity we suppose that K decouples RM. Note that /6.152/ can be written as

$$G_{ck}/z_1, z_2/ = \frac{1}{d/z_1, z_2/} \Big[ B_k/1-a_1-\cdots-a_{n-d_k-1}/ + C_k/\bar{A}+BK/^{d_k+1} B/1-a_1-\cdots$$
$$-a_{n-d_k-2}/ + \cdots + C_k/\bar{A}+BK/^{n-2}B/1-a_1/ + C_k/\bar{A}+BK/^{n-1}B \Big] H \qquad /6.162/$$

For decoupled RM $G_{ck}/z_1, z_2/$ is diagonal and nonsingular. Therefore, $G_{ck}/z_1, z_2/$ should have only a nonzero element in the kth position. The polynomials $1-a_1-\cdots-a_{n-d_k-1}$, $1-a_1-\cdots-a_{n-d_k-2}, \ldots, 1-a_1$, 1, which are linearly independent, are the coefficients of the rows $B_k^*$, $C_k/\bar{A}+BK/^{d_k+1}B, \ldots, C_k/\bar{A}+BK/^{n-2}B$ and $C_k/\bar{A}+BK/^{n-1}$. We conclude that rows of /6.160/ should to linearly dependent and rank $T_k = 1$.
To prove sufficiency let us assume that /6.161/ is satisfied.
From /6.148/ and /6.150/ it follows that

$$C_k/\bar{A}+BK/^{d_k}B = C_k \bar{A}^{d_k} B = B_k^* \neq 0 \qquad /6.163/$$

Therefore the matrix /6.160/ takes the form

$$T_k = \begin{bmatrix} k_{n-1}B_k^* \\ k_{n-2}B_k^* \\ \cdots \\ k_{d_k+1}B_k^* \\ B_k^* \\ 0 \end{bmatrix} \qquad /6.164/$$

where $k_i$ /i = $d_k+1,\ldots,n-1$/ are scalars.

Using /6.164/ and /6.159/ we can write /6.162/ as

$$G_{c_k}/z_1,z_2/ = \frac{1}{d/z_1,z_2/} p/z_1,z_2/ B_k^* H = \frac{p/z_1,z_2/}{d/z_1,z_2/} e_k \qquad /6.165/$$

where
$$p/z_1,z_2/ = k_{n-1}+k_{n-2}/1-a_1/+\ldots+k_{d_k+1}/1-a_1-\ldots-a_{n-d_k-2}/ +$$
$$+/1-a_1-\ldots-a_{n-d_k-1}/$$

$e_k$ = /0...0$d_k$ 0.../ and $d_k$ is the kth element of the diagonal matrix D. Thus the matrices F and /6.159/ decouple RM. □

Let us define the matrix K as follows

$$K = /B^*/^{-1} \left[ \sum_{i=0}^{d_{max}} M_i C\bar{A}^i - A^*\bar{A} \right] \qquad /6.166/$$

where

$$M_i = \begin{bmatrix} m_{i1} & 0 & \cdots & 0 \\ 0 & m_{i2} & \cdots & 0 \\ \cdots & \cdots & \cdots & \cdots \\ 0 & 0 & & m_{im} \end{bmatrix} \qquad /6.167a/$$

and

$$m_{ij} = 0 \quad \text{for} \quad i = d_j+1,\ldots,d_{max} \qquad /6.167b/$$

$$d_{max} = \max d_i \qquad /6.167c/$$

**Theorem 6.12**

The matrix /6.166/ satisfied the condition /6.161/.

## Proof

From /6.166/ and $B^* = A^* B$ it follows that

$$\sum_{i=0}^{d_{max}} M_i CA^i = A\bar{A} + BK = A/\bar{A}+BK/ \qquad /6.168/$$

Using /6.148/ and /6.168/ we can write

$$C_k/\bar{A}+BK/^{d_k+1} B = C_k \bar{A}^{d_k}/\bar{A}+BK/B = \sum_{i=0}^{d_{max}} m_{ik} C_k \bar{A}^i B = m_{d_k,k} B_k^* \qquad /6.169/$$

Furthermore, using /6.148/, /6.167b/, /6.168/ and /6.169/ we obtain

$$C_k/\bar{A}+BK/^{d_k+2} B = C_k \bar{A}^{d_k}/\bar{A}+BK/^2 B = \sum_{i=0}^{d_{max}} m_{ik} C_k \bar{A}^i/\bar{A}+BK/B =$$

$$= m_{d_k-1,k} B_k^* + m_{d_k,k}^2 B_k^* \qquad /6.170/$$

Similarly we get

$$C_k/\bar{A}+BK/^{d_k+j+1} = \sum_{i=0}^{d_{max}} m_{ik} C_k \bar{A}^i/\bar{A}+BK/^j B = \sum_{i=0}^{j} m_{d_k-j+1,k} C_k$$

$$/\bar{A}+BK/^{d_k+i} B \quad \text{for } j = 0,1,\ldots,n-d_k-1 \qquad /6.171/$$

For /6.169/-/6.171/ it follows that the matrix /6.166/ satisfied the condition /6.161/

Note that in /6.166/ there are $m + \sum_{i=1}^{m} d_i$ unspecified nonzero elements of the matrices $M_i$. The unspecified elements can be used in design procedure for meeting other additional requirements, for example in eigenvalue assignment.

## Problems

**6.1.** Show that a full order observer for RM exists if

/i/ $A_{12}C_2^+C_2 = A_{12}$

/ii/ the pair $//I-C_2C_2^+/C_1, A_{11}-A_{12}C_2^gC_1/$ is detectable

/iii/ the pair $/A_{22},C_2/$ is detectable

where $C_2^g$ denotes the generalized inverse matrix of $C_2$, satisfying the condition $C_2C_2^gC_2 = C_2$.

Hint: Show that the state error vector $e = z-x$ satisfies the equation $e' = Fe$, where $F = A-HC$. Next show that if /i/-/iii/ are satisfied then H may be chosen so that $F^{i,j} \to 0$ as $i,j \to \infty$.

**6.2.** Consider RM described by /6.1/ and /6.2/ with the given boundary conditions /6.3/ and

$$w/i,j/ = [M_1 M_2] \begin{bmatrix} x^h/i,j/ \\ x^v/i,j/ \end{bmatrix} \quad /a/$$

Show that the 2-D system described by /6.4/ and

$$\hat{w}/i,j/ = [L_1 L_2] \begin{bmatrix} z^h/i,j/ \\ z^v/i,j/ \end{bmatrix} + K\, y/i,j/$$

is an asymptotic observer of /a/, i.e. $\lim_{i,j \to \infty} \hat{w}/i,j/ = \lim_{i,j \to \infty} w/i,j/$, if the following relations hold

$$T_1 A_{11} = F_{11}T_1 + H_1C_1, \quad T_1 A_{12} = F_{12}T_2 + H_1C_2$$

$$T_2 A_{21} = F_{21}T_1 + H_2C_1, \quad T_2 A_{22} = F_{22}T_2 + H_2C_2$$

$$G_1 = T_1 B_1, \quad G_2 = T_2 B_2, \quad M_1 = L_1 T_1 + KC_1,$$

$$M_2 = L_2 T_2 + KC_2$$

where

$$T = \begin{bmatrix} T_1 & 0 \\ 0 & T_2 \end{bmatrix} \quad /6.a/$$

Hint: Use a similar way as in proof of theorem 6.1.

6.3. Show that the closed-loop system consisting of the system /6.1/, /6.2/, the observer /6.4/, /6.5/ and the state feedback law $u = K_f x$ is described by the equation

$$\begin{bmatrix} x' \\ e' \end{bmatrix} = \begin{bmatrix} A+BK_f & BK_f L \\ 0 & F \end{bmatrix} \begin{bmatrix} x \\ e \end{bmatrix}$$

Hint: Use /6.1/, /6.14/ and /6.12/

6.4. Show that for square matrices P and Q

$$/P+Q/^{i,j} = P^{i,j} + P^{i-1,j}Q^{1,0} + P^{i,j-1}Q^{0,1} + \ldots + P^{0,1}Q^{i,j-1} +$$
$$+ P^{1,0}Q^{i-1,j} + Q^{i,j} \quad \text{for } i,j \geqslant 0$$

Hint: By induction using $P^{i,j} = P^{1,0}P^{i-1,j} + P^{0,1}P^{i,j-1}$ and $Q^{1,0}Q^{i-1,j} + Q^{0,1}Q^{i,j-1}$.

6.5. Let $/p_k, q_k/$ and $/\bar{p}_k, \bar{q}_k/$ be pairs of integers defined by

$$/p_k, q_k/ = \min\{/s,t/: C_k A^s, {}^t B^{1,0} \neq 0, \; /0,0/ \leqslant /s,t/ < /n_1, n_2/\}$$

$$/\bar{p}_k, \bar{q}_k/ = \min\{/s,t/: C_k A^s, {}^t B^{0,1} \neq 0, \; /0,0/ \leqslant /s,t/ < /n_1, n_2/\}$$

where by minimum over the set /s,t/ we mean the minimum order $\varrho$ of the pair /s,t/ which is defined by $\varrho = s+t$.

Show that RM is decouplable if the matrix $B^*$, where its kth row $B_k^*$ is given by

$$B_k^* \begin{cases} = C_k A^{p_k, q_k} B^{1,0} & \text{for } p_k + q_k < \bar{p}_k + \bar{q}_k \\ = C_k A^{p_k, q_k} B^{0,1} & \text{for } \bar{p}_k + \bar{q}_k < p_k + q_k \end{cases}$$

is nonsingular and the following conditions are satisfied:
If $p_k + q_k < \bar{p}_k + \bar{q}_k$, then

$$C_k A^{i,j} = 0 \quad \text{for } i+j = p_k + q_k + 1 \text{ except } /i,j/ = /p_k + 1, q_k/$$

and if $\bar{p}_k + \bar{q}_k < p_k + q_k$, then

$$C_k A^{i,j} = 0 \quad \text{for } i+j = \bar{p}_k + \bar{q}_k + 1 \text{ except } /i,j/ = /\bar{p}_k, \bar{q}_q + 1/$$

Hint: Use $H = /B^*/^{-1}$ and $K = -/B^*/^{-1}A^*$, where the kth row $A_k^*$ of the matrix $A^*$ is given by

$$A_k^* \begin{cases} = C_k A^{p_k+1,q_k} & \text{if } p_k+q_k < \bar{p}_k+\bar{q}_k \\ = C_k A^{p_k,q_k+1} & \text{if } \bar{p}_k+\bar{q}_k < p_k+q_k \end{cases}$$

6.6. For RM with

$$A = \begin{bmatrix} 0 & 1 & 0 \\ 1 & -3 & 1 \\ \hdashline -1 & 0 & -1 \end{bmatrix}, \quad B = \begin{bmatrix} 0 & 0 \\ 1 & 0 \\ \hdashline 0 & 1 \end{bmatrix}, \quad C = \begin{bmatrix} 1 & 0 & 0 \\ 0 & 0 & 1 \end{bmatrix}$$

find K and H such that

$$G_c/z_1,z_2/ = C\left[Z-A-BK\right]^{-1}BH = \begin{bmatrix} \frac{1}{z_1} & 0 \\ 0 & \frac{1}{z_2} \end{bmatrix}$$

Answer:

$$K = \begin{bmatrix} -1 & 3 & -1 \\ 1 & 0 & 1 \end{bmatrix}, \quad H = \begin{bmatrix} 1 & 0 \\ 0 & 1 \end{bmatrix}$$

Hint: Use $K = -/B^*/^{-1}A^*$ and $H = /B^*/^{-1}$.

6.7. Show that if the matrices K and H decouple RM with the matrices A,B,C, then the matrices $\bar{K} = KT$ and $\bar{H} = H$ decouple RM with $\bar{A} = TAT^{-1}$, $\bar{B} = TB$ and $\bar{C} = CT^{-1}$ where T is given by /6.a/.

Hint: Show that $\bar{C}\left[Z-\bar{A}-\bar{B}\bar{K}\right]^{-1}\bar{B}\bar{H} = C\left[Z-A-BK\right]^{-1}BH$.

## References

[1] T.Hinamoto, F.W.Fairman, J.Shimonishi: Stabilization of 2D filters using 2D observers. Int.J.Systems Sci. vol.13, No.2, 1982, pp. 177-191

[2] S.Kawaji: Minimal order state observer for two-dimensional systems. Reprint of 9th IFAC World Congress, Budapest, July 2-6, 1984, vol. IX, pp. 153-158

[3] B.Mertzios: Decoupling of 2-D systems by dynamic state feedback controllers. Reprint of 9th IFAC World Congress, Budapest, July 2-6, 1984, vol. VIII, pp. 144-149

[4] B.Mertzios, P.N.Paraskevopoulos: On the input-output decoupling of 2-D systems by state feedback. Journal of the Franklin Institute, vol. 314, No.1, July 1982, pp. 55-76

[5] P.N.Paraskevopoulos: Exact model-matching of 2-D system via state feedback. Journal of the Franklin Institute, vol. 308, No.5, Nov.1979, pp. 475-486

[6] P.N.Paraskevopoulos: Transfer function matrix synthesis of two-dimensional system. IEEE Trans.Autom.Control, vol.AC-25, No.2, April 1980, pp. 321-324

[7] P.N.Paraskevopoulos, O.I.Kosmidou: Dynamic compensation for exact model-matching of two-dimensional systems. Int.J.Systems Sci. vol. 11, No.10, 1980, pp. 1163-1175

[8] P.N.Paraskevopoulos, P.Stavroulakis: Decoupling of linear multi-variable two-dimensional system via state feedback. IEEE Proc., vol 129, Pt.D, No.1, Jan. 1982, pp. 15-20

[9] T.G.Pimenides, S.G.Tzafestas: Feedback decoupling-controller design of 3-D systems in state space. Mathematics and Computers in Simulation XXIV /1982/, pp.341-352. North Holland Publishing Company

[10] M.Sebek: Model matching of 2-D multi-input multi-output systems. Reprints of 9th IFAC World Congress, Budapest, July 2-6, 1984, vol. IX, pp. 148-152

[11] P.Stavroulakis, P.N.Paraskevopoulos: Low- sensitivity observer-compensator design for two-dimensional digital systems. IEE Proc. vol. 129, Pt.D, No.5, Sept.1982, pp. 193-200

[12] P.Stavroulakis, P.N.Paraskevopoulos: Reduced order feedback law implementation for 2-D digital systems. Signal Processing: Theories and Applications. M. Kunt and F de Coulon /ed./, North Holland Publishing Company EURASIP 1980, pp.403-408

[13] P.Stavroulakis, P.N.Paraskevopoulos: Reduced order feedback law implementation for 2-D digital systems. Int. J.Systems Sci., vol. 12, No.5, 1981, pp. 525-537

[14] S.G.Tzafestas, T.G.Pimenides: Exact model-matching control of three-dimensional systems using state and output feedback. Int.J.Systems Sci., 1982, vol. 13, No. 11, pp. 1171-1187

[15] S.G.Tzafestas, T.G.Pimenides: State observer design for 3-dimensional systems. Proc MELECON'83, Athens, May 24-26,1983, pp. 182-186

[16] M.Morf, B.C.Lévy, S.Y. Kung : New results in 2-D systems theory, Part I: 2-D polynomial matrices factorization and coprimeness. Proc. IEEE, vol. 65, No.6, June 1977, pp.861-872.

# 7. DEADBEAT CONTROL AND DEADBEAT SERVO PROBLEM

## 7.1. POLYNOMIAL DESIGN OF DEADBEAT CONTROL LAWS

### 7.1.1. Problem formulation

Consider the Roesser's model /RM/ described by the equations

$$x' = Ax + Bu \qquad /7.1/$$
$$y = Cx \qquad /7.2/$$

where

$$x' = \begin{bmatrix} x^h/i+1,j/ \\ x^v/i,j+1/ \end{bmatrix}, \quad x = \begin{bmatrix} x^h/i,j/ \\ x^v/i,j/ \end{bmatrix}$$

$$A = \begin{bmatrix} A_{11} & A_{12} \\ A_{21} & A_{22} \end{bmatrix}, \quad B = \begin{bmatrix} B_1 \\ B_2 \end{bmatrix}, \quad C = \begin{bmatrix} C_1 & C_2 \end{bmatrix}$$

$x^h/i,j/ \in R^{n_1}$ is the horizontal state vector,

$x^v/i,j/ \in R^{n_2}$ is vertical state vector,

$u = u/i,j/ \in R^m$ is the input vector,

$y = y/i,j/ \in R^l$ is the output vector,

$A_{ij}$, $B_i$, $C_i$ are real constant matrices of appropriate dimensions.

The boundary conditions of RM are given by

$$x^h/0,j/, \quad j = 0,1,2,\ldots \text{ and } x^v/i,0/, \quad i = 0,1,2,\ldots \qquad /7.3/$$

It is assumed that the boundary conditions are finite, i.e. there exist positive integers $I_o$ and $J_o$ such that

$$x^h/0,j/ = 0 \text{ for } j > J_o \text{ and } x^v/i,0/ = 0 \text{ for } i > I_o \qquad /7.3a/$$

Using 2-D Z transformation to /7.1/ we obtain

$$\bar{A}X = \bar{B}U + \bar{C} \qquad /7.4/$$

where

$$\bar{A} = \begin{bmatrix} I_{n_1} - d_1 A_{11}, & -d_1 A_{12} \\ -d_2 A_{21}, & I_{n_2} - d_2 A_{22} \end{bmatrix} \in R^{n \times n}[d_1, d_2],$$

$$\bar{B} = \begin{bmatrix} d_1 B_1 \\ d_2 B_2 \end{bmatrix} \in R^{n \times m}[d_1, d_2] \qquad n = n_1 + n_2$$

/7.4a/

$\bar{C} \in R^{n \times 1}[d_1, d_2]$ depends on the boundary conditions /7.3/,

$R^{k \times l}[d_1, d_2]$ denotes the set of kxl polynomial matrices in $d_1 = z_1^{-1}$ and $d_2 = z_2^{-1}$, X, U are 2-D Z transforms of x/i,j/ and u/i,j/, respectively.

Let us consider a 2-D linear dynamic regulator of the form

$$U = -Q P^{-1} X \qquad /7.5/$$

where

$$Q \in R^{m \times n}[d_1, d_2], \quad P \in R^{n \times n}[d_1, d_2]$$

It is desired to find the regulator which transfers any finite boundary conditions /7.3a/ to zero for minimal i and j.

Thus, the problem can be stated as follows. Given the matrices A and B, find polynomial matrices P and Q such that X is a polynomial vector of least degrees possible in $d_1$ and $d_2$.

## Problem solution

Substitution of /7.5/ into /7.4/ yields

$$/\bar{A} P + \bar{B} Q / P^{-1} X = \bar{C}$$

or

$$X = P / \bar{A} P + \bar{B} Q /^{-1} \bar{C} \qquad /7.6/$$

and

$$U = -Q / \bar{A} P + \bar{B} Q /^{-1} \bar{C} \qquad /7.7/$$

Note that X and U are polynomial vectors for any $\bar{C}$ if and only if

$$\bar{A} P + \bar{B} Q = I_n \qquad /7.8/$$

In this case from /7.6/ and /7.7/ we have

$$X = P\bar{C}, \quad U = -Q\bar{C} \qquad /7.9/$$

The equation /7.8/ has a solution if and only if the matrices $\bar{A}$, $\bar{B}$ are zero left coprime /ZLC/ /see Appendix/.

Note that X is a polynomial vector of least degrres possible in $d_1$ and $d_2$ for any $\bar{C}$ if each column of P is of least degrees possible in $d_1$ and $d_2$.

Therefore we have established the following

### Theorem 7.1

The problem has a solution if and only if $\bar{A}$ and $\bar{B}$ are ZLC.

If $\bar{A}$ and $\bar{B}$ are ZLC, then matrices P, Q can be found by the use of the following

### Algorithm 7.1

**Step 1.** Using /7.4a/ find $\bar{A}$ and $\bar{B}$

**Step 2.** Carry out the reduction /using elementary column operations/

$$\begin{bmatrix} \bar{A} & \bar{B} \\ \hline I_n & 0 \\ \hline 0 & I_m \end{bmatrix} \rightarrow \begin{bmatrix} I_n & 0 \\ \hline U_1 & U_2 \\ \hline U_3 & U_4 \end{bmatrix}$$

and find the general solution

$$\begin{bmatrix} P \\ \hline Q \end{bmatrix} = \begin{bmatrix} U_1 & U_2 \\ \hline U_3 & U_4 \end{bmatrix} \begin{bmatrix} I_n \\ \hline T \end{bmatrix}$$

to the equation /7.8/ where:

T is an arbitrary polynomial matrix.

**Remark:** The algorithm given in Appendix can be also used for finding P and Q.

**Step 3.** Choose T so that each column of P is of least possible degrees in $d_1$ and $d_2$.

## Example 7.1

Given RM with

$$A = \begin{bmatrix} 0 & 1 \\ -1 & 2 \end{bmatrix}, \qquad B = \begin{bmatrix} 0 \\ 1 \end{bmatrix}$$

find P and Q of the regulator. In this case $n_1 = n_2 = 1$ and $m = 1$.

### Step 1

$$\bar{A} = \begin{bmatrix} I_{n_1} - d_1 A_{11}, & -d_1 A_{12} \\ -d_2 A_{21}, & I_{n_2} - d_2 A_{22} \end{bmatrix} = \begin{bmatrix} 1, & -d_1 \\ d_2, & 1 - 2d_2 \end{bmatrix}$$

$$\bar{B} = \begin{bmatrix} d_1 B_1 \\ d_2 B_2 \end{bmatrix} = \begin{bmatrix} 0 \\ d_2 \end{bmatrix}$$

/7.10/

It is easy to check that /7.10/ are ZLC.

### Step 2

Using elementary column operations we carry out the reduction

$$\begin{bmatrix} \bar{A} & | & \bar{B} \\ \hline I_n & | & 0 \\ \hline 0 & | & I_m \end{bmatrix} = \left[ \begin{array}{cc|c} 1 & -d_1 & 0 \\ d_2 & 1-2d_2 & d_2 \\ \hline 1 & 0 & 0 \\ 0 & 1 & 0 \\ \hline 0 & 0 & 1 \end{array} \right] \rightarrow \left[ \begin{array}{cc|c} 1 & 0 & 0 \\ 0 & 1 & 0 \\ \hline 1 & d_1 & -d_1 d_2 \\ 0 & 1 & -d_2 \\ \hline -1 & 2-d_1 & 1+d_1 d_2 - 2d_2 \end{array} \right]$$

Hence the general solution to the equation

$$\begin{bmatrix} 1 & -d_1 \\ d_2 & 1-2d_2 \end{bmatrix} P + \begin{bmatrix} 0 \\ d_2 \end{bmatrix} Q = \begin{bmatrix} 1 & 0 \\ 0 & 1 \end{bmatrix}$$

has the form

$$P = \begin{bmatrix} 1 - d_1 d_2 t_1, & d_1 - d_1 d_2 t_2 \\ -d_2 t_1, & 1 - d_2 t_2 \end{bmatrix}$$

$$Q = \begin{bmatrix} -1 + /1 + d_1 d_2 - 2d_2/t_1, & 2-d_1 + /1 + d_1 d_2 - 2d_2/t_2 \end{bmatrix}$$

where $t_1, t_2$ are arbitrary polynomials in $d_1$ and $d_2$.

Step 3. It is easy to see that each column of P is of least degrees possible in $d_1$ and $d_2$ for $t_1 = t_2 = 0$. Then the desired matrices are

$$P = \begin{bmatrix} 1 & d_1 \\ 0 & 1 \end{bmatrix}, \quad Q = \begin{bmatrix} -1, & 2-d_1 \end{bmatrix}$$

## 7.2. OUTPUT DEADBEAT CONTROL PROBLEM

### 7.2.1. Problem formulation

Consider RM described by the equations /7.1/ and /7.2/ with given finite boundary conditions /7.3a/.

It is desired to find an input sequence $u/i,j/$ of a duration $/N_1,N_2/$, that is

$$u/i,j/ \begin{cases} \neq 0 & \text{for } /0,0/ \leq /i,j/ < /N_1,N_2/ \\ = 0 & \text{for } /i,j/ \geq /N_1,N_2/ \end{cases} \qquad /7.11/$$

such that the output $y/i,j/$ of RM reaches zero after a finite number of steps and remains zero thereafter, i.e.

$$y/i,j/ = 0 \quad \text{for } /i,j/ \geq /N_1,N_2/ \qquad /7.12/$$

We assumed that

$$x^h/0,j/ = 0 \text{ for } j \geq N_2 \quad \text{and} \quad x^v/i,0/ = 0 \text{ for } i \geq N_1 \qquad /7.13/$$

Thus, the output deadbeat control problem can be stated as follows. Given the matrices A, B, C of RM and finite boundary conditions satisfying /7.13/, find a pair $/N_1,N_2/$ of positive integers and an input sequence /7.11/ such that /7.12/ holds.

### 7.2.2. Problem solution

Two methods of solving the problem will be presented.

**Method 1.** Using 2-D Z transformation to /7.1/ and /7.2/ it can be shown /in a similar way as in 2.1/ that

$$Y = G/d_1,d_2/U + H/d_1,d_2/ \qquad /7.14/$$

where

$$G/d_1,d_2/ = C \begin{bmatrix} I_{n_1}-d_1A_{11}, & -d_1A_{12} \\ -d_2A_{21}, & I_{n_2}-d_2A_{22} \end{bmatrix}^{-1} \begin{bmatrix} d_1B_1 \\ d_2B_2 \end{bmatrix} =$$

$$= \sum_{i=0}^{\infty} \sum_{j=0}^{\infty} g_{ij} \, d_1^i \, d_2^j \quad /g_{00} = 0/ \qquad /7.15/$$

$$H/d_1,d_2/ = C \begin{bmatrix} I_{n_1}-d_1A_{11}, & -d_1A_{12} \\ -d_2A_{21}, & I_{n_2}-d_2A_{22} \end{bmatrix}^{-1} X_o = \sum_{i=0}^{\infty} \sum_{j=0}^{\infty} h_{ij} d_1^i d_2^j \qquad /7.16/$$

$X_o = X_o/d_1,d_2/$ is the polynomial vector in $d_1 = z_1^{-1}$ and $d_2 = z_2^{-1}$, which depends on the given boundary conditions,

and

$Y = Y/d_1,d_2/$, $U = U/d_1,d_2/$ are 2-D Z transforms of $y/i,j/$ and $u/i,j/$, respectively, i.e.

$$Y = \sum_{i=0}^{\infty} \sum_{j=0}^{\infty} y/i,j/d_1^i d_2^j \qquad /7.17a/$$

$$U = \sum_{i=0}^{\infty} \sum_{j=0}^{\infty} u/i,j/d_1^i d_2^j \qquad /7.17b/$$

Substitution of /7.15/, /7.16/ and /7.17/ into /7.14/ yields

$$\sum_{i=0}^{\infty} \sum_{j=0}^{\infty} y/i,j/d_1^i d_2^j = \sum_{i=0}^{\infty} \sum_{j=0}^{\infty} \sum_{k=0}^{i} \sum_{l=0}^{j} g_{i-k,j-l} u/k,l/ d_1^i d_2^j +$$

$$+ \sum_{i=0}^{\infty} \sum_{j=0}^{\infty} h_{ij} d_1^i d_2^j \qquad /7.18/$$

Equating the coefficients of equal powers of $d_1$ and $d_2$ we obtain

$$y/i,j/ = \sum_{k=0}^{i} \sum_{l=0}^{j} g_{i-k,j-l} \, u/k,l/ + h_{ij} \qquad /7.19/$$

for $i,j = 0,1,2,\ldots$

Taking into account /7.11/ and /7.12/ from /7.19/ for $i = N_1, N_1+1, \ldots, N_1+n_1$ and $j = N_2, N_2+1, \ldots, N_2+n_2$ we obtain the set of

$$\left[/N_1+n_1+1//N_2+n_2+1/ - /N_1+1//N_2+1/-N_2n_1-N_1n_2+1/\right]1$$

linear equations in $\left[/N_1+1//N_2+1/-1\right]$m unknowns $u/i,j/$, $/0,0/ \leq i,j / N_1,N_2/$

of the form

$$\sum_{/0,0/\leq/i,j/</N_1,N_2/} g_{N_1-i,N_2-j}\, u/i,j/ = -h_{N_1,N_2}$$

$$\sum_{/0,0/\leq/i,j/</N_1,N_2/} g_{N_1-i+1,N_2-j}\, u/i,j/ = -h_{N_1+1,N_2}$$

$$\sum_{/0,0/\leq/i,j/</N_1,N_2/} g_{N_a-i,N_2-j+1}\, u/i,j/ = -h_{N_1,N_2+1}$$

. . . . . . . . . . . . . . . . . . . . . . . .

$$\sum_{/0,0/\leq/i,j/</N_1,N_2/} g_{N_1+n_1-i,N_2+n_2-j}\, u/i,j/ = -h_{N_1+n_1,N_2+n_2}$$

or

$$gu = h \qquad /7.20/$$

where

$$g = \begin{bmatrix} g_{N_1 N_2} & g_{N_1-1,N_2} & g_{N_1,N_2-1} & \cdots & g_{10} & g_{01} \\ g_{N_1+1,N_2} & g_{N_1,N_2} & g_{N_1+1,N_2-1} & \cdots & g_{20} & g_{11} \\ g_{N_1,N_2+1} & g_{N_1-1,N_2+1} & g_{N_1 N_2} & \cdots & g_{11} & g_{02} \\ \cdots & \cdots & \cdots & \cdots & \cdots & \cdots \\ g_{N_1+n_1,N_2+n_2} & g_{N_1+n_1-1,N_2+n_2} & \cdots & & g_{n_1+1,n_2} & g_{n_1,n_2+1} \end{bmatrix}$$

$$/7.20a/$$

$$u = \begin{bmatrix} u/0,0/ \\ u/1,0/ \\ u/0,1/ \\ \cdots \\ u/N_1-1,N_2/ \\ u/N_1,N_2-1/ \end{bmatrix}, \quad h = -\begin{bmatrix} h_{N_1 N_2} \\ h_{N_1+1,N_2} \\ h_{N_1,N_2+1} \\ \cdots \\ h_{N_1+n_1,N_2+n_2} \end{bmatrix}$$

From Kronecker-Capelli's theorem it follows that the equation /7.20/ has a solution if and only if

$$\text{rank } g = \text{rank}\left[g \mid h\right] \qquad /7.21/$$

It can be easily shown that if /7.11/ holds and

$$y/i,j/ = 0 \text{ for } /N_1,N_2/ \leqslant /i,j/ \leqslant /N_1+n_1,N_2+n_2/$$

then $y/i,j/ = 0$, for $/i,j/ \geqslant /N_1,N_2/$.

Thus, the following theorem has been proved.

## Theorem 7.2

The output deadbeat control problem has a solution if and only if the condition /7.21/ is satisfied.

Note that, if g is of full row rank, then a necessary condition for the existence of a solution to the problem is

$$\left[/N_1+1//N_2+1/-1\right]m > \left[/N_1+n_1+1//N_2+n_2+1/ - /N_1+1//N_2+1/ -N_2n_1-N_1n_2+1\right]1 \qquad /7.22/$$

If the condition /7.21/ is satisfied the following algorithm can be used for finding a solution to the problem.

## Algorithm 7.2

**Step 1** Choose $N_1$ and $N_2$ such that the condition /7.21/ /in particular case /7.22// is satisfied.

**Step 2** Using /7.15/ and /7.16/ find $G/d_1,d_2/$, $H/d_1,d_2/$ and

$g_{10}, g_{01}, g_{11}, \ldots, g_{N_1+n_1,N_2+n_2}, h_{N_1,N_2}, h_{N_1+1,N_2}, h_{N_1,N_2+1},$

$\ldots, h_{N_1+n_1,N_2+n_2}$.

**Step 3** Write the equation /7.20/ and find its solution u.

## Example 7.2

Given RM with

$$A = \begin{bmatrix} 0 & 1 \\ -1 & 2 \end{bmatrix}, \quad B = \begin{bmatrix} 0 \\ 1 \end{bmatrix}, \quad C = \begin{bmatrix} 1 & 0 \end{bmatrix}$$

and
$$x^h/0,j/ = 0, \quad j \geq 0, \quad x^v/i,0/ \begin{cases} =1, & i=0 \\ =0, & i>0 \end{cases}$$

find a solution to the problem.

In this case $n_1 = n_2 = 1$ and $l = m = 1$.

**Step 1**

It is easy to check that $N_1 = N_2 = 1$ satisfy the condition /7.21/, since rank $g$ = rank $\begin{bmatrix} g \mid h \end{bmatrix} = 3$.

**Step 2**

$$G/d_1,d_2/ = C \begin{bmatrix} I_{n_1} - d_1 A_{11}, & -d_1 A_{12} \\ -d_2 A_{21}, & I_{n_2} - d_2 A_{22} \end{bmatrix}^{-1} \begin{bmatrix} d_1 B_1 \\ d_2 B_2 \end{bmatrix} =$$

$$= \begin{bmatrix} 1 & 0 \end{bmatrix} \begin{bmatrix} 1, & -d_1 \\ d_2, & 1-2d_2 \end{bmatrix}^{-1} \begin{bmatrix} 0 \\ d_2 \end{bmatrix} = \frac{d_1 d_2}{d_1 d_2 - 2d_2 + 1} =$$

$$= d_1 d_2 + 2d_1 d_2^2 - d_1^2 d_2^2 + \ldots$$

and

$$H/d_1,d_2/ = C \begin{bmatrix} I_{n_1} - d_1 A_{11}, & -d_1 A_{12} \\ -d_2 A_{21}, & I_{n_2} - d_2 A_{22} \end{bmatrix}^{-1} X_o =$$

$$= \begin{bmatrix} 1 & 0 \end{bmatrix} \begin{bmatrix} 1, & -d_1 \\ d_2, & 1-2d_2 \end{bmatrix}^{-1} \begin{bmatrix} 0 \\ 1 \end{bmatrix} = \frac{d_1}{d_1 d_2 - 2d_2 + 1} =$$

$$= d_1 + 2d_1 d_2 - d_1^2 d_2 + 4d_1 d_2^2 - 4d_1^2 d_2^2 \ldots$$

**Step 3**

In this case the equation /7.20/ has the form

$$\begin{bmatrix} 1 & 0 & 0 \\ 0 & 1 & 0 \\ 2 & 0 & 1 \\ -1 & 2 & 0 \end{bmatrix} \begin{bmatrix} u/0,0/ \\ u/1,0/ \\ u/0,1/ \end{bmatrix} = \begin{bmatrix} -2 \\ 1 \\ -4 \\ 4 \end{bmatrix} \qquad /7.23/$$

and its solution is

$u/0,0/ = -2$, $u/1,0/ = 1$, otherwise $u/i,j/ = 0$.

## Method 2

The general response formula for RM with /7.11/ and /7.13/ is given by

$$y/i,j/ = \sum_{l=0}^{N_2} CA^{i,j-l} \begin{bmatrix} x^h/0,l/ \\ 0 \end{bmatrix} + \sum CA^{i-k,j} \begin{bmatrix} 0 \\ x^v/k,0/ \end{bmatrix} +$$

$$+ \sum_{/0,0/ \leq /k,l/ < /N_1,N_2/} CM_{i-k,j-l}\, u/k,l/ \quad \text{for } /i,j/ \geq /N_1,N_2/ \qquad /7.23/$$

where

$$M_{i-k,j-l} = A^{i-k-1,j-l}\begin{bmatrix} B_1 \\ 0 \end{bmatrix} + A^{i-k,j-l-1}\begin{bmatrix} 0 \\ B_2 \end{bmatrix} \qquad /7.24/$$

From /7.23/ for $i = N_1, N_1+1, \ldots, N_1+n_1$ and $j = N_2, N_2+1, \ldots, N_2+n_2$ we obtain the set of $[/N_1+n_1+1//N_2+n_2+1/ - /N_1+1//N_2+1/ - N_1 n_2 - N_2 n_1 + 1]$ 1 linear equations in $/N_1+1//N_2+1/-1$ in unknowns $u/i,j/$, $/0,0/ \leq /i,j/ <$ $/N_1,N_2/$ of the form

$$M\, u = Q \qquad /7.25/$$

where

$$M = \begin{bmatrix} CM_{N_1 N_2} & CM_{N_1-1,N_2} & CM_{N_1,N_2-1} & \cdots & CM_{01} \\ CM_{N_1+1,N_2} & CM_{N_1 N_2} & CM_{N_1+1,N_2-1} & \cdots & CM_{11} \\ \cdots & \cdots & \cdots & \cdots & \cdots \\ CM_{N_1+n_1,N_2+n_2} & CM_{N_1+n_1-1,N_2+n_2} & CM_{N_1+n_2,N_2+n_2-1} & \cdots & CM_{n_1,n_2+1} \end{bmatrix}$$

/7.26/

$$u = \begin{bmatrix} u/0,0/ \\ u/1,0/ \\ u/0,1/ \\ \cdots \\ u/N_1,N_2-1/ \end{bmatrix}, \qquad Q = \begin{bmatrix} Q_{N_1 N_2} \\ Q_{N_1+1,N_2} \\ \cdots \\ Q_{N_1+n_1,N_2+n_2} \end{bmatrix}$$

$$Q_{i,j} = -\sum_{l=0}^{N_2} CA^{i,j-l} \begin{bmatrix} x^h/0,1/ \\ 0 \end{bmatrix} - \sum_{k=0}^{N_1} CA^{i-k,j} \begin{bmatrix} 0 \\ x^v/k,0/ \end{bmatrix}$$

From Kronecker-Capelli's theorem it follows that the equation /7.25/ has a solution if and only if

$$\text{rank } M = \text{rank} \begin{bmatrix} M & | & Q \end{bmatrix} \qquad /7.27/$$

Using the theorem 2.1 and /7.23/-/7.26/ it can be shown that if /7.11/ holds and $y/i,j/ = 0$ for $/N_1,N_2/ \leq /i,j/ \leq /N_1+n_1,N_2+n_2/$, then $y/i,j/ = 0$ for $/i,j/ \geq /N_1,N_2/$.

Thus, the following theorem has been proved.

### Theorem 7.3

The output deadbeat control problem has a solution if and only if the condition /7.27/ is satisfied.

If M is of full row rank, then the problem has a solution only if /7.22/ holds.

A solution to the problem can be found by the used of the following.

### Algorithm 7.3

**Step 1**   Choose positive integers $N_1$ and $N_2$ such that the condition /7.27/ is satisfied.

**Step 2**   Using /7.24/ and /7.26/, find M and Q.

**Step 3**   Write the equation /7.25/ and find its solution u.

**Remark:**   It is easy to show /see Problem 7.3/ that $M = g$ and $Q = h$.

### Example 7.3

Using Method 2 solve the same problem as in Example 7.2.

**Step 1**   In a similar way as in Example 7.2 we choose $N_1 = N_2 = 1$.

**Step 2**   Using /7.24/ and /7.26/ we obtain

$$M = \begin{bmatrix} CM_{11} & CM_{01} & CM_{10} \\ CM_{21} & CM_{11} & CM_{20} \\ CM_{12} & CM_{02} & CM_{11} \\ CM_{22} & CM_{12} & CM_{21} \end{bmatrix} = \begin{bmatrix} 1 & 0 & 0 \\ 0 & 1 & 0 \\ 2 & 0 & 1 \\ -1 & 2 & 0 \end{bmatrix}$$

$$Q = \begin{bmatrix} Q_{11} \\ Q_{21} \\ Q_{12} \\ Q_{22} \end{bmatrix} = \begin{bmatrix} 2 \\ 1 \\ -4 \\ 4 \end{bmatrix}$$

**Step 3**

The equation /7.25/ has the same form as /7.23/ in Example 7.2. Therefore

$u/0,0/ = -2$, $u/1,0/ = 1$, otherwise $u/i,j/ = 0$.

## 7.3. OUTPUT DEADBEAT CONTROL OF CLOSED-LOOP SYSTEMS
### 7.3.1. Problem formulation

Consider RM described by the equations /7.1/ and /7.2/ with boundary conditions /7.3/. The closed-loop system with the control law

$$u = Kx + Hr \qquad /7.28/$$

is described by the equations

$$x' = /A + Bk/x + BHr \qquad /7.29/$$

$$y = Cx \qquad /7.30/$$

where

$$K = [K_1 K_2], \quad K_1 \in R^{m \times n_1}, \quad K_2 \in R^{m \times n_2}, \quad H \in R^{m \times 1}$$

and $r = r/i,j/ \in R^1$ is the reference input.

The problem can be stated as follows [12]. Given A, B, C of RM, find K and H such that

$$y/i,j/ = 1/i,j/ \quad \text{for} \quad i \geq I, \quad j \geq J \qquad /7.31/$$

for same positive integers $I, J : I \leq n_1, J \leq n_2$, where

$1/i,j/$ is the 2-D unit step function: $1/i,j/ \begin{cases} =1 & \text{for } i,j \geq 0 \\ = 0 & \text{for } i,j < 0 \end{cases}$

### 7.3.2. Problem solution

The 2-D Z transform of $r/i,j/ = 1/i,j/$ has the form

$$R/z_1,z_2/ = \sum_{i=0}^{\infty} \sum_{j=0}^{\infty} 1/i,j/z_1^{-i}z_2^{-j} = \frac{z_1 z_2}{/z_1-1//z_2-1/} \quad \text{for } |z_1|>1, |z_2|>1$$

The transfer function of the closed-loop system /7.29/, /7.30/ is given by

$$G_C/z_1,z_2/ = C[Z-A-BK]^{-1} BH \qquad /7.32/$$

<u>Theorem 7.4</u>

The condition /7.31/ is satisfied if and only if

$$G_C/z_1,z_2/ = z_1^{-I}z_2^{-J} \sum_{i=0}^{I} \sum_{j=0}^{J} h_{ij} z_1^{I-i} z_2^{J-j} \qquad /7.33a/$$

and

$$\sum_{i=0}^{I} \sum_{j=0}^{J} h_{ij} = 1 \qquad /h_{oo} = 0/ \qquad /7.33b/$$

<u>Proof</u>

Taking into account the shift property of the operators $z_1^{-1}$ and $z_2^{-1}$ applied to $1/i,j/$ we can write

$$Y/z_1,z_2/ = [h_{10}z_1^{-1}+h_{01}z_2^{-1}+h_{11}z_1^{-1}z_2^{-1}+ \ldots +h_{IJ}z_1^{-I}z_2^{-J}]R/z_1,z_2/ \qquad /7.34/$$

It is easy to check that if /7.33/ is satisfied, then /7.31/ holds and

$$y/k,l/ = \sum_{i=0}^{k} \sum_{j=0}^{l} h_{ij} \quad \text{for } k<I, l<J$$

The formula /7.33a/ follows from /7.34/.  □

The matrix K is chosen so that

$$\det[Z-A-BK] = z_1^{n_1} z_2^{n_2} \qquad /7.35/$$

This can be done by matching of the coefficients or by the use one of the methods presented in 5.5.

The matrix H is chosen so that /7.33/ are satisfied. Note that a necessary condition for eigenvalue assignment is

$$m \geq 1 + \frac{n_1 n_2}{n_1+n_2}$$

It follows from the method of coefficients matching. The number of entries of K /which is equal to $m/n_1+n_2//$ must be greater or equal to the number $n_1 n_2 + n_1 + n_2$ of coefficients of the closed-loop characteristic polynomial.

If the conditions /7.33/ are satisfied, then K and H can be found by the use of the following

### Algorithm 7.4

**Step 1** Using the coefficient matching method or one of the methods presented in 5.5, find K so that /7.35/ is satisfied.

**Step 2** Find
$$A_C = A + BK$$

**Step 3** Using /7.32/ find $G_C/z_1, z_2/$.

**Step 4** Choose H so that /7.33/ are satisfied.

### Example 7.4

Solve the problem for RM with

$$A = \begin{bmatrix} A_{11} & A_{12} \\ A_{21} & A_{22} \end{bmatrix} = \left[\begin{array}{cc|cc} -1 & 6 & 1 & 0 \\ -1 & 8 & 1 & -2 \\ \hline 2 & -6 & -3 & 3 \\ 1 & -3 & 1 & -1 \end{array}\right], \quad B = \begin{bmatrix} B_1 \\ B_2 \end{bmatrix} = \left[\begin{array}{cc} 1 & -1 \\ 1 & 0 \\ \hline 0 & 2 \\ 0 & 1 \end{array}\right],$$

$$C = \begin{bmatrix} C_1 & C_2 \end{bmatrix} = \left[\begin{array}{cc|cc} 1 & 0 & 0 & 1 \end{array}\right]$$

**Step 1** Using the method we obtain

$$K = \begin{bmatrix} 5/4 & 37/4 & 0 & 0 \\ -1 & 3 & 4 & -4 \end{bmatrix}$$

**Step 2**

$$A_C = A + BK = \left[\begin{array}{cc|cc} 5/4 & -\frac{25}{4} & -3 & 4 \\ \frac{1}{2} & -\frac{5}{4} & 1 & -2 \\ \hline 0 & 0 & 5 & -5 \\ 0 & 0 & 5 & -5 \end{array}\right]$$

**Step 3** Using /7.32/ we obtain for $H = [h_1, h_2]$

$$G_C/z_1,z_2/ = \frac{/h_1 - h_2/z_1 z_2^2 + h_2 z_1^2 z_2 - /5h_1 + \frac{5}{4} h_2/z_2^2 + 5h_2 z_1^2 - 2h_2 z_1 z_2 - \frac{10}{4} h_2 z_2}{z_1^2 z_2^2} +$$

$$+ \frac{5h_2 z_1 + 150 h_2}{z_1^2 z_2^2}$$

**Step 4** It is easy to check that /7.33b/ is satisfied if

$$-4h_1 + \frac{167}{4} h_2 = 1. \text{ For } h_2 = 0 \text{ we get } h_1 = -\frac{1}{4} \text{ and}$$

$$H = \begin{bmatrix} -\frac{1}{4} & 0 \end{bmatrix}$$

Hence

$$G_C/z_1,z_2/ = -\frac{1}{4} z_1^{-1} + \frac{5}{4} z_1^{-2}$$

and $I = 2$, $J = 0$.

## 7.4. DEADBEAT CONTROL OF OPEN-LOOP SYSTEM

### 7.4.1. Problem formulation

Consider a plant /for example RM/ described by the equation

$$y = A^{-1} B u + A^{-1} C_o \qquad /7.36/$$

where $y \in R^{l \times 1}/d_1,d_2/$, $u \in R^{m \times 1}/d_1,d_2/$ are 2-D Z transforms of the output vector $y/i,j/$ and the input vector $u/i,j/$, respectively and $A \in R^{l \times l}[d_1,d_2]$, $B \in R^{l \times m}[d_1,d_2]$, $C_o \in R^{l \times 1}[d_1,d_2]$ are polynomial matrices in $d_1 = z_1^{-1}$, $d_2 = z_2^{-1}$. It is assumed that $A$ is invertible. $A^{-1}B$ is the transfer function matrix of the plant and $A^{-1}C_o$ represents the effect of non-zero boundary conditions on the plant output.
Let the 2-D Z transforms of a reference input vector $r/i,j/$ be given by the equation

$$r = F^{-1}G \qquad /7.37/$$

where $F \in R^{l \times l}[d_1,d_2]$ is invertible and $G \in R^{l \times 1}[d_1,d_2]$.
Note that $r/i,j/$ can be considered as a free motion of a reference generator. Varying $G$ in /7.37/ /boundary conditions of the generator/

we may generate a whole class of reference input vectors.
It is desired to find $u/i,j/$ such that the tracking error $e/i,j/ = r/i,j/-y/i,j/$ vanishes for all $i \geq N_1$, $j \geq N_2$ where $N_1, N_2$ are some positive integers. Thus, the problem can be stated as follows [18]. Given A, B, $C_o$, F and G, find $u/i,j/$ such that e is a polynomial vector of least possible degree in $d_1$ and $d_2$.

## 7.4.2. Problem solution

### Theorem 7.5.

The problem has a solution if

/i/ F is a right divisor of A, i.e. there exists a matrix $A_o \in R^{l \times l} [d_1, d_2]$ such that
$$A = A_o F \qquad /7.38/$$

/ii/ A, B are zero left coprime /ZLC/.

### Proof

From /7.36/-/7.38/ we have
$$e = r-y = F^{-1}G - A^{-1}Bu - A^{-1}C_o$$
and
$$Ae + Bu = C \qquad /7.39/$$
where
$$C = AF^{-1}G - C_o = A_o G - C_o \qquad /7.40/$$

Thus, the desired u can be found as a minimal order solution with respect to X of the equation
$$AX + BY = C \qquad /7.41/$$
where $X = e$ and $Y = u$.

By theorem A.5.3 there exist polynomial matrices $\bar{X}$, $\bar{Y}$ such that
$$A\bar{X} + B\bar{Y} = I \qquad /7.42/$$
if and only if A, B are ZLC.

Postmultiplying /7.42/ by C we obtain

$$A\bar{X}C + B\bar{Y}C = C$$

and
$$X = \bar{X}C, \quad Y = \bar{Y}C$$

Therefore, the equation /7.41/ has a solution for any C if and only if A, B are ZLC. □

If the assumptions /i/, /ii/ of Theorem 7.5 are satisfied then a solution to the problem can be found by the use of the following

## Algorithm 7.5

**Step 1** Using /7.40/ find C.

**Step 2** Carry out the reduction /using elementary column operations/

$$\begin{bmatrix} A & | & B \\ \hline I_1 & | & 0 \\ \hline 0 & | & I_m \end{bmatrix} \rightarrow \begin{bmatrix} I_1 & | & 0 \\ \hline U_1 & | & U_2 \\ \hline U_3 & | & U_4 \end{bmatrix}$$

and find the general solution

$$\begin{bmatrix} X \\ Y \end{bmatrix} = \begin{bmatrix} U_1 & U_2 \\ U_3 & U_4 \end{bmatrix} \begin{bmatrix} C \\ T \end{bmatrix}$$

to the equation /7.41/, where T is an arbitrary polynomial matrix.

**Remark:** The algorithm A.7.2 given in Appendix can be also used for finding X and Y.

**Step 3** Choose T so that X is of least degrees possible in $d_1$ and $d_2$ and find $u/i,j/$.

## Example 7.5

Given

$$A = F = \begin{bmatrix} 1 & d_1 d_2 \\ 0 & d_2 \end{bmatrix}, \quad B = \begin{bmatrix} d_1 \\ 1 \end{bmatrix}, \quad C_0 = \begin{bmatrix} d_2 \\ d_1^2 d_2 \end{bmatrix} \text{ and } G = \begin{bmatrix} 1+d_2 \\ d_1^2 d_2 + d_1 \end{bmatrix}$$

find a solution $u/i,j/$ to the problem.

In this case $l = 2$, $m = 1$, $A_o = I_2$ and it is easy to see that the conditions /i/, /ii/ of theorem 7.5 are satisfied.

**Step 1**

$$C = A_o G - C_o = G - C_o = \begin{bmatrix} 1 \\ d_1 \end{bmatrix}$$

**Step 2** Using elementary column operations we carry out the reduction

$$\begin{bmatrix} A & B \\ \hline I_1 & 0 \\ 0 & I_m \end{bmatrix} = \begin{bmatrix} 1 & d_1 d_2 & d_1 \\ 0 & d_2 & 1 \\ \hline 1 & 0 & 0 \\ 0 & 1 & 0 \\ \hline 0 & 0 & 1 \end{bmatrix} \longrightarrow \begin{bmatrix} 1 & 0 & 0 \\ 0 & 1 & 0 \\ \hline 1 & -d_1 & 0 \\ 0 & 0 & 1 \\ \hline 0 & 1 & -d_2 \end{bmatrix}$$

Hence the general solution to the equation

$$\begin{bmatrix} 1 & d_1 d_2 \\ 0 & d_2 \end{bmatrix} X + \begin{bmatrix} d_1 \\ 1 \end{bmatrix} Y = \begin{bmatrix} 1 \\ d_1 \end{bmatrix}$$

has the form

$$\begin{bmatrix} X \\ \hline Y \end{bmatrix} = \begin{bmatrix} 1 & -d_1 & 0 \\ 0 & 0 & 1 \\ \hline 0 & 1 & -d_2 \end{bmatrix} \begin{bmatrix} 1 \\ d_1 \\ \hline t \end{bmatrix}$$

where t is an arbitrary polynomial in $d_1$ and $d_2$.

**Step 3** For t = 0 we have

$$X = \begin{bmatrix} 1 & -d_1^2 \\ 0 & \end{bmatrix}, \qquad Y = \begin{bmatrix} d_1 \end{bmatrix}$$

and

$u/0,0/ = 0, \quad u/1,0/ = 1, \quad u/0,1/ = 0, \quad u/i,j/ = 0$

for i,j > 1.

The above considerations can be extended for n-D linear system [18].

## 7.5. DEADBEAT SERVO PROBLEM FOR SINGLE-INPUT SINGLE-OUTPUT SYSTEMS

### 7.5.1. Problem formulation

Consider a 2-D linear plant described by the equation

$$y = \frac{B}{A} u + \frac{C}{A} \qquad /7.43/$$

where $y = y/d_1,d_2/$, $u = u/d_1,d_2/$ are the 2-D Z transforms of the output $y/i,j/$ and the input $u/i,j/$ and A, B and C are polynomials in $d_1 = z_1^{-1}$ and $d_2 = z_2^{-1}$.

It is assumed that A and B are zero-coprime polynomials /i.e. they have no zero in common/ such that $A/0,0/ \neq 0$ and $B/0,0/ = 0$. $\frac{B}{A}$ is the transfer function of the plant and $\frac{C}{A}$ represents the effect of boundary conditions on the plant output.

Let the 2-D Z transform $r = r/d_1,d_2/$ of a reference input $r/i,j/$ be given by

$$r = \frac{G}{F} \qquad /7.44/$$

where G and F are factor-coprime polynomials in $d_1$ and $d_2$. Note that r can be considered as a free motion of a reference generator. Varying G in /7.44/ /boundary conditions of the generator/ we can generate a whole class of reference inputs.

It is desired to find the control law

$$Pu = -Qy + Rr \qquad /7.45/$$

such that the tracking error $e = r-y$ and the input u vanish for minimal i and j and for any boundary conditions of the plant and reference generator, where P, Q and R are factor-coprime polynomials in $d_1$ and $d_2$. The problem can be stated as follows [2]. Given A, B and F, find P, Q and R such that e and u are polynomials of least possible degrees in $d_1$ and $d_2$ for any C and G.

### 7.5.2. Problem solution

Substitution of /7.43/ and /7.44/ into /7.45/ yields

$$u = \frac{AR}{AP+BQ}\frac{G}{P} - \frac{AQ}{AP+BQ}\frac{C}{A} \qquad /7.46/$$

and

$$e = \left(1 - \frac{BR}{AP+BQ}\right)\frac{G}{F} - \frac{AP}{AP+BQ}\frac{C}{A} \qquad /7.47/$$

From /7.46/ it follows that u is a polynomial for any C and G if and only if

$$AP + BQ = 1 \qquad /7.48/$$

and

$$\frac{A}{F} = A_o \qquad /7.49/$$

where $A_o$ is a polynomial in $d_1$ and $d_2$.

From /7.47/ it follows that e is a polynomial if and only if F is a divisor of 1-BR, i.e. if and only if there exists a polynomial S such that 1-BR = FS or

$$FS + BR = 1 \qquad /7.50/$$

If /7.48/-/7.50/ are satisfied, then /7.46/ and /7.47/ take the form

$$u = A_o RG - QC, \qquad e = SG - PC,$$

or

$$\begin{bmatrix} u \\ e \end{bmatrix} = \begin{bmatrix} A_o R & -Q \\ S & -P \end{bmatrix} \begin{bmatrix} G \\ C \end{bmatrix} \qquad /7.51/$$

### Theorem 7.6.

The problem has a solution if and only if

/i/ F is a divisor of A,

/ii/ A and B are zero-coprime polynomials.

## Proof

Note that u and e are polynomials if and only if /i/ holds and /7.48/, /7.50/ are satisfied. The equation /7.48/ has a solution if and only if /ii/ holds /see Appendix/. Is it easy to show that the zero coprimeness of A, B implies the zero coprimeness of F, B. Thus, the equation /7.50/ has a solution if /ii/ holds. □

It can easily be shown /see Appendix/ that, if $P_o$, $Q_o$ and $S_o$, $R_o$ are particular solutions to /7.48/ and /7.50/, then the general solutions to /7.48/ and /7.50/ have the form

$$P = P_o + Bt_1, \quad Q = Q_o - At_1$$
$$S = S_o + Bt_2, \quad R = R_o - Ft_2 \qquad /7.52/$$

where $t_1$ and $t_2$ are arbitrary polynomials in $d_1$ and $d_2$.

Since $\begin{bmatrix} u \\ e \end{bmatrix}$ is to be of least possible degrees in $d_1$ and $d_2$ for any C and G, we must choose $t_1$ and $t_2$ so that the degrees of every column of the matrix

$$\begin{bmatrix} A_o R & -Q \\ S & -P \end{bmatrix} = \begin{bmatrix} A_o R_o - A_o F t_2, & -Q_o + A t_1 \\ S_o + B t_2, & -P_o - B t_1 \end{bmatrix} \qquad /7.53/$$

are minimal.

If the conditions /i/, /ii/ of theorem 7.6 are satisfied, then P, Q and R can be found by the use of the following

### Algorithm 7.6

**Step 1** Using elementary column operations, carry out the reductions

$$\begin{bmatrix} A & B \\ 1 & 0 \\ 0 & 1 \end{bmatrix} \longrightarrow \begin{bmatrix} 1 & 0 \\ U_1 & U_2 \\ U_3 & U_4 \end{bmatrix}$$

and

and

$$\begin{bmatrix} F & B \\ 1 & 0 \\ 0 & 1 \end{bmatrix} \rightarrow \begin{bmatrix} 1 & 0 \\ V_1 & V_2 \\ V_3 & V_4 \end{bmatrix}$$

Then $P_o = U_1$, $Q_o = U_3$ and $S_o = V_1$, $R_o = V_3$.

**Remark:**

The same result can be obtained by equating the coefficients at like powers of $d_1$ and $d_2$ in /7.48/ and /7.50/ or by the use of one of the algorithms given in Appendix.

**Step 2**  Using /7.49/ find $A_o$.

**Step 3**  Choose the polynomials $t_1$ and $t_2$ so that the degrees in $d_1$ and $d_2$ of every column of /7.53/ are minimal.

**Example 7.6.**

Given

$$A = 1 + /1+d_1/d_2 + /1+2d_1/d_2^2 + /1+d_1/d_2^3, \quad B = d_2 \qquad /7.54a/$$

and

$$F = 1 + d_1 d_2 + /1+d_1/d_2^2 \qquad /7.54b/$$

find P, Q and R.

In this case it is easy to see that A and B are zero-coprime and F is a divisor of A.

**Step 1**  Using elementary column operations we carry out the reductions

$$\begin{bmatrix} 1+/1+d_1/d_2+/1+2d_1/d_2^2+/1+d_1/d_2^3, & d_2 \\ 1 & 0 \\ 0 & 1 \end{bmatrix} \rightarrow \begin{bmatrix} 1 & 0 \\ 1 & * \\ -/1+d_1/-/1+2d_1/d_2-/1+d_1/d_2^2 & * \end{bmatrix}$$

and

$$\begin{bmatrix} 1+d_1 d_2+/1+d_1/d_2^2, & d_2 \\ 1 & 0 \\ 0 & 1 \end{bmatrix} \rightarrow \begin{bmatrix} 1 & 0 \\ 1 & * \\ -d_1-/1+d_1/d_2 & * \end{bmatrix}$$

where * denotes polynomials in $d_1$ and $d_2$.

Then
$$P_o = 1, \quad Q_o = -/1+d_1/-/1+2d_1/d_2\ -/1+d_1/d_2^2 \quad /7.55a/$$
and
$$S_o = 1, \quad R_o = -d_1-/1+d_1/d_2 \quad /7.55b/$$

**Step 2** For /7.54/ we have
$$A_o = \frac{A}{F} = 1+d_2 \quad /7.56/$$

**Step 3** Taking into account /7.55/ and /7.56/ we obtain
$$\begin{bmatrix} A_o R & -Q \\ S & -P \end{bmatrix} = \begin{bmatrix} -/1+d_2/M, & t_1+/1+d_1/N_1+/1+2d_1/N_2 \\ 1+d_2 t_2, & -1-d_2 t_1 \end{bmatrix} \quad /7.57/$$

where
$$M = d_1+t_2+d_1 d_2 t_2+/1+d_1/\!/d_2+d_2^2 t_2/$$
$$N_1 = 1+d_2 t_1+d_2^2+d_2^3 t_1, \quad N_2 = d_2+d_2^3 t_1$$

It is easy to see that the degrees of every column of /7.57/ are minimal for $t_1 = t_2 = 0$.

Hence the desired solution is $P = P_o$, $Q = Q_o$ and $R = R_o$.

## 7.6. DEADBEAT SERVO PROBLEM FOR MULTIVARIABLE LINEAR SYSTEM

### 7.6.1. Problem formulation

Consider a 2-D linear multivariable plant described by the equation
$$y = A^{-1}Bu + A^{-1}C \quad /7.58/$$
where $y \in R^{1\times 1}/d_1,d_2/$ is the ouput sequence, $u \in R^{m\times 1}/d_1,d_2/$ is the input sequence. $A \in R^{1\times 1}[d_1,d_2]$, $B \in R^{1\times m}[d_1,d_2]$, $C \in R^{1\times 1}[d_1,d_2]$.
It is assumed that A is invertible and A, B are ZLC /see Appendix/.
$A^{-1}B$ is the transfer function matrix of the plant and $A^{-1}C$ represents the effect of boundary conditions on the plant output.

Let a class of reference sequences $r \in R^{1\times 1}/d_1,d_2/$ be given by the equation

$$r = F^{-1}G \qquad /7.59/$$

where $F \in R^{1\times 1}[d_1,d_2]$, $G \in R^{1\times 1}[d_1,d_2]$. It is assumed that F is invertible and F, G are factor left coprime. Note that r can be considered as a free motion of a reference generator. Varying G in /7.59//boundary conditions of the generator/ we can generate a whole class of reference sequences.

It is desired to find a 2-D linear controller described by the equation

$$Pu = -Qy + Rr + S \qquad /7.60/$$

such that the tracking error $e = r-y$ and u vanish for minimal i and j and for any boundary conditions of the plant, the reference generator and the controller, where $P \in R^{m\times m}[d_1,d_2]$, $Q \in R^{m\times 1}[d_1,d_2]$, $R \in R^{m\times 1}[d_1,d_2]$ and $S \in R^{m\times 1}[d_1,d_2]$.

The problem can be stated as follows [9].

Given A, B and F, find P, Q and R so that the tracking error e and u are polynomial vectors of least possible degrees in $d_1$ and $d_2$ for any C, G and S.

## 7.6.2. Problem solution

### Theorem 7.7.

The problem has a solution if

/i/ F is a divisor of A, i.e. $A = A_o F$ for some $A_o \in R^{1\times 1}[d_1,d_2]$

/ii/ A and B are ZLC.

### Proof

By theorem A.5.4 there exist $A_2 \in R^{m\times m}[d_1,d_2]$, $B_2 \in R^{1\times m}[d_1,d_2]$, $\bar{P} \in R^{m\times m}[d_1,d_2]$, $\bar{Q} \in R^{m\times 1}[d_1,d_2]$, $\bar{P}_2 \in R^{1\times 1}[d_1,d_2]$ and $\bar{Q}_2 \in R^{m\times 1}[d_1,d_2]$ such that

$$\begin{bmatrix} A & B \\ \bar{Q} & -\bar{P} \end{bmatrix} \begin{bmatrix} \bar{P}_2 & B_2 \\ \bar{Q}_2 & -A_2 \end{bmatrix} = \begin{bmatrix} I_1 & 0 \\ 0 & I_m \end{bmatrix} \qquad /7.61/$$

if and only if /ii/ is satisfied.

If /i/ holds, then premultiplying /7.61/ by $\begin{bmatrix} A_o^{-1} & 0 \\ 0 & I_m \end{bmatrix}$ and postmultiplying by $\begin{bmatrix} A_o & 0 \\ 0 & I_m \end{bmatrix}$ we obtain

$$\begin{bmatrix} F & A_o^{-1}B \\ \bar{Q} & -\bar{P} \end{bmatrix} \begin{bmatrix} \bar{P}_2 A_o & B_2 \\ \bar{Q}_2 A_o & -A_2 \end{bmatrix} = \begin{bmatrix} I_1 & 0 \\ 0 & I_m \end{bmatrix} \qquad /7.62/$$

From /7.58/-/7.60/ for the closed-loop system we have

$$u = A_2 [PA_2 + QB_2]^{-1} /RF^{-1}G + S - QA^{-1}C/ \qquad /7.63/$$

$$e = /I_1 - B_2[PA_2+QB_2]^{-1}R/F^{-1}G - /I_1 - B_2[PA_2+QB_2]^{-1}Q/A^{-1}C +$$
$$- B_2[PA_2+QB_2]^{-1}S \qquad /7.64/$$

where $B_2 A_2^{-2}$ is a zero right coprime factorization of $A^{-1}B$.

If
$$P = \bar{P} - KB, \quad Q = \bar{Q} + KA, \quad R = \bar{Q} + LF \qquad /7.65/$$

then from /7.61/-/7.64/ we obtain

$$u = /\bar{Q}_2 A_o + A_2 L/G + A_2 S - /\bar{Q}_2 + A_2 K/C \qquad /7.66/$$

$$e = /\bar{P}_2 A_o - B_2 L/G - B_2 S - /\bar{P}_2 - B_2 K/C \qquad /7.67/$$

Note that /7.66/ and /7.67/ are polynomial vectors for any C, G, S and every $K \in R^{m \times l}[d_1,d_2]$, $L \in R^{l \times l}[d_1,d_2]$. □

### Theorem 7.8.

If the plant /7.58/ is strictly causal /rank A/0,0/ = rank A/$d_1,d_2$/ and B/0,0/ = 0/, then the controller given by /7.65/ is causal.

### Proof

From $AB_2 = BA_2$ it follows that B/0,0/ = 0 implies $B_2$/0,0/ = 0, since det A/0,0/ ≠ 0. Thus, for $d_1 = d_2 = 0$ from $QB_2 + PA_2 = I_m$, we have

$$P/0,0/A_2/0,0/ = I_m \text{ and } \det P/0,0/ \neq 0$$

Therefore, the controller /7.65/ is causal. □

From /7.66/ and /7.67/ it follows that if $r = 0$ then for any boundary conditions specified by polynomial matrices C and S the system output y and the controller output u are polynomial vectors, thus the closed-loop system is stable.

If the conditions /i/, /ii/ of theorem 7.7 are satisfied, then P, Q and R can be found by the use of the following

### Algorithm 7.7.

**Step 1** Using elementary column operations, carry out the reduction

$$\begin{bmatrix} A & B \\ I_1 & 0 \\ 0 & I_m \end{bmatrix} \rightarrow \begin{bmatrix} I_1 & 0 \\ U_1 & U_2 \\ U_3 & U_4 \end{bmatrix}$$

Then $A_2 = U_4$ and $B_2 = -U_2$.

**Step 2** Using elementary row operations, carry out the reduction

$$\begin{bmatrix} A_2 & I_m & 0 \\ B_2 & 0 & I_1 \end{bmatrix} \rightarrow \begin{bmatrix} I_m & V_1 & V_2 \\ 0 & V_3 & V_4 \end{bmatrix}$$

Then a solution to the equation

$$\bar{P} A_2 + \bar{Q} B_2 = I_m$$

is given by $\bar{P} = V_1$, $\bar{Q} = V_2$.

**Remark:** The algorithm A.7.2 can also be used for finding $\bar{P}$, $\bar{Q}$.

**Step 3** Find the general solution in the parametric form /7.65/.

**Step 4** Choose polynomial matrices K, L so that each column of the matrices

$$\bar{Q}_2 A_o + A_2 L, \quad \bar{Q}_2 + A_2 K, \quad \bar{P}_2 A_o - B_2 L \text{ and } \bar{P}_2 - B_2 K$$

is of least possible degrees in $d_1$ and $d_2$ and find the desired P, Q and R.

**Example 7.7.**

Solve the problem for

$$A = F = \begin{bmatrix} d_1 & -1 \\ 1 & d_2 \end{bmatrix} \quad \text{and} \quad B = \begin{bmatrix} 0 \\ d_2 \end{bmatrix} \qquad /7.68/$$

It is easy to check that the matrices /7.68/ are ZLC and $A_o = I_2$.

**Step 1** Using elementary column operations, we carry out the reduction

$$\left[\begin{array}{c|c} A & B \\ \hline I_1 & 0 \\ \hline 0 & I_m \end{array}\right] = \left[\begin{array}{cc|c} d_1 & -1 & 0 \\ 1 & d_2 & d_2 \\ \hline 1 & 0 & 0 \\ 0 & 1 & 0 \\ \hline 0 & 0 & 1 \end{array}\right] \rightarrow \left[\begin{array}{cc|c} 1 & 0 & 0 \\ 0 & 1 & 0 \\ \hline 0 & 1 & -d_2 \\ -1 & d_1 & -d_1 d_2 \\ \hline 1 & -d_1 & 1+d_1 d_2 \end{array}\right]$$

Then $\quad A_2 = \begin{bmatrix} 1+d_1 d_2 \end{bmatrix} \quad \text{and} \quad B_2 = \begin{bmatrix} d_2 \\ d_1 d_2 \end{bmatrix}$

**Step 2** Using elementary row operations, we carry out the reduction

$$\left[\begin{array}{c|c|c} A_2 & I_m & 0 \\ \hline B_2 & 0 & I_1 \end{array}\right] = \left[\begin{array}{c|cc|c} 1+d_1 d_2 & 1 & 0 & 0 \\ \hline d_2 & 0 & 1 & 0 \\ d_1 d_2 & 0 & 0 & 1 \end{array}\right] \rightarrow \left[\begin{array}{c|cc|c} 1 & 1 & 0 & -1 \\ \hline 0 & -d_2 & 1 & d_2 \\ 0 & 0 & -d_1 & 1 \end{array}\right]$$

Then a solution to the equation

$$\overline{P}\begin{bmatrix} 1+d_1 d_2 \end{bmatrix} + \overline{Q}\begin{bmatrix} d_2 \\ d_1 d_2 \end{bmatrix} = 1$$

has the form

$$\overline{P} = 1 \quad \text{and} \quad \overline{Q} = \begin{bmatrix} 0 & -1 \end{bmatrix}$$

**Step 3** The general solution is

$$P = \begin{bmatrix} 1-k_1 d_2 \end{bmatrix}, \quad Q = \begin{bmatrix} k_1 d_1 + k_2, & -1-k_1+k_2 d_2 \end{bmatrix}$$

and

$$R = \begin{bmatrix} l_1 d_1 + l_2, & -1-l_1+l_2 d_2 \end{bmatrix}$$

where $k_1, k_2, l_1$ and $l_2$ are arbitrary polynomials in $d_1$ and $d_2$.

**Step 4** Taking into account that

$$\bar{P}_2 = I_2 \quad \text{and} \quad \bar{Q}_2 = \begin{bmatrix} 0 & -1 \end{bmatrix}$$

we obtain

$$\bar{Q}_2 A_0 + A_2 L = \begin{bmatrix} /1+d_1 d_2/l_1, & -1+/1+d_1 d_2/l_2 \end{bmatrix}$$

$$\bar{Q}_2 + A_2 K = \begin{bmatrix} /1+d_1 d_2/k_1, & -1+/1+d_1 d_2/k_2 \end{bmatrix}$$

$$\bar{P}_2 A_0 - B_2 L = \begin{bmatrix} 1-d_2 l_1 & -d_2 l_2 \\ -d_1 d_2 l_1 & 1-d_1 d_2 l_2 \end{bmatrix} \qquad /7.69/$$

$$P_2 - B_2 K = \begin{bmatrix} 1-d_2 k_1 & -d_2 k_2 \\ -d_1 d_2 k_1 & 1-d_1 d_2 k_2 \end{bmatrix}$$

It is easy to see that each column of /7.69/ is of least possible degrees in $d_1$ and $d_2$ for $k_1 = k_2 = l_1 = l_2 = 0$.
Hence, the desired solution is

$$P = \bar{P}, \qquad Q = R = \bar{Q}$$

## PROBLEMS

1. Given the 3-D plant with

$$A = \begin{bmatrix} 1 & 0 \\ d_1 & 1 \end{bmatrix}, \quad B = \begin{bmatrix} d_2 \\ d_3 \end{bmatrix}, \quad C_o = \begin{bmatrix} d_2 \\ d_1+d_3 \end{bmatrix}$$

and the reference input with

$$F = \begin{bmatrix} 1 & 0 \\ d_1 & 1 \end{bmatrix}, \quad G = \begin{bmatrix} d_1+d_2 \\ d_1+2d_3 \end{bmatrix}$$

find the input vector $u/i,j,k/$ such that the tracking error $e/i,j,k/$ vanishes for minimal $i+j+k$ and $i$.

Hint: Use a simular procedure as in Algorithm 7.5.

Answer: $u/1,0,0/ = -1$, $u/0,0,1/ = 1$, $u/2,0,0/ = -1$, $u/2,1,0/ = 1$ otherwise $u/i,j,k/ = 0$.

2. Show for RM described by /7.1/, /7.2/ that if $u/i,j/ = 0$, $/i,j/ \geqslant /N_1,N_2/$ and $y/i,j/ = 0$, $/N_1,N_2/ \leqslant /i,j/ \leqslant /N_1+n_1,N_2+n_2/$, then $y/i,j/ = 0$ for $/i,j/ \geqslant /N_1,N_2/$.

Hint: Use the relation /1.54/.

3. Show that the matrices M and Q defined by /7.26/ are equal to the matrices g and h defined by /7.20a/, respectively.

Hint: Compare the relations /7.19/ and /7.23/.

4. Show that the deadbeat servo problem for n-D multivariable linear system has a solution if [10]:

/i/ F is a right divisor of A,

/ii/ A, B are unimodular left-coprime, i.e. there exists unimodular matrix U /det $U \in R\backslash 0$/ such that $[A \mid B] U = [I \mid 0]$.

Hint: Use a similar manner as in proof of theorem 7.7.

5. Given a 2-D system described by the equation

$$y/i,j/ + 2y/i-1,j/ + 2y/i,j-1/ + y/i,j-1/ =$$
$$= u/i,j/ + 3u/i-1,j/ + 3u/i,j-1/$$

find a pair $/N_1,N_2/$ and an input sequence $u/i,j/$ for $/0,0/ \leq /i,j/ < /N_1,N_2/$ such that $y/i,j/ = 0$ for $/i,j/ \geq /1,1/$.

Hint. Find the transfer function of the system and use the algorithm 7.2.

Answer: $u/0,0/ = u_0$, $u/1,0/ = u/0,1/ = 2u_0$, $u/1,1/ = u_0$, where $u_0$ arbitrary.

6. Given $A \in R^{n \times n}[d_1,d_2]$ and $B \in R^{n \times m}[d_1,d_2]$, find $A_1 \in R^{m \times m}[d_1,d_2]$ and $B_1 \in R^{n \times m}[d_1,d_2]$ such that $AB = B_1 A_1$ and the pair $/A_1,B/$ is zero right coprime /ZRC/.

Hint: Choose

$$A_1 = \begin{bmatrix} 1 & a_{12} & \cdots & a_{1m} \\ 0 & 1 & \cdots & a_{2m} \\ \vdots & & & \vdots \\ 0 & 0 & \cdots & 1 \end{bmatrix}$$

and note that the pair $/A_1,B/$ is ZRC for any B and $A_1^{-1}$ is a polynomial matrix.

Answer: $B_1 = ABA_1^{-1}$.

7. Let $AU_1 + BU_3 = I$. Find U such that $[A \vdots B] U = [I \vdots 0]$

Hint: Note that $[A \vdots B] \begin{bmatrix} U_1 B \\ U_3 B & -I \end{bmatrix} = 0$

Answer: $U = \begin{bmatrix} U_1, & U_1 B \\ U_3, & U_3 B - I \end{bmatrix}$

## REFERENCES

[1] B.Eichstaedt: Multivariable closed-loop deadbeat control: a polynomial-matrix approach. Automatica, vol. 18, No.5, 1982, pp. 589-593

[2] Fu-Yih Shih: Design algorithms for digital control systems with deadbeat unit step response. IEE Proc. vol. 130, No.3, 1983, pp. 119-127

[3] R.Isermann: Digital control systems. Springer-Verlag, Berlin 1981

[4] D.Jordan, J.Korn: Deadbeat algorithms for multivariable process control. IEE Trans. AC-25, 1980, pp. 486-491

[5] T.Kaczorek: Deadbeat control in multivariable nonlinear time-varying systems with constraints of control inputs. Int.J.Control, vol. 38, No.2, 1983, pp. 449-458

[6] T.Kaczorek: Deadbeat control of 2-D linear systems. Bull.Acad. Polon. Sci., Ser. sci.techn., vol. 32, No.5-6, 1984 /in press/

[7] T.Kaczorek: Deadbeat control of multi-input-output n-D linear systems. Prof. of Symposium on Large Scale Systems, July 11-15, 1983, Warsaw, Poland /in press/

[8] T.Kaczorek: Deadbeat servo problem for 2-dimensional linear systems. Int.J.Control, vol. 37, No.6, 1983, pp. 1349-1353

[9] T.Kaczorek: Dead-beat servo problem for 2-D multivariable systems. Proc. Third International Conference on Systems Engineering. Sept. 5-7, 1984, Dayton, USA /in press/

[10] T.Kaczorek: Dead-beat servo problem for n-D multi-input-output linear systems. Preprinto IFAC 9th World Congress, Budapest July 2-6, 1984, vol. IX, pp. 159-162

[11] T.Kaczorek: Solving of 2-D polynomial matrix equations. Proc. III Int. Conference on Functional Differential Systems and Related Topics. Zielona Góra 1983

[12] T.Kaczorek, M.Kocięcki: Deadbeat control in 2-D linear systems. Bull. Acad. Polon. Sci., Ser.sci.techn., vol. 32, No.11-12, 1984, /in press/

[13] V.Kučera: A dead-beat servo problem. Int. J.Control, vol. 32, No.1, 1980, pp. 107-114

[14] V.Kučera: Discrete Linear Control. The Polynomial Equation Approach. John Wiley, Chichester 1979

[15] B.Leden: Output deadbeat control - A geometric approach. Int. J. Control, vol. 26, 1977, pp. 493-507

[16] M.Šebek: Multivariable dead-beat servo problem. Kibernetika, vol. 16, No.5, 1980, pp. 442-453

[17] H.Seraji: Deadbeat control of discrete-time systems using output feedback. Int. J. Control, vol. 21, 1975, pp. 213-223

[18] S.G. Tzafestas, N.J.Theodorou: Open-loop deadbeat control of multidimensional system. Paper presented on IMACS European Meeting, DIGITECH'84, July 9-12, 1984, Patras, Greece.

APPENDIX

## 1. Function of 2-D matrix

Let us assume that the matrix

$$A = \begin{bmatrix} A_{11} & A_{12} \\ A_{21} & A_{22} \end{bmatrix} \qquad A_{ii} \in R^{n_i \times n_i} \quad /i = 1,2/ \qquad /A.1.1/$$

has a separable /factorable/ characteristic polynomial in the form

$$p/z_1 z_2/ = \det \begin{bmatrix} I_{n_1} z_1 - A_{11}, & -A_{12} \\ -A_{21}, & I_{n_2} z_2 - A_{22} \end{bmatrix} = p_1/z_1/p_2/z_2/ \qquad /A.1.2/$$

where

$$p_i/z_i/ = /z_i - z_{i1}/^{m_{i1}} /z_i - z_{i2}/^{m_{i2}} \ldots /z_i - z_{iN_i}/^{m_{iN_i}} \qquad /A.1.3/$$

$$\text{for } i = 1,2$$

$N_i \leq n_i$.

The set of 2-D eigenvalues of the matrix A is defined as follows

$$\{/z_{1i}, z_{2j}/ \;:\; /0,0/ < /i,j/ \leq /N_1, N_2/\} \qquad /A.1.4/$$

where $z_{10} = z_{20} = 0$.

For the polynomial

$$w/z_1, z_2/ = \sum_{i=0}^{M_1} \sum_{j=0}^{M_2} b_{ij} z_1^i z_2^j \qquad /A.1.5/$$

we define w/A/ as the matrix

$$w/A/ := \sum_{i=0}^{M_1} \sum_{j=0}^{M_2} b_{ij} A^{i,j} \qquad /A.1.6/$$

The notion of 2-D matrix polynomial is used to define an arbitrary function of the matrix A.

### Definition A.1.1

Let $f/z_1, z_2/$ be an analytic function in an open set containing the 2-D eigenvalues $/z_{1i}, z_{2j}/$ for $/0,0/ < /i,j/ \leq /N_1, N_2/$ of the matrix A.

Let $w/z_1,z_2/$ be a polynomial such that

$$w^{/r,s/}/z_{1i},z_{2j}/ = f^{/r,s/}/z_{1i},z_{2j}/ \quad \text{for } /0,0/</i,j/\leq/N_1,N_2/ \qquad /A.1.7/$$

and $/0,0/\leq/r,s/\leq/m_{1i}-1,m_{2j}-1/$

where

$$f^{/r,s/}/z_1,z_2/ = \frac{\partial^{r+s} f/z_1,z_2/}{\partial z_1^r \partial z_2^s} \qquad /A.1.8/$$

Then we define the function $f$ of the matrix $A$ as $f/A/ = w/A/$.
We introduce the following notation

$$w_{i0}/z_i/ = \begin{cases} p_i/z_i/ & \text{if 0 is a root of the polynomial } p_i/z_i/ \\ z_i p_i/z_i/ & \text{if 0 is not a root of the polynomial } p_i/z_i/ \end{cases} \quad /i=1,2/$$

Let us expand the expression $/w_{i0}/z_1//^{-1}$ in partial fractions, i.e.

$$/w_{i0}/z_i//^{-1} = \sum_{k=0}^{N_i} \sum_{l=0}^{m_{ik}-1} \frac{c_{ikl}}{/z_i - z_{ik}/^{l+1}} =$$

$$= \sum_{k=0}^{N_i} \frac{c_{ik}/z_i/}{/z_i - z_{ik}/^{m_{ik}}} \qquad /i = 1,2/ \qquad /A.1.9/$$

where $c_{ik}/z_i/$ are polynomials of the degree less than $m_{ik}$.

Let us define

$$v_{i,j}/z_1,z_2/ := \frac{w_{10}/z_1/w_{20}/z_2/c_{1i}/z_1/c_{2j}/z_2/}{/z_1-z_{1i}/^{m_{1i}}/z_2-z_{2j}/^{m_{2j}}} \qquad /A.1.10/$$

for $/0,0/</i,j/\leq/N_1,N_2/$

<u>Lemma A.1.1.</u>

The polynomials /A.1.10/ satisfy the following conditions

$$\sum_{/0,0/</i,j/\leq/N_1,N_2/} v_{i,j}/z_1,z_2/ = 1 \qquad \forall\, z_1 \in R \text{ and } z_2 \in R \qquad /A.1.11/$$

$$v_{i,j}/z_{1i},z_{2j}/ = 1 \quad \text{for } /0,0/</i,j/\leqslant/N_1,N_2/ \qquad /A.1.12/$$

$$v_{i,j}^{/r,s/}/z_{1i},z_{2j}/ = 0 \quad \text{for } /0,0/</i,j/\leqslant/N_1,N_2/ \text{ and}$$
$$/0,0/</r,s/\leqslant/m_{1i}-1,m_{2j}-1/ \qquad /A.1.13/$$

$$v_{h,k}^{/r,s/}/z_{1i},z_{2j}/ = 0 \quad \text{for } /0,0/</i,j/\leqslant/N_1,N_2/ \text{ and}$$
$$/0,0/</r,s/\leqslant/m_{1i}-1,m_{2j}-1/ \text{ and}$$
$$/0,0/</h,k/\leqslant/N_1,N_2/ \qquad /A.1.14/$$

<u>Proof</u>

From /A.1.9/ we have

$$\sum_{i=0}^{N_1} \frac{w_{10}/z_1/\ c_{1i}/z_1/}{/z_1-z_{1i}/^{m_{1i}}} = 1$$

and $\qquad\qquad\qquad\qquad\qquad\qquad\qquad\qquad\qquad\qquad\qquad\qquad\qquad\qquad$ /A.1.15/

$$\sum_{j=0}^{N_2} \frac{w_{20}/z_2/\ c_{2j}/z_2/}{/z_2-z_{2j}/^{m_{2j}}} = 1$$

Multipying by sides the expressions /A.1.15/ and taking into account /A.1.10/ we obtain /A.1.11/.

From /A.1.9/ it follows that

$$/w_{10}/z_i//^{-1} = \frac{c_{i0}/z_i/d_0/z_i/+c_{i1}/z_i/d_1/z_i/+\ldots+c_{iN_i}/z_i/d_{N_i}/z_i}{/z_i-z_{i0}/^{m_{i0}}\ /z_i-z_{i1}/^{m_{i1}}\ldots/z_i-z_{iN_i}/^{m_{iN_1}}}$$

$$/i = 1,2/$$

where

$$d_j/z_i/ = \prod_{\substack{k=0 \\ k\neq j}}^{N_i} /z_i-z_{ik}/^{m_{ik}} \qquad /i = 1,2; j = 0,1,\ldots,N_i/$$

Hence

$$\left.\frac{w_{i0}/z_i/c_{ik}/z_i/}{/z_i-z_{ik}/^{m_{ik}}}\right|_{z_i=z_{ik}} = 1 \quad /i=1,2;\ k=0,1,\ldots,N_i/ \qquad /A.1.16/$$

since $\quad d_j/z_{ik}/ = 0 \quad$ for $\quad k \neq j$ /A.1.17/

The relation /A.1.12/ follows immediately from /A.1.16/.
Differentiating /A.1.10/ and taking into account the above relations we obtain /A.1.13/.
The relation /A.1.14/ can be proved in a similar way. □

## Theorem A.1.1 [12]

If $f/z_1, z_2/$ is an analytic function in an open set containing the 2-D eigenvalues $/z_{1i}, z_{2j}/$ for $/0,0/</i,j/\leqslant/N_1,N_2/$ of the matrix A, then

$$f/A/ = \sum_{/0,0/</i,j/\leqslant/N_1,N_2/} \sum_{/0,0/\leqslant/r,s/\leqslant/m_{1i}-1,m_{2j}-1/} Z_{i,j,r,s} \; f^{/r,s/}/z_{1i},z_{2j}/ \qquad /A.1.18/$$

where

$$Z_{i,j,r,s} = q_{i,j,r,s}/A/ \qquad /.1.19/$$

and $q_{i,j,r,s}/z_1,z_2/$ is a residual polynomial obtained via the 2-D division algorithm /theorem 1.3/ from the polynomial

$$v_{i,j,r,s}/z_1,z_2/ = \frac{w_{10}/z_1/ w_{20}/z_2/ c_{1i}/z_1/ c_{2j}/z_2/}{r!\,s!} /z_1 - z_{1i}/^{r-m_{1i}} /z_2 - z_{2j}/^{s-m_{2j}}$$

for $/0,0/</i,j/\leqslant/N_1,N_2/$ and $/0,0/\leqslant/r,s/\leqslant/m_{1i}-1,m_{2j}-1/$

/A.1.20/

## Proof

Consider for the function $f/z_1,z_2/$ the interpolating polynomial

$$w/z_1,z_2/ = \sum_{/0,0/</i,j/\leqslant/N_1,N_2/} \sum_{/0,0/\leqslant/r,s/\leqslant/m_{1i}-1,m_{2j}-1/} f^{/r,s/}/z_{1i},z_{2j}/ \; v_{i,j,r,s}/z_1,z_2/$$

/A.1.21/

where

$$v_{i,j,r,s}/z_1,z_2/ = \frac{1}{r!\,s!}\, v_{i,j}/z_1,z_2//z_1-z_{1i}/^r/z_2-z_{2j}/^s \qquad /A.1.22/$$

Differentialing /A.1.22/ and taking into account /A.1.12/, /A.1.13/ and /A.1.14/ we obtain

$$v_{i,j,r,s}^{/r,s/}/z_{1k},z_{2l}/ = \begin{cases} 1 & \text{for } k = i \text{ and } l = j \\ 0 & \text{otherwise} \end{cases} \qquad /A.1.23/$$

From /A.1.21/ and /A.1.23/ it follows that

$$w^{/r,s/}/z_{1i},z_{2j}/ = f^{/r,s/}/z_{1i},z_{2j}/ \qquad /A.1.24/$$

$$\text{for } /0,0/\leqslant/i,j/\leqslant/N_1,N_2/ \text{ and } /0,0/\leqslant/r,s/\leqslant/m_{1i}-1,m_{2j}-1/$$

Hence by the definition A.1.1, the 2-D divison algorithm and the equation /1.48/ we obtain /A.1.1/ and /A.1.19/

In particular case when the 2-D eigenvalues $/z_{1i},z_{2j}/$ of the matrix A are distinct $|/m_{i1} = m_{i2} = \ldots = m_{iN_i} = 1$ and $N_i = n_i$ for $i = 1,2/$ from /A.1.9/ and /A.1.10/ we obtain

$$c_{ik} = \prod_{\substack{l=0 \\ l\neq k}}^{n_1} \frac{1}{/z_{ik}-z_{il}/} \qquad \text{for } i = 1,2 \text{ and } k = 0,1,\ldots,n_1 \qquad /A.1.25/$$

and

$$v_{i,j}/z_1,z_2/ = \prod_{\substack{k=0 \\ k\neq i}}^{n_1} \frac{z_1-z_{1k}}{z_{1i}-z_{1k}} \prod_{\substack{l=0 \\ l\neq j}}^{n_2} \frac{z_2-z_{21}}{z_{2j}-z_{21}} \qquad /A.1.26/$$

$$\text{for } /0,0/\leqslant/i,j/\leqslant/n_1,n_2/$$

Note that $\deg v_{i,j}/z_1,z_2/ < n_k$, $k = 1,2$ $//n_1,n_2/\geqslant/1,1//$ and by the 2-D divison algorithm $v_{i,j}/z_1,z_2/ = q_{i,j}/z_1,z_2/$ for $/0,0/\leqslant/i,j/\leqslant/n_1,n_2/$.

We have proved the following

## Corollary A.1.1 [8]

If $f/z_1,z_2/$ is an analytic function in an open set containing the 2-D eigenvalues $/z_{1i},z_{2j}/$ for $/0,0/</i,j/\leq/n_1,n_2/$ of the matrix A and the characteristic polynomial $p/z_1,z_2/$ has the form

$$p/z_1,z_2/ = /z_1-z_{11}//z_1-z_{12}/\ldots/z_1-z_{1n_1}//z_2-z_{21}//z_2-z_{22}/\ldots/z_2-z_{2n_2}/$$

$$/z_{10} = z_{20} = 0/$$

then

$$f/A/ = \sum_{/0,0/</i,j/\leq/n_1,n_2} Z_{i,j}\, f/z_{1i},z_{2j}/ \qquad /A.1.27/$$

where
$Z_{i,j} = v_{i,j}/A/$ and the polynomial $v_{i,j}/z_1,z_2/$ is given by /A.1.26/.

From above considerations the following algorithm for finding $f/A/$ follows

### Algorithm A.1.1

**Step 1** For the given matrix A find the 2-D characteristic polynomial and the set /A.1.4/ of 2-D eigenvalues $/z_{1i},z_{2j}/$ of the matrix A.

**Step 2** Using /A.1.20/ find the polynomials $v_{i,j,r,s}/z_1,z_2/$ for $/0,0/</i,j/\leq/N_1,N_2/$ and $/0,0/\leq/r,s/\leq/m_{1i}-1,m_{2j}-1/$.

**Step 3** Using the 2-D division algorithm /theorem 1.3/ find the residual polynomial $q_{i,j,r,s}/z_1,z_2/$.

**Step 4** Using /A.1.19/ fing $Z_{i,j,r,s}$ for $/0,0/</i,j/\leq/N_1,N_2/$ and $/0,0/\leq/r,s/\leq/m_{1i}-1,m_{2j}-1/$.

**Step 5** Using /A.1.18/ find $f/A/$.

### Example A.1.1.

Find $A^{i,j}$ for the matrix

$$A = \begin{bmatrix} 1 & 2 & | & 0 \\ 0 & 1 & | & 3 \\ \hline 0 & 0 & | & 2 \end{bmatrix} \qquad /A.1.28/$$

**Step 1**  The 2-D characteristic polynomial of the matrix /A.1.28/ has the form

$$p/z_1,z_2/ = \begin{vmatrix} z_1-1 & -2 & 0 \\ 0 & z_1-1 & -3 \\ 0 & 0 & z_2-2 \end{vmatrix} = /z_1-1/^2/z_2-2/ \qquad /A.1.29/$$

Hence $p_1/z_1/ = /z_1-1/^2$, $p_2/z_2/ \; /z_2-2/$ and $N_1 = 1$, $N_2 = 1$, $m_{10} = 1$, $m_{11} = 2$, $m_{20} = 1$, $m_{21} = 1$.

The set of 2-D eigenvalues of the matrix /A.1.28/ is

$$\{/z_{1i},z_{2j}/ : /0,0/ \leqslant /i,j/ \leqslant /N_1,N_2/\} = \{/1,0/, /0,2/, /1,2/\}$$

**Step 2**  In this case $w_{10}/z_1/ = z_1/z_1-1/^2$, $w_{20}/z_2/ = z_2/z_2-2/$ and

$$\frac{1}{w_{10}/z_1/} = \frac{1}{z_1/z_1-1/^2} = \frac{1}{z_1} + \frac{2-z_1}{/z_1-1/^2}$$

$$\frac{1}{w_{20}/z_2/} = \frac{1}{z_2/z_2-2/} = -\frac{1}{2z_2} + \frac{1}{2/z_2-2/}$$

Hence

$$c_{10} = 1, \quad c_{11}/z_1/ = 2-z_1, \quad c_{20} = -\frac{1}{2}, \quad c_{21} = \frac{1}{2}$$

Using /A.1.20/ we calculate

$$v_{0,1,0,0}/z_1,z_2/ = \frac{w_{10}/z_1/w_{20}/z_2/c_{10}/z_1/c_{21}/z_2/}{/z_1-z_{10}/^{m_{10}} /z_2-z_{21}/^{m_{21}}} = \frac{1}{2}/z_1-1/^2 z_2$$

$$v_{1,0,0,0}/z_1,z_2/ = \frac{w_{10}/z_1/w_{20}/z_2/c_{11}/z_1/c_{20}/z_2/}{/z_1-z_{11}/^{m_{11}} /z_2-z_{20}/^{m_{20}}} =$$

$$= \frac{1}{2} z_1/z_1-2//z_2-2/$$

$$v_{1,1,0,0}/z_1,z_2/ = \frac{w_{10}/z_1/w_{20}/z_2/c_{11}/z_1/c_{21}/z_2/}{/z_1-z_{11}/^{m_{11}} /z_2-z_{21}/^{m_{21}}} = -\frac{1}{2} z_1 z_2 /z_1-2/$$

$$v_{1,0,1,0}/z_1,z_2/ = \frac{w_{10}/z_1/w_{20}/z_2/c_{11}/z_1/c_{20}/z_2/}{/z_1-z_{11}/^{m_{11}-1} /z_2-z_{20}/^{m_{20}}} = \frac{1}{2} z_1/z_1-1//z_1-2//z_2-2/$$

$$v_{1,1,1,0}/z_1,z_2/ = \frac{w_{10}/z_1/w_{20}/z_2/c_{11}/z_1/c_{21}/z_2/}{/z_1-z_{11}/^{m_{11}-1} /z_2-z_{21}/^{m_{21}}} = -\frac{1}{2} z_1 z_2 /z_1-1//z_1-2/$$

**Step 3** Using /A.1.29/ we evaluate the residual polynomials

$$q_{0,1,0,0}/z_1,z_2/ = z_1^2 - 2z_1 + 1$$

$$q_{1,0,0,0}/z_1,z_2/ = -\frac{1}{2} z_2 + 1$$

$$q_{1,1,0,0}/z_1,z_2/ = -z_1^2 + 2z_1 + \frac{1}{2} z_2 - 1$$

$$q_{1,0,1,0}/z_1,z_2/ = -\frac{1}{2} z_1 z_2 + z_1 + \frac{1}{2} z_2 - 1$$

$$q_{1,1,1,0} z_1,z_2/ = -z_1^3 + 3z_1^2 + \frac{1}{2} z_1 z_2 - 3z_1 - \frac{1}{2} z_2 + 1$$

**Step 4** Using /A.1.19/ we obtain

$$Z_{0,1,0,0} = A^{2,0} - 2A^{1,0} + I = \begin{bmatrix} 0 & 0 & 6 \\ 0 & 0 & -3 \\ 0 & 0 & 1 \end{bmatrix}$$

$$Z_{1,0,0,0} = -\frac{1}{2} A^{0,1} + I = \begin{bmatrix} 1 & 0 & 0 \\ 0 & 1 & 0 \\ 0 & 0 & 0 \end{bmatrix}$$

$$Z_{1,1,0,0} = -A^{2,0} + 2A^{1,0} + \frac{1}{2} A^{0,1} - I = \begin{bmatrix} 0 & 0 & -6 \\ 0 & 0 & 3 \\ 0 & 0 & 0 \end{bmatrix}$$

$$Z_{1,0,1,0} = -\frac{1}{2} A^{1,1} + A^{1,0} + \frac{1}{2} A^{0,1} - I = \begin{bmatrix} 0 & 2 & 0 \\ 0 & 0 & 0 \\ 0 & 0 & 0 \end{bmatrix}$$

$$Z_{1,1,1,0} = -A^{3,0} + 3A^{2,0} + \frac{1}{2} A^{1,1} - 3A^{1,0} - \frac{1}{2} A^{0,1} + I = \begin{bmatrix} 0 & 0 & 6 \\ 0 & 0 & 0 \\ 0 & 0 & 0 \end{bmatrix}$$

**Step 5**  From /A.1.18/ we obtain the desired matrix

$$A^{i,j} = Z_{0,1,0,0}\, z_{10}^{i} z_{21}^{j} + Z_{1,0,0,0}\, z_{11}^{i} z_{20}^{j} + Z_{1,1,0,0}\, z_{11}^{i} z_{21}^{j} +$$

$$+ Z_{1,0,1,0}\, i z_{11}^{i-1} z_{20}^{j} + Z_{1,1,1,0}\, i z_{11}^{i-1} z_{21}^{j} = \begin{bmatrix} 0^j, & 2i0^j, & 6/i+0^i-1/2^j \\ 0, & 0^j, & 3/1-0^i/2^j \\ 0, & 0, & 0^i 2^j \end{bmatrix}$$

for $i,j \geqslant 0$ and

$$A^{i,j} = \begin{bmatrix} 0 & 0 & 6/i-1/2^j \\ 0 & 0 & 3;2^j \\ 0 & 0 & 0 \end{bmatrix} \quad \text{for } i,j > 0$$

## 2. Two-dimensional Z transformation

**Definition A.2.1.**

Two-dimensional /2-D/ Z transform $F/z_1,z_2/$ of a discrete 2-D function $f/i,j/$ satisfying the condition $f/i,j/ = 0$ for $i<0$ or $j<0$ is defined by the formula

$$F/z_1,z_2/ = Z\big[f/i,j/\big] := \sum_{i=0}^{\infty} \sum_{j=0}^{\infty} f/i,j/\, z_1^{-i} z_2^{-j} \qquad /A.2.1/$$

**Theorem A.2.1.**

If $F/z_1,z_2/ = Z\big[f/i,j/\big]$, then

$$Z\big[f/i+k,j+l/\big] = z_1^k z_2^l\, F/z_1,z_2/ - \sum_{i=0}^{k-1} \sum_{j=0}^{l-1} f/i,j/ z_1^{k-i} z_2^{l-j} +$$

$$- \sum_{i=0}^{k-1} \sum_{j=0}^{\infty} f/i,j/ z_1^{k-i} z_2^{l-j} - \sum_{i=k}^{\infty} \sum_{j=0}^{l-1} f/i,j/ z_1^{k-i} z_2^{l-j} \qquad /A.2.2/$$

**Proof**  Using /A.2.1/ we obtain

$$Z\big[f/i+k,j+l/\big] = \sum_{i=0}^{\infty} \sum_{j=0}^{\infty} f/i+k,j+l/\, z_1^{-i} z_2^{-j} =$$

$$= \sum_{t=k}^{\infty} \sum_{s=1}^{\infty} f/t,s/ z_1^{k-t} z_2^{l-s} = \sum_{t=0}^{\infty} \sum_{s=0}^{\infty} f/t,s/ z_1^{k-t} z_2^{l-s} +$$

$$- \sum_{t=0}^{k-1} \sum_{s=0}^{l-1} f/t,s/ z_1^{k-t} z_2^{l-s} - \sum_{i=0}^{k-1} \sum_{j=1}^{\infty} f/i,j/ z_1^{k-i} z_2^{l-j} +$$

$$- \sum_{i=k}^{\infty} \sum_{j=0}^{l-1} f/i,j/ z_1^{k-i} z_2^{l-j} = z_1^k z_2^l \, F/z_1, z_2/ +$$

$$- \sum_{t=0}^{k-1} \sum_{s=0}^{l-1} f/t,s/ z_1^{k-t} z_2^{l-s} - \sum_{i=0}^{k-1} \sum_{j=1}^{\infty} f/i,j/ z_1^{k-i} z_2^{l-j} +$$

$$- \sum_{i=k}^{\infty} \sum_{j=0}^{l-1} f/i,j/ z_1^{k-i} z_2^{l-j}$$

In particular case for $k > 0$, $l = 0$ we obtain

$$Z\left[f/i+k,j/\right] = z_1^k \, F/z_1, z_2/ - \sum_{i=0}^{k-1} \sum_{j=0}^{\infty} f/i,j/ z_1^{k-i} z_2^{-j} \qquad /A.2.3/$$

and for $k = 1$, $l = 0$ and $k = 0$, $l = 1$

$$Z\left[f/i+1,j/\right] = z_1 \, F/z_1, z_2/ - z_1 \, F/0, z_2/ \qquad /A.2.4/$$
$$Z\left[f/i,j+1/\right] = z_2 \, F/z_1, z_2/ - z_2 \, F/z_1, 0/ \qquad /A.2.4/$$

Note that if $f/i,0/ = 0$ for $i > 0$ and $f/0,j/ = 0$ for $j > 0$ we have

$$Z\left[f/i+1,j/\right] = z_1 \, F/z_1, z_2/ - z_1 \, f/0,0/$$
$$Z\left[f/i,j+1/\right] = z_2 \, F/z_1, z_2/ - z_2 \, f/0,0/ \qquad /A.2.4'/$$

In a similar way we can prove the following

<u>Theorem A.2.2</u>

If
$$F/z_1, z_2/ = Z\left[f/i,j/\right],$$

then
$$Z[f/i-k,j-l/] = z_1^{-k} z_2^{-l} \, F/z_1,z_2/ \qquad /A.2.5/$$

**Definition A.2.2.**

The 2-D discrete function $f/i,j/$ defined by the formula

$$f/i,j/ = f_1/i,j/ * f_2/i,j/ = \sum_{k=0}^{i} \sum_{l=0}^{j} f_1/i-k,j-l/ \, f_2/k,l/ \qquad /A.2.6/$$

is called the 2-D convolution of functions $f_1/i,j/$ and $f_2/i,j/$

**Theorem A.2.3.**

If
$$F_1/z_1,z_2/ = Z[f_1/i,j/] \quad \text{and} \quad F_2/z_1,z_2/ = Z[f_2/i,j/]$$
then
$$Z[f_1/i,j/ * f_2/i,j/] = F_1/z_1,z_2/ \, F_2/z_1,z_2/ \qquad /A.2.7/$$

**Proof**

Taking into account that $f_1/i-k,j-l/ = 0$ for $k>i$ or $l>j$ and using /A.2.1/, /A.2.4/ and /A.2.3/ we obtain

$$Z[f_1/i,j/ * f_2/i,j/] = Z\left[\sum_{k=0}^{i} \sum_{l=0}^{j} f_1/i-k,j-l/ \, f_2/k,l/\right] =$$

$$= Z\left[\sum_{k=0}^{\infty} \sum_{l=0}^{\infty} f_1/i-k,j-l/f_2/k,l/\right] = \sum_{k=0}^{\infty} \sum_{l=0}^{\infty} Z[f_1/i-k,j-l/] f_2/k,l/ =$$

$$= \sum_{k=0}^{\infty} \sum_{l=0}^{\infty} F_1/z_1,z_2/ f_2/k,l/ z_1^{-k} z_2^{-l} = F_1/z_1,z_2/ \, F_2/z_1,z_2/. \qquad \square$$

**Theorem A.2.4.**

If
$$A = \begin{bmatrix} A_1 & A_{12} \\ A_{21} & A_{22} \end{bmatrix} \qquad A_{ii} \in R^{n_i \times n_i} \quad /i=1,2/$$

then

$$Z\left[A^{i,j}\right] = \begin{bmatrix} I_{n_1} & -z_1^{-1}A_{11}, & -z_1^{-1}A_{12} \\ -z_2^{-1}A_{21} & , & I_{n_2}-z_2^{-1}A_{22} \end{bmatrix}^{-1} \qquad /A.2.8/$$

**Proof** Consider the equation

$$\begin{bmatrix} x^h/i+1,j/ \\ x^v/i,j+1/ \end{bmatrix} = A \begin{bmatrix} x^h/i,j/ \\ x^v/i,j/ \end{bmatrix} \qquad /A.2.9/$$

with the boundary conditions $x^h/0,j/ = 0$ for $j > 0$ and $x^v/i,0/ = 0$ for $i > 0$.

From /1.17/ it follows that

$$\begin{bmatrix} x^h/i,j/ \\ x^v/i,j/ \end{bmatrix} = A^{i,j} \begin{bmatrix} x^h/0,0/ \\ x^v/0,0/ \end{bmatrix} \qquad /A.2.10/$$

Using /A.2.4'/ from /A.2.9/ we obtain

$$\begin{bmatrix} z_1 x^h/z_1,z_2/ \\ z_2 x^v/z_1,z_2/ \end{bmatrix} = A \begin{bmatrix} x^h/z_1,z_2/ \\ x^v/z_1,z_2/ \end{bmatrix} + \begin{bmatrix} z_1 x^h/0,0/ \\ z_2 x^v/0,0/ \end{bmatrix}$$

and

$$\begin{bmatrix} x^h/z_1,z_2/ \\ x^v/z_1,z_2/ \end{bmatrix} = \begin{bmatrix} I_{n_1} - z_1^{-1}A_{11}, & -z_1^{-1}A_{12} \\ -z_2^{-1}A_{21} & , & I_{n_2}-z_2^{-1}A_{22} \end{bmatrix}^{-1} \begin{bmatrix} x^h/0,0/ \\ x^v/0,0/ \end{bmatrix} \qquad /A.2.11/$$

The formula /A.2.8/ follows from comparison of /A.2.10/ and /A.2.11/. □
Note that using /A.2.7/ and /A.2.8/ from /A.2.6/ we can obtain /1.15/.

## 3. Euclidean algorithm, Hermite and Smith forms of 2-D polynomial matrices

### 3.1. Euclidean algorithm

We shall consider a 2-D polynomial

$$a/z_n,z_2/ = \sum_{i=0}^{n_2} a_i/z_1/z_2^i = \sum_{j=0}^{n_1} a_j/z_2/z_1^j$$

as an element of $F/z_1/[z_2]$, i.e. as a polynomial in $z_2$ with coefficients $a_i/z_1/$ in the field $F/z_1/$ or as an element of $F/z_2/[z_1]$.

Let

$$a/z_1,z_2/ = \sum_{i=0}^{n} a_i/z_1/z_2^i \qquad /A.3.1a/$$

and

$$b/z_1,z_2/ = \sum_{i=0}^{m} b_i/z_1/z_2^i \qquad /A.3.1b/$$

with

$$\deg_{z_2} a/z_1,z_2/ = n \leq \deg_{z_2} b/z_1,z_2/ = m$$

Note that we can work over the field $F/z_1/$ in a similar manner as in the 1-D case over the real field. Using this approach we can find two polynomials $\bar{q}/z_1,z_2/, \bar{r}/z_1,z_2/ \in F/z_1/[z_2]$ such that

$$a/z_1,z_2/ = \bar{q}/z_1,z_2/ \, b/z_1,z_2/ + \bar{r}/z_1,z_2/ \qquad /A.3.2/$$

where

$$\deg_{z_2} \bar{r}/z_1,z_2/ < \deg_{z_2} b/z_1,z_2/$$

Let

$$\bar{q}/z_1,z_2/ = \frac{q_0/z_1/}{p_0/z_1/} z_2^{n-m} + \ldots + \frac{q_{n-m-1}/z_1/}{p_{n-m-1}/z_1/} z_2 + \frac{q_{n-m}/z_1/}{p_{n-m}/z_1/}$$

and let $p/z_1/$ be the least common multiple of all the $p_i/z_1/$, i.e.

$$p/z_1/ = L\,M\,C\left[p_0/z_1/,p_1/z_1/,\ldots,p_{n-m}/z_1/\right]$$

Taking into account that

$$\bar{q}(z_1,z_2) = \frac{q(z_1,z_2)}{p(z_1)}$$

and multiplying /A.3.2/ by $p(z_1)$ we obtain

$$p(z_1)\,a(z_1,z_2) = q(z_1,z_2)\,b(z_1,z_2) + r(z_1,z_2) \qquad /A.3.3/$$

where $r(z_1,z_2) = p(z_1)\bar{r}(z_1,z_2)$.

Note that $\deg_{z_2} r(z_1,z_2) < m$ and that $p(z_1)$ and $q(z_1,z_2)$ are coprime.

Symmetrically, a 2-D Euclidean division algorithm can also be defined in $F[z_2][z_1]$. Thus we can find $\hat{p}(z_2)$, $\hat{q}(z_1,z_2)$ and $\hat{r}(z_1,z_2)$ such that

$$\hat{p}(z_2)\,a(z_1,z_2) = \hat{q}(z_1,z_2)\,b(z_1,z_2) + \hat{r}(z_2,z_2) \qquad /A.3.4/$$

when $\deg_{z_1} \hat{r}(z_1,z_2) < \deg_{z_1} b(z_1,z_2)$ and $\hat{p}(z_2)$ and $\hat{q}(z_1,z_2)$ are coprime.

### Example A.3.1.

Given

$$a(z_1,z_2) = z_1 z_2^2 + z_2 + z_1$$

$$b(z_1,z_2) = z_1 z_2 + z_1$$

find $p(z_1)$, $q(z_1,z_2)$ and $r(z_1,z_2)$.

It is easy to check that

$$a(z_1,z_2) = (z_2 + \frac{1-z_1}{z_1})\,(b(z_1,z_2) + 2z_1 - 1$$

Hence

$$\bar{q}(z_1,z_2) = z_2 + \frac{1-z_1}{z_1} = \frac{q(z_1,z_2)}{p(z_1)}, \qquad \bar{r}(z_1,z_2) = 2z_1 - 1$$

and $r(z_1,z_2) = p(z_1)\bar{r}(z_1,z_2) = z_1(2z_1-1)$

where

$$q(z_1,z_2) = z_1 z_2 - z_1 - 1$$

$$p(z_1) = z_1$$

Thus
$$z_1/z_1z_2^2+z_2+z_1/ = /z_1z_2-z_1+1//z_1z_2+z_1/+z_1/2z_1-1/$$

## 3.2. Hermite form of 2-D polynomial matrices

We are interested in finding the Hermite form for a polynomial matrix $A/z_1,z_2/ \in F^{m \times n}[z_1,z_2]$ with respect to /w.r.t/ $F[z_1][z_2]$. Without loss of generality we can assume that $A/z_1,z_2/$ is of full rank. In a similar manner as in the 1-D case over the real field we can work here over the field $F/z_1/$. Using this approach we can find an m×m unimodular matrix $\bar{U}/z_1,z_2/ \in F^{m \times m}/z_1/[z_2]/$ i.e. a polynomial matrix in $z_2$ with coefficients in the field $F/z_1/$ and with nonzero det $\bar{U}/z_1,z_2/ \in F/z_1//$ such that
if m ⩾ n

$$\bar{U}/z_1,z_2/A/z_1,z_2/ = \begin{bmatrix} \bar{a}_{11} & \bar{a}_{12} & \cdots & \bar{a}_{1n} \\ 0 & \bar{a}_{22} & \cdots & \bar{a}_{2n} \\ \cdot & \cdot & \cdots & \cdot \\ 0 & 0 & \cdots & \bar{a}_{nn} \\ 0 & 0 & \cdots & 0 \\ \cdot & \cdot & \cdots & \cdot \\ 0 & 0 & \cdots & 0 \end{bmatrix} \qquad /A.3.5a/$$

or if n ⩾ m

$$\bar{U}/z_1,z_2/A/z_1,z_2/ = \begin{bmatrix} \bar{a}_{11} & \bar{a}_{12} & \cdots & \bar{a}_{1m} & \cdots & \bar{a}_{1n} \\ 0 & \bar{a}_{22} & \cdots & \bar{a}_{2m} & \cdots & \bar{a}_{2n} \\ \cdot & \cdot & \cdots & \cdot & \cdots & \cdot \\ 0 & 0 & \cdots & \bar{a}_{mm} & \cdots & \bar{a}_{mn} \end{bmatrix} \qquad /A.3.5b/$$

where $\deg_{z_2} \bar{a}_{ii} > \deg_{z_2} \bar{a}_{ki}$ for $k \neq i$.

Let $p_i/z_1/$ be the least common denominator of all entriess of ith row of $\bar{U}/z_1,z_2/$. Then premultiplying /A.3.5/ by diag $[p_1/z_1/, p_2/z_1/, \ldots, p_m/z_1/]$ we obtain if $m \geqslant n$

$$U/z_1,z_2/ \; A/z_1,z_2/ = \begin{bmatrix} \hat{a}_{11} & \hat{a}_{12} & \cdots & \hat{a}_{1n} \\ 0 & \hat{a}_{22} & \cdots & \hat{a}_{2n} \\ \vdots & \vdots & & \vdots \\ 0 & 0 & \cdots & \hat{a}_{nn} \\ 0 & 0 & \cdots & 0 \\ \vdots & \vdots & & \vdots \\ 0 & 0 & \cdots & 0 \end{bmatrix} \qquad /A.3.6a/$$

or if $n \; m$

$$U/z_1,z_2/ \; A/z_1,z_2/ = \begin{bmatrix} \hat{a}_{11} & \hat{a}_{12} & \cdots & \hat{a}_{1m} & \cdots & \hat{a}_{1n} \\ 0 & \hat{a}_{22} & \cdots & \hat{a}_{2m} & \cdots & \hat{a}_{2n} \\ \vdots & \vdots & & \vdots & & \vdots \\ 0 & 0 & \cdots & \hat{a}_{mm} & \cdots & \hat{a}_{mn} \end{bmatrix} \qquad /A.3.6b/$$

where

$$U/z_1,z_2/ = \text{diag}\left[p_1/z_1/, \; p_2/z_1/,\ldots,p_m/z_1/\right] \overline{U}/z_1,z_2/ \in F^{m \times m}[z_1,z_2]$$

and

$$\hat{a}_{ij} = p_i/z_1/\bar{a}_{ij} \qquad \begin{aligned} i &= 1,2,\ldots,m \\ j &= 1,2,\ldots,n \end{aligned}$$

are polynomials in $z_1$ and $z_2$.

Note that $\deg_{z_2} \hat{a}_{ii} > \deg_{z_2} \hat{a}_{ki}$ for $k \neq i$.

Symmetrically, we can find the Hermite form of $A/z_1,z_2/$ w.r.t $F[z_2][z_1]$.

<u>Example A.3.2.</u>

Find the Hermite form /w.r.t $F[z_1][z_2]$/ of

$$A/z_1,z_2/ = \begin{bmatrix} z_2^2 & z_1 z_2 \\ z_2 & z_1 \\ z_1 & z_2 \end{bmatrix} \qquad /A.3.7/$$

Using elementary row operations we carry out the reduction

$$\left[\begin{array}{cc|ccc} z_2^2 & z_1 z_2 & 1 & 0 & 0 \\ z_2 & z_1 & 0 & 1 & 0 \\ z_1 & z_2 & 0 & 0 & 1 \end{array}\right] \rightarrow \left[\begin{array}{cc|ccc} z_1 & z_2 & 0 & 0 & 1 \\ 0 & z_1 - \dfrac{z_2^2}{z_1} & 0 & 1 & -\dfrac{z_2}{z_1} \\ 0 & 0 & 1 & -z_2 & 0 \end{array}\right]$$

Hence

$$\begin{bmatrix} 0 & 0 & 1 \\ 0 & 1 & -\frac{z_2}{z_1} \\ 1 & -z_2 & 0 \end{bmatrix} \begin{bmatrix} z_2^2 & z_1 z_2 \\ z_2 & z_1 \\ z_1 & z_2 \end{bmatrix} = \begin{bmatrix} z_1 & z_2 \\ 0 & z_1 - \frac{z_2^2}{z_1} \\ 0 & 0 \end{bmatrix} \qquad /A.3.8/$$

and

$$\bar{U}/z_1,z_2/ = \begin{bmatrix} 0 & 0 & 1 \\ 0 & 1 & -\frac{z_2}{z_1} \\ 1 & -z_2 & 0 \end{bmatrix}$$

Premultiplying /A.3.8/ by

$$\operatorname{diag}\left[p_1/z_1/, p_2/z_1/, p_3/z_1\right] = \begin{bmatrix} 1 & 0 & 0 \\ 0 & z_1 & 0 \\ 0 & 0 & 1 \end{bmatrix}$$

we obtain

$$\begin{bmatrix} 0 & 0 & 1 \\ 0 & z_1 & -z_2 \\ 1 & -z_2 & 0 \end{bmatrix} \begin{bmatrix} z_2^2 & z_1 z_2 \\ z_2 & z_1 \\ z_1 & z_2 \end{bmatrix} = \begin{bmatrix} z_1 & z_2 \\ 0 & z_1^2 - z_2^2 \\ 0 & 0 \end{bmatrix}$$

### 3.3. Smith form of D-2 polynomial matrices

We are interested in finding the Smith form for a polynomial matrix $A/z_1,z_2/ \in F^{m \times n}[z_1,z_2]$ w.r.t $F[z_1][z_2]$. Without loss of generality we can assume that $A/z_1,z_2/$ is of full rank. In a similar manner as in the 1-D case over the real field we can work here over the field $F/z_1/$. Using this approach we can find two unimodular matrices $\bar{U}/z_1,z_2/ \in F^{m \times m}/z_1/[z_2]$, $\bar{V}/z_1,z_2/ \in F^{n \times n}/z_1/[z_2]$ /with nonzero det $\bar{U}/z_1,z_2/$, det $\bar{V}/z_1,z_2/ \in F/z_1/$ such that
if $m \geqslant n$

$$\bar{U}/z_1,z_2/A/z_1,z_2/\bar{V}/z_1,z_2/ = \begin{bmatrix} \tilde{a}_{11} & 0 & \cdots & 0 \\ 0 & \tilde{a}_{22} & \cdots & 0 \\ \vdots & & \ddots & \vdots \\ 0 & 0 & \cdots & \tilde{a}_{nn} \\ 0 & 0 & \cdots & 0 \\ \vdots & \vdots & & \vdots \\ 0 & 0 & \cdots & 0 \end{bmatrix} \qquad /A.3.9a/$$

or if $n \geqslant m$

$$\bar{U}(z_1,z_2) A(z_1,z_2) \bar{V}(z_1,z_2) = \begin{bmatrix} \tilde{a}_{11} & 0 & \cdots & 0 & \cdots & 0 \\ 0 & \tilde{a}_{22} & \cdots & 0 & \cdots & 0 \\ \cdots & \cdots & \cdots & \cdots & \cdots & \cdots \\ 0 & 0 & \cdots & \tilde{a}_{mm} & \cdots & 0 \end{bmatrix} \quad /A.3.9b/$$

where $\tilde{a}_{ii}$ divides $\tilde{a}_{kk}$ for $k>i$, i.e. $\tilde{a}_{ii} \mid \tilde{a}_{kk}$.

Let $p_i(z_1)/q_i(z_1)/$ be the least common denominator of all entries of ith row of $\bar{U}(z_1,z_2)$ /column of $\bar{V}(z_1,z_2)$/. Then premultipying /A.3.9/ by $\mathrm{diag}\left[p_1(z_1), p_2(z_1), \ldots, p_m(z_1)\right]$ and postmultiplying by $\mathrm{diag}\left[q_1(z_1), q_2(z_1), \ldots, q_n(z_1)\right]$ we obtain the 2-D Smith form of $A(z_1,z_2)$:

if $m \geqslant n$

$$U(z_1,z_2) A(z_1,z_2) V(z_1,z_2) = \begin{bmatrix} a_{11} & 0 & \cdots & 0 \\ 0 & a_{22} & \cdots & 0 \\ \cdots & \cdots & \cdots & \cdots \\ 0 & 0 & \cdots & a_{nn} \\ 0 & 0 & \cdots & 0 \\ \cdots & \cdots & \cdots & \cdots \\ 0 & 0 & \cdots & 0 \end{bmatrix} \quad /A.3.10a/$$

or if $n \geqslant m$

$$U(z_1,z_2) A(z_1,z_2) V(z_1,z_2) = \begin{bmatrix} a_{11} & 0 & \cdots & 0 & \cdots & 0 \\ 0 & a_{22} & \cdots & 0 & \cdots & 0 \\ \cdots & \cdots & \cdots & \cdots & \cdots & \cdots \\ 0 & 0 & \cdots & a_{mm} & \cdots & 0 \end{bmatrix} \quad /A.3.10b/$$

where

$$U(z_1,z_2) = \mathrm{diag}\left[p_1(z_1), p_2(z_1), \ldots, p_m(z_1)\right]\bar{U}(z_1,z_2) \in F^{m\times m}[z_1,z_2]$$

$$V(z_1,z_2) = \bar{V}(z_1,z_2)\mathrm{diag}\left[q_1(z_1), q_2(z_1), \ldots, q_n(z_1)\right] \in F^{n\times n}[z_1,z_2]$$

$$a_{ij} = p_i(z_1)\, \tilde{a}_{ij} q_j(z_1)$$

are polynomials in $z_1$ and $z_2$ and $a_{ii} \mid a_{kk}$ for $k>i$. Symmetrically, we can find the Smith form of $A(z_1,z_2)$ w.r.t $F[z_2][z_1]$.

## Example A.3.3.

Find the Smith form w.r.t $F[z_1][z_2]$ of the matrix /A.3.7/.
Using the result obtained in Example A.3.2 we carry out the reduction

$$\begin{bmatrix} z_1 & z_2 \\ 0 & z_1 - \frac{z_2^2}{z_1} \\ \hline 0 & 0 \\ 1 & 0 \\ 0 & 1 \end{bmatrix} \rightarrow \begin{bmatrix} z_1 & 0 \\ 0 & z_1 - \frac{z_2^2}{z_1} \\ \hline 0 & 0 \\ 1 & -\frac{z_2}{z_1} \\ 0 & 1 \end{bmatrix}$$

Hence

$$\begin{bmatrix} 0 & 0 & 1 \\ 0 & 1 & -\frac{z_2}{z_1} \\ 1 & -z_2 & 0 \end{bmatrix} \begin{bmatrix} z_1^2 & z_1 z_2 \\ z_2 & z_1 \\ z_1 & z_2 \end{bmatrix} \begin{bmatrix} 1 & -\frac{z_2}{z_1} \\ 0 & 1 \end{bmatrix} = \begin{bmatrix} z_1 & 0 \\ 0 & z_1 - \frac{z_2^2}{z_1} \\ 0 & 0 \end{bmatrix} \quad /A.3.11/$$

and

$$\bar{U}/z_1,z_2/ = \begin{bmatrix} 0 & 0 & 1 \\ 0 & 1 & -\frac{z_2}{z_1} \\ 1 & -z_2 & 0 \end{bmatrix}, \quad \bar{V}/z_1,z_2/ = \begin{bmatrix} 1 & -\frac{z_2}{z_1} \\ 0 & 1 \end{bmatrix}$$

Premultiplying /A.3.11/ by

$$\text{diag}\left[p_1/z_1/, p_2/z_1/, p_3/z_1/\right] = \begin{bmatrix} 1 & 0 & 0 \\ 0 & z_1 & 0 \\ 0 & 0 & 1 \end{bmatrix}$$

and postmultiplying by

$$\text{diag}\left[q_1/z_1/, q_2/z_1/\right] = \begin{bmatrix} 1 & 0 \\ 0 & z_1 \end{bmatrix}$$

we obtain

$$\begin{bmatrix} 0 & 0 & 1 \\ 0 & z_1 & -z_2 \\ 1 & -z_2 & 0 \end{bmatrix} \begin{bmatrix} z_2^2 & z_1 z_2 \\ z_2 & z_1 \\ z_1 & z_2 \end{bmatrix} \begin{bmatrix} 1 & -z_2 \\ 0 & z_1 \end{bmatrix} = \begin{bmatrix} z_1 & 0 \\ 0 & z_1^3 - z_1 z_2^2 \\ 0 & 0 \end{bmatrix}$$

Matrices $A/z_1,z_2/ \in F^{m \times n}[z_1,z_2]$, $B/z_1,z_2/ \in F^{m \times n}$ are said to be equivalent over $F[z_1,z_2]$ if and only if there exist unimodular matrices $U/z_1,z_2/ \in F^{m \times m}[z_1,z_2]$, $V/z_1,z_2/ \in F^{n \times m}[z_1,z_2]$ such that

$$A/z_1,z_2/ = U/z_1,z_2/ \; B/z_1,z_2/ \; V/z_1 z_2/$$

where $\det U/z_1,z_2/$, $\det V/z_1,z_2/ \in F\setminus\{0\}$.
It was shown in [14] that if there exist equal Smith forms of $A/z_1,z_2/$ w.r.t $R[z_1][z_2]$ and w.r.t $R[z_2][z_1]$, then $A/z_1,z_2/$ is equivalent to its Smith form over $R[z_1,z_2]$.

## 4. Factorization of 2-D polynomial matrices

### 4.1. Primitive factorization

**Definition A.4.1.**

A matrix $A/z_1,z_2/ \in F^{m \times n}[z_1,z_2]/m \leq n/$ is said to be primitive in $F[z_1][z_2]$ if and only if $A/z_1^o,z_2/$ is of full rank for all fixed $z_1^o$.
By full rank, we mean that there is an $m \times m$ submatrix whose determinant is not zero element of $F/z_2/$.
For example the matrix

$$A/z_1,z_2/ = \begin{bmatrix} 1 & z_2^2 & -z_2 \\ 0 & 1 & z_1^2 \end{bmatrix}$$

is primitive in $F[z_1][z_2]$ /and also in $F[z_2][z_1]$/, since it is of full rank for all $z_1$ /and also for all $z_2$/.
Note that in the scalar case the polynomial

$$a/z_1,z_2/ = \sum_{i=0}^{n} a_i/z_1/z_2^i$$

is primitive if and only if $a_i/z_1/$ for $i = 0,1,\ldots,n$ have no non-trivial common factor.

## Lemma A.4.1. [15]

Let $A = A/z_1, z_2/ \in F^{m \times n}[z_1, z_2]$ be a given full rank matrix with

$$A = BC = DE \qquad /A.4.1/$$

for some $B \in F^{m \times m}[z_1, z_2]$, $C \in F^{m \times n}[z_1, z_2]$, $D \in F^{m \times p}[z_1, z_2]$,

$E \in F^{p \times n}[z_1, z_2]$, where $p \leq n$.

If E is primitive and $\det B \in F/z_1/$, then $H = B^{-1}D$ is a polynomial matrix and

$$C = HE \qquad /A.4.2/$$

## Proof

The lemma will be proved by contradiction. Suppose that H is a rational /nonpolynomial/ matrix, then there is at least one row, say the first one, which is nonpolynomial. Since F has been assumed algebraically closed, this means there exists some $z_1^o$ and $g/z_1/$ such that

$$H_1 = \left[ \frac{h_{11}/z_1,z_2/}{g/z_1//z_1-z_1^o/}, \frac{h_{12}/z_1,z_2/}{g/z_1//z_1-z_1^o/}, \ldots, \frac{h_{1p}/z_1,z_2/}{g/z_1//z_1-z_1^o/} \right]$$

where $h_{1i}/z_1,z_2/$ are not all identically zero.

Note that there exists an unimodular matrix $V \in F^{p \times p}[z_2]$ such that

$$H_1 V = \left[ \frac{\bar{h}_{11}/z_1,z_2/}{g/z_1//z_1-z_1^o/}, \frac{\bar{h}_{12}/z_1,z_2/}{g/z_1//z_1-z_1^o/}, \ldots, \frac{\bar{h}_{1p}/z_1,z_2/}{g/z_1//z_1-z_1^o/} \right]$$

with

$$\bar{h}_{11}/z_1^o,z_2/ \neq 0$$

and $\qquad /A.4.3/$

$$\bar{h}_{1i}/z_1^o,z_2/ = 0 \quad \text{for } i = 2,3,\ldots,p.$$

Now we have from /A.4.2/ that

$$C = HVV^{-1}E = \bar{H}\bar{E} \qquad /A.4.4/$$

where $\bar{H} = HV$ and $\bar{E} = V^{-1}E$.

Since $V^{-1}$ is unimodular, $V^{-1}E$ is also primitive.

Let $C_1/z_1,z_2/$ be the first row of $C$ and $\bar{E}_i/z_1,z_2/$ be the ith row of $\bar{E}$. Then /A.4.4/ yields

$$C_1/z_1,z_2/ = \sum_{i=1}^{p} \frac{\bar{h}_{1i}}{g/z_1//z_1-z_1^o/} \bar{E}_i/z_1,z_2/$$

or

$$/z_1-z_1^o/g/z_1/\, C/z_1,z_2/ = \sum_{i=1}^{p} \bar{h}_{1i}\, \bar{E}_i/z_1,z_2/$$

Taking into account /A.4.3/ and substituting $z_1 = z_1^o$ we obtain

$$\bar{h}_{11}/z_1^o,z_2/\, \bar{E}_1/z_1^o,z_2/ = 0$$

and $\bar{E}_1/z_1^o,z_2/ = 0$, which implies that $\bar{E}_1/z_1,z_2/$ contains the factor $/z_1-z_1^o/$. This contradicts the assumption that $\bar{E}/z_1,z_2/$ and therefore $E/z_1,z_2/$ is primitive. Thus we conclude that $H = B^{-1}D$ is a polynomial matrix. □

<u>Theorem A.4.1.</u> [15]

Let $A/z_1,z_2/ \in F^{m \times n}[z_1,z_2]/m \leq n/$ be a given full rank matrix, then there exist a unique $\bar{A}/z_2,z_2/$ /modulo a right unimodular matrix/ and a unique $A^*/z_1,z_2/$ /modulo a left unimodular matrix/ such that

$$A/z_1,z_2/ = \bar{A}/z_1,z_2/\, A^*/z_1,z_2/ \qquad \text{/A.4.5/}$$

where $\det \bar{A}/z_1,z_2/ = \bar{a}/z_2/ \in F[z_2]$ and $A^*/z_1,z_2/$ is primitive in $F[z_2][z_1]$.

<u>Proof</u>

We shall prove the theorem by the following constructive algorithm.

**Step 1** Find all the roots $z_{2i}$, which make $A/z_1,z_2/$ not full rank. This can be done by taking a nonsingular minor and calculating its determinant, say $d/z_1,z_2/$. Then we can factor $d/z_1,z_2/ = \bar{d}/z_1/\ d^*/z_1,z_2/$ and obtain our candidate roots from the set of roots of $\bar{d}/z_1/ = 0$. By checking each candidate root we can pick out those making $A/z_1,z_2/$ not full rank and obtain the set of $z_{2i}$'s.

**Step 2** Since $A/z_1,z_{2i}/$ has not full rank, the Hermite form of $A/z_1,z_{2i}/$ has its last row identically equal to zero. Thus we can find a unimodular matrix $V_1/z_1/$ such that

$$V_1/z_1/A/z_1,z_{2i}/ = \begin{bmatrix} a_{11} & \cdots & a_{1n} \\ \cdot & \cdot\cdot\cdot\cdot\cdot & \cdot \\ a_{m-1,1} & \cdots & a_{m-1,n} \\ 0 & \cdots & 0 \end{bmatrix}$$

or

$$V_1/z_1/A/z_1,z_2/ = \begin{bmatrix} 1 & 0 & \cdots & 0 & 0 \\ 0 & 1 & \cdots & 0 & 0 \\ \cdot & \cdot & \cdot\cdot\cdot\cdot\cdot & \cdot & \cdot \\ 0 & 0. & \cdots & 1 & 0 \\ 0 & 0 & \cdots & 0 & /z_2-z_{21}/ \end{bmatrix} \hat{A}/z_1,z_2/$$

for some $\hat{A}/z_1,z_2/$.

Hence

$$A/z_1,z_2/ = V_1^{-1}/z_1/ \begin{bmatrix} 1 & 0 & \cdots & 0 & 0 \\ 0 & 1 & \cdots & 0 & 0 \\ 0 & 0 & \cdots & 1 & 0 \\ 0 & 0 & \cdots & 0 & /z_2-z_{21}/ \end{bmatrix} \hat{A}/z_1,z_2/ =$$

$$= \bar{A}_1/z_1,z_2/\ \hat{A}/z_1,z_2/$$

where

$$\bar{A}_1/z_1,z_2/ = V_1^{-1}/z_1/ \begin{bmatrix} 1 & 0 & \cdots & 0 & 0 \\ 0 & 1 & \cdots & 0 & 0 \\ 0 & 0 & \cdots & 1 & 0 \\ 0 & 0 & \cdots & 0 & /z_2-z_{21}/ \end{bmatrix}$$

Continuing the same procedure on $\hat{A}/z_1,z_2/$ we shall obtain a final result of the form

$$A/z_1,z_2/ = \bar{A}_1/z_1,z_2/\bar{A}_2/z_1,z_2/\ldots\bar{A}_k/z_1,z_2/A^*/z_1,z_2/ =$$

$$= \bar{A}/z_1,z_2/\, A^*/z_1,z_2/$$

where

$$\bar{A}/z_1,z_2/ = \prod_{i=1}^{k} \bar{A}_i/z_1,z_2/$$

with

$$\det \bar{A}/z_1,z_2/ = \prod_{i=1}^{k} \det \bar{A}_i/z_1,z_2/ = \bar{a}/z_2/$$

and $A^*/z_1,z_2/$ is primitive /otherwise the procedure could be still continued/.

To prove uniqueness let us assume

$$A/z_1,z_2/ = \bar{A}/z_1,z_2/\, A^*/z_1,z_2/ = \bar{A}'/z_1,z_2/\, A_1^*/z_1,z_2/$$

By Lemma A.4.1, taking $p = m \leq n$, we have $H = \bar{A}^{-1}/z_1,z_2/\, \bar{A}'/z_1,z_2/$ is a polynomial matrix. Note that $H^{-1} = \bar{A}'^{-1}/z_1,z_2/\bar{A}/z_1,z_2/$ is also a polynomial matrix. Thus $\bar{A}/z_1,z_2/$ and $\bar{A}'/z_1,z_2/$ are related by a unimodular matrix. □

Example A.4.1.

Find a primitive factorization /A.4.5/ of the matrix

$$A/z_1,z_2/ = \begin{bmatrix} -z_1+z_2 & z_1+z_2 \\ 2z_1+z_2 & z_2 \end{bmatrix}$$

Since $z_1 = 0$ is a root, the corresponding unimodular matrix is

$$V = \begin{bmatrix} 1 & 0 \\ 1 & -1 \end{bmatrix}$$

and

$$V\, A/z_1,z_2/ = \begin{bmatrix} -z_1+z_2 , & z_1+z_2 \\ -3z_1 , & z_1 \end{bmatrix} = \begin{bmatrix} 1 & 0 \\ 0 & z_1 \end{bmatrix} \begin{bmatrix} -z_1+z_2, & z_1+z_2 \\ -3 , & 1 \end{bmatrix}$$

Hence we have the desired factorization

$$A/z_1,z_2/ = V^{-1} \begin{bmatrix} 1 & 0 \\ 0 & z_1 \end{bmatrix} \begin{bmatrix} -z_1+z_2, & z_1+z_2 \\ -3 , & 1 \end{bmatrix} = \begin{bmatrix} 1 & 0 \\ 1 & -z_1 \end{bmatrix} \begin{bmatrix} -z_1+z_2, & z_1+z_2 \\ -3 , & 1 \end{bmatrix}$$

Symmetrically, we can find a unique $\hat{A}/z_1,z_2/$ /modulo a left unimodular matrix/ and a unique $\hat{A}^*/z_1,z_2/$ /modulo a right unimodular matrix/ such that

$$A/z_1,z_2/ = \hat{A}^*/z_1,z_2/\, \hat{A}/z_1,z_2/$$

where $\det \hat{A}/z_1,z_2/ = \hat{a}/z_2/ \in F[z_2]$ and $\hat{A}^*/z_1,z_2/$ is primitive in $F[z_1][z_2]$.

## 4.2. General factorization

Consider an arbitrary square matrix $A/z_1,z_2/ \in F^{m \times m}[z_1,z_2]$.

**Lemma A.4.2.** [15]

If $\det A/z_1,z_2/ = \prod_{i=1}^{k} \bar{a}_i/z_2/ a^*/z_1,z_2/$ where $a^*/z_1,z_2/$ is a primitive polynomial w.r.t $F[z_2][z_1]$, then $A/z_1,z_2/$ can be factored as follows

$$A/z_1,z_2/ = \prod_{i=1}^{k} \bar{A}_i/z_1,z_2/\, A^*/z_1,z_2/ \qquad /A.4.6/$$

with

$$\det \bar{A}_i/z_1,z_2/ = \bar{a}_i/z_2/$$

and

$$\det A^*/z_1,z_2/ = a^*/z_1,z_2/$$

**Proof** Following the primitive factorization algorithm w.r.t $F[z_2][z_1]$ we extract the roots of $\bar{a}_1/z_1,z_2/$ first, then the roots of $\bar{a}_2/z_1,z_2/$ and so on. The corresponding matrices are $\bar{A}_1/z_1,z_2/, \ldots, \bar{A}_k/z_1,z_2/$ and $A^*/z_1,z_2/$ is the residual matrix. □

**Theorem A.4.2.** [15]

If
$$\det A/z_1,z_2/ = \prod_{i=1}^{k} a_i/z_1,z_2/$$

then $A/z_1,z_2/$ can be factored such that

$$A/z_1,z_2/ = \prod_{i=1}^{k} A_i/z_1,z_2/ \qquad /A.4.7/$$

where
$$\det A_i/z_1,z_2/ = a_i/z_1,z_2/ \quad /i=1,2,\ldots,k/$$

and $a_i/z_1,z_2/$ are arbitrary polynomials.

**Proof** We assume first that $A/z_1,z_2/$ is a primitive polynomial matrix w.r.t $F[z_2][z_1]$ so that $\det A/z_1,z_2/ = a/z_1,z_2/$ is primitive and all the $a_i/z_1,z_2/$'s are primitive.

The Smith form of $A = A/z_1,z_2/$ is $S = UAV$, where $S$ is diagonal and $\det U = u/z_2/$, $\det V = v/z_2/$. Let $S = \bar{S}S^*$ be the primitive factorization of $S$. Since $S$ is primitive and diagonal $\det S = \det A/z_1,z_2/ = \prod_{i=1}^{k} a_i/z_1,z_2/$ and

$$S^* = \prod_{i=1}^{k} S_i^*$$

with
$$\det S_i^* = a_i/z_1,z_2/$$

Thus
$$S = UAV = \bar{S} S^* = \bar{S} \prod_{i=1}^{k} S_i^*$$

and by Lemma A.4.1

$$AV = \Lambda \prod_{i=1}^{k} S_i^*$$

where $\Lambda = U^{-1}S$ is a polynomial matrix.

Now, if we set $\Lambda_o = \Lambda$, and if $\Lambda_{i-1}$ is unimodular in $z_1$, then

$$M_i = \Lambda_{i-1} S_i^* = A_i \Lambda_i \quad \text{for } i = 1,\ldots,k \qquad /A.4.8/$$

are, respectively, the left and right primitive factorization of $M_i$, so that

$$\det A_i = \det S_i^* = a_i/z_1,z_2/$$

and $\Lambda_i$ is unimodular in $z_1$. Hence

$$AV = (\prod_{i=1}^{k} A_i) \Lambda_k$$

and by Lemma A.4.1

$$A = (\prod_{i=1}^{k} A_i) W$$

where $W = {}_k V^{-1}$ is a unimodular matrix in $z_1$ and $z_2$.

If A is not primitive w.r.t $F[z_2][z_1]$, let

$$a_i/z_1,z_2/ = \bar{a}_i/z_1,z_2/ \, a_i^*/z_1,z_2/ \qquad /i = 1,2,\ldots,k/$$

where $a_i^*/z_1,z_2/$ is primitive.

Then

$$a/z_1,z_2/ = \bar{a}/z_1,z_2/ \, a^*/z_1,z_2/$$

where

$$\bar{a}/z_1,z_2/ = \prod_{i=1}^{k} \bar{a}_i/z_1,z_2/$$

and

$$a^*/z_1,z_2/ = \prod_{i=1}^{k} a_i^*/z_1,z_2/$$

Using Lemma A.4.2 we can factor $A/z_1,z_2/$ as follows

$$A/z_1,z_2/ = \prod_{i=1}^{k} \bar{A}_i/z_1,z_2/ \; A^*/z_1,z_2/ \qquad /A.4.9/$$

where $A/z_1,z_2/$ is primitive and

$$\det \bar{A}_i/z_1,z_2/ = \bar{a}_i/z_1,z_2/ \qquad /i=1,2,\ldots,k/$$

Now, using the first part of the proof $A^*/z_1,z_2/$ can be factored as

$$A^*/z_1,z_2/ = \prod_{i=1}^{k} A_i^*/z_1,z_2/ \qquad /A.4.10/$$

where

$$\det A_i^*/z_1,z_2/ = a_i^*/z_1,z_2/$$

Substitution of /A.4.10/ into /A.4.9/ yields

$$A/z_1,z_2/ = \left( \prod_{i=1}^{k} \bar{A}_i/z_1,z_2/ \right) \left( \prod_{i=1}^{k} A_i^*/z_1,z_2/ \right)$$

By interchanging left and right primitive factorization like in /A.4.8/ we can interleave the $\bar{A}_i/z_1,z_2/$ and $A_i^*/z_1,z_2/$ so that /A.4.7/ holds. □

This factorization is in general not unique.

### Example A.4.2.

Find two factorizations of the matrix

$$A/z_1,z_2/ = \begin{bmatrix} 3z_1 z_2 & 0 \\ z_1 z_2^2 & 2z_1^2 z_2 \end{bmatrix}$$

It is easy to check that

$$A/z_1,z_2/ = A_1/z_1,z_2/ A_2/z_1,z_2/ = \bar{A}_1/z_1,z_2/ \bar{A}_2/z_1,z_2/$$

and $\det A/z_1,z_2/ = \det A_1/z_1,z_2/ \det A_2/z_1,z_2/ =$

$$= \det \bar{A}_1/z_1,z_2/ \det \bar{A}_2/z_1,z_2/$$

where

$$A_1/z_1,z_2/ = \begin{bmatrix} 1 & z_1 \\ -z_2 & z_1z_2 \end{bmatrix}, \quad A_2/z_1,z_2/ = \begin{bmatrix} z_1z_2 & -z^2 \\ 2z_2 & z_1 \end{bmatrix}$$

$$\bar{A}_1/z_1,z_2/ = \begin{bmatrix} 1 & 0 \\ \frac{1}{3}/z_2-z_1/ & z_1^2 z_2 \end{bmatrix}, \quad \bar{A}_2/z_1,z_2/ = \begin{bmatrix} 3z_1z_2 & 0 \\ 1 & 2 \end{bmatrix}$$

It was shown in [18] that for $n \geqslant 3$ it is not always possible for a given $A = A/z_1,\ldots,z_n/$ with det $A = d/z_1,\ldots,z_n/ = d_1/z_1,\ldots,z_n/ \cdot d_2/z_1,\ldots,z_n/$ to find two n-D polynomial matrices $A_1$ and $A_2$ such that det $A_i = d_i/z_1,\ldots,z_n/$, $i=1,2$ and $A = A_1 A_2$.

## 5. Coprimeness of 2-D polynomials and polynomial matrices

### 5.1. Definitions

**Definition A.5.1.**

A 2-D polynomial matrix $C = C/z_1,z_2/$ /$B = B/z_1,z_2//$ is called a right /left/ divisor /RD /LD// of $A = A/z_1,z_2/$ if there exists a matrix B /C/ such that

$$A = BC$$

and A is called a left /right/ multiple of C /B/.

**Definition A.5.2.**

A 2-D polynomial matrix $R = R/z_1,z_2/$ /$L = L/z_1,z_2//$ is called a common right /left/ divisor /CRD /CLD// of $A = A/z_1,z_2/$ and $B = B/z_1,z_2/$ if there exist two 2-D polynomial matrices $A_1 = A_1/z_1,z_2/$, $B_1 = B_1/z_1,z_2/$ /$A_2 = A_2/z_1,z_2/$, $B_2 = B_2/z_1,z_2//$ such that $A = A_1 R$ and $B = B_1 R$ /$A = LA_2$, and $B = LB_2/$.

**Definition A.5.3.**

A 2-D polynomial matrix $D = D/z_1,z_2/$ /$E = E/z_1,z_2//$ is called common right /left/ multiple /CRM /CLM// of $A = A/z_1,z_2/$ and $B = B/z_1,z_2/$

if there exist two 2-D polynomial matrices $C_1 = C_1/z_1,z_2/$, $C_2 = C_2/z_1,z_2/$ $/F_1 = F_1/z_1,z_2/$, $F_2 = F_2/z_1,z_2//$ such that $D = AC_1$ and $D = BC_2$ $/E = F_1A$ and $E = F_2B/$.

#### Definition A.5.4.

A 2-D polynomial matrix $P = P/z_1,z_2/$ $/Q = Q/z_1,z_2//$ is called geatest common right /left/ divisor /GCRD /GCLD// of $A = A/z_1,z_2/$ and $B = B/z_1,z_2/$ if:

/i/ P /Q/ is a CRD /CLD/ of A and B,

/ii/ P /Q/ is left /right/ multiple of every CRD /CLD/ of A and B.

#### Definition A.5.5.

A 2-D polynomial matrix $G = G/z_1,z_2/$ $/H = H/z_1,z_2//$ is called a least common right /left/ multiple /LCRM /LCLM// of $A = A/z_1,z_2/$ and $B = B/z_1,z_2/$ if

/i/ G /H/ is CRM /CLM/ of A and B,

/ii/ G /H/ is left /right/ divisor of every CRM /CLM/ of A and B.

#### Definition A.5.6.

2-D polynomial matrices $A = A/z_1,z_2/$, $B = B/z_1,z_2/$ are called factor right /left/ coprime /FRC /FLC// w.r.t $F[z_1,z_2]$ if their GCRD /GCLD/ is a unimodular matrix $U = U/z_1,z_2//\det U \in R \setminus \{0\}/$.

#### Definition A.5.7.

Polynomial matrices $A \in F^{p \times m}[z_1,z_2]$, $B \in F^{q \times m}[z_1,z_2]$ $/p+q \geqslant m \geqslant 1/$ are called zero right coprime /ZRC/ if there exists no a pair $/z_1,z_2/$ which is a zero of all mxm minors of the matrix $\begin{bmatrix} A \\ B \end{bmatrix}$.

Polynomial matrices $A \in F^{m \times p}[z_1,z_2]$, $B \in F^{m \times q}[z_1,z_2]$ are called zero left coprime /ZLC/ if the transposed matrices $A^T$, $B^T$ and ZRC.

## Definition A.5.8.

Polynomial matrices $A \in F^{p \times m}[z_1,z_2]$, $B \in F^{q \times m}[z_1,z_2]$ /$p+q \geq m \geq 1$/ are called minor right coprime /MRC/ if all the m×m minors of the matrix $\begin{bmatrix} A \\ B \end{bmatrix}$ are coprime /relatively prime/, i.e. their greatest common polynomial divisor is a nonzero constant. It can be shown [18] that for 1-D case the definitions A.5.6, A.5.7 and A.5.8 are equivalent and for 2-D case the definitions A.5.6 and A.5.8 are equivalent.

Note that the polynomials $A/z_1,z_2/ = z_1$ and $B/z_1,z_2/ = z_2$ possess the common zero $z_1 = 0$, $z_2 = 0$ but they are coprime. Thus, for 2-D case the definitions A.5.6 and A.5.7 are not equivalent. In general case for n-D systems the zero coprimeness implies the minor coprimeness and the minor coprimeness implies the factor coprimeness [18].

### 5.2. Coprimeness of 2-D polynomials

Consider a 2-D polynomial of the form

$$a/z_1,z_2/ = \sum_{i=0}^{n_2} a_i/z_1/z_2^i = \sum_{j=0}^{n_1} a_j/z_2/z_1^j \qquad /A.5.1/$$

The polynomial /A.5.1/ is called primitive if and only if $a_i/z_1/$, $i=1,2,\ldots,n_1$ and $a_j/z_2/$, $j=1,2,\ldots,n_2$ are coprime.

The polynomial /A.5.1/ can be written as

$$a/z_1,z_2/ = \alpha/z_1/ \beta/z_2/ \; \bar{a}/z_1,z_2/ \qquad /A.5.1a/$$

where $\alpha/z_1/$, $\beta/z_2/$ are the so-called contents of $a/z_1,z_2/$ and $\bar{a}/z_1,z_2/$ is a primitive polynomial.

Our problem can be stated as follows: Given two 2-D /primitive/ polynomials $\bar{a}/z_1,z_2/$, $\bar{b}/z_1,z_2/$, find necessary and sufficient conditions and tests for the /factor/ coprimeness.

Consider 2-D primitive polynomials

$$\bar{a}/z_1,z_2/ = a_{n_1}/z_2/ \; \hat{a}/z_1,z_2/ = a_{n_2}/z_1/ \; \tilde{a}/z_1,z_2/$$
$$\bar{b}/z_1,z_2/ = b_{m_1}/z_2/ \; \hat{b}/z_1,z_2/ = b_{m_2}/z_1/ \; \tilde{b}/z_1,z_2/$$
/A.5.2/

where

$\hat{a}/z_1,z_2/$, $\hat{b}/z_1,z_2//\tilde{a}/z_1,z_2/$, $\tilde{b}/z_1,z_2//$ are monic polynomials in $z_1$ /$z_2$/ with rational coefficients in $z_2$ /$z_1$/.

$$\hat{a}/z_1,z_2/ = z_1^{n_1} + \hat{a}_{n_1-1} z_1^{n_1-1} + \ldots + \hat{a}_1 z_1 + \hat{a}_0 \quad /\hat{a}_i = \hat{a}_i/z_2/, i=0,1,\ldots,n_1-1/$$
$$\tilde{a}/z_1,z_2/ = z_2^{n_2} + \tilde{a}_{n_2-1} z_2^{n_2-1} + \ldots + \tilde{a}_1 z_2 + \tilde{a}_0 \quad /\tilde{a}_i = \tilde{a}_i/z_1/, i=0,1,\ldots,n_2-1/$$
/A.5.3/
$$\hat{b}/z_1,z_2/ = z_1^{m_1} + \hat{b}_{m_1-1} z_1^{m_1-1} + \ldots + \hat{b}_1 z_1 + \hat{b}_0 \quad /\hat{b}_i = \hat{b}_i/z_2/, i=0,1,\ldots,m_1-1/$$
$$\tilde{b}/z_1,z_2/ = z_2^{m_2} + \tilde{b}_{m_2-1} z_2^{m_2-1} + \ldots + \tilde{b}_1 z_2 + \tilde{b}_0 \quad /\tilde{b}_i = \tilde{b}_i/z_1/, i=0,1,\ldots,m_2-1/$$

Let associate with /A.5.3/ the companion matrices

$$\hat{A} = \begin{bmatrix} 0 & 1 & 0 & \cdots & 0 \\ 0 & 0 & 1 & \cdots & 0 \\ \vdots & & & & \vdots \\ 0 & 0 & 0 & \cdots & 1 \\ -\hat{a}_0 & -\hat{a}_1 & \cdots & & -\hat{a}_{n_1-1} \end{bmatrix} \quad \tilde{A} = \begin{bmatrix} 0 & 1 & 0 & \cdots & 0 \\ 0 & 0 & 1 & \cdots & 0 \\ \vdots & & & & \vdots \\ 0 & 0 & 0 & \cdots & 1 \\ -\tilde{a}_0 & -\tilde{a}_1 & \cdots & & -\tilde{a}_{n_2-1} \end{bmatrix}$$

/A.5.4/

and

$$\hat{B} = \begin{bmatrix} 0 & 1 & 0 & \cdots & 0 \\ 0 & 0 & 1 & \cdots & 0 \\ \vdots & & & & \vdots \\ 0 & 0 & 0 & \cdots & 1 \\ -\hat{b}_0 & -\hat{b}_1 & \cdots & & -\hat{b}_{m_1-1} \end{bmatrix} \quad \tilde{B} = \begin{bmatrix} 0 & 1 & 0 & \cdots & 0 \\ 0 & 0 & 1 & \cdots & 0 \\ \vdots & & & & \vdots \\ 0 & 0 & 0 & \cdots & 1 \\ -\tilde{b}_0 & -\tilde{b}_1 & \cdots & & -\tilde{b}_{m_2-1} \end{bmatrix}$$

<u>Remark</u>

Instaed of /A.5.4/ we can also assume the matrices $\hat{A}$, $\tilde{A}$, $\hat{B}$ and $\tilde{B}$ the following forms

$$\begin{bmatrix} 0 & 0 & \cdots & 0 & -a_0 \\ 1 & 0 & \cdots & 0 & -a_1 \\ 0 & 1 & \cdots & 0 & -a_2 \\ \cdot & \cdot & & \cdot & \cdot \\ 0 & 0 & \cdots & 1 & -a_{n-1} \end{bmatrix}, \begin{bmatrix} -a_{n-1} & -a_{n-2} & \cdots & -a_1 & -a_0 \\ 1 & 0 & \cdots & 0 & 0 \\ 0 & 1 & \cdots & 0 & 0 \\ \cdot & \cdot & & \cdot & \cdot \\ 0 & 0 & \cdots & 1 & 0 \end{bmatrix}$$

and

$$\begin{bmatrix} -a_{n-1} & 1 & 0 & \cdots & 0 \\ -a_{n-2} & 0 & 1 & \cdots & 0 \\ \vdots & \vdots & \vdots & & \vdots \\ -a_1 & 0 & 0 & \cdots & 1 \\ -a_0 & 0 & 0 & \cdots & 0 \end{bmatrix}$$

It is easy to check that

$$\det\left[I_{n_1}z_1 - \hat{A}\right] = \hat{a}/z_1, z_2/ \,, \quad \det\left[I_{n_2}z_2 - \tilde{A}\right] = \tilde{a}/z_1, z_2/ \qquad /A.5.5/$$

and                                                                                                     /A.5.5/

$$\det\left[I_{m_1}z_1 - \hat{B}\right] = \hat{b}/z_1, z_2/ \,, \quad \det\left[I_{m_2}z_2 - \tilde{B}\right] = \tilde{b}/z_1, z_2/$$

### Theorem A.5.1

2-D primitive polynomials /A.5.2/ are /factor/ coprime if and only if

$$\det\left[\hat{B}^{n_1} + \hat{a}_{n_1-1}\hat{B}^{n_1-1} + \cdots + \hat{a}_1\hat{B} + \hat{a}_0 I_{m_1}\right] \neq 0$$

and                                                                                                     /A.5.6a/

$$\det\left[\tilde{B}^{n_2} + \tilde{a}_{n_2-1}\tilde{B}^{n_2-1} + \cdots + \tilde{a}_1\tilde{B} + \tilde{a}_0 I_{m_2}\right] \neq 0$$

or

$$\det\left[\hat{A}^{m_1} + \hat{b}_{m_1-1}\hat{A}^{m_1-1} + \cdots + \hat{b}_1\hat{A} + \hat{b}_0 I_{n_1}\right] \neq 0$$

and                                                                                                     /A.5.6b/

$$\det\left[\tilde{A}^{m_2} + \tilde{b}_{m_2-1}\tilde{A}^{m_2-1} + \cdots + \tilde{b}_1\tilde{A} + \tilde{b}_0 I_{n_2}\right] \neq 0$$

### Proof

Let $\lambda_1, \lambda_2, \ldots, \lambda_{m_1} /\lambda_i = \lambda_i/z_2//$ be eigenvalues of $\hat{B}$ and

$$a/\lambda/ = \lambda^{n_1} + \hat{a}_{n_1-1}\lambda^{n_1-1} + \cdots + \hat{a}_1\lambda + \hat{a}_0 \qquad /A.5.7/$$

It is well known that $a/\lambda_i/$, $i = 1, \ldots, m_1$, are the eigenvalues of $a/B/$.
From the relation

$$\det a/\hat{B}/ = \prod_{i=1}^{n_1} a/\lambda_i/$$

it follows that $\det a/\hat{B}/ = 0$ if and only if at least one of $\lambda_1, \lambda_2, \ldots \lambda_{m_1}$ is a root of /A.5.7/. Thus, /A.5.2/ are /factor/ coprime if and only if /A.5.6a/ or /A.5.6b/ is satisfied. □

The theorem A.5.1 can be extended for n-D polynomials [2].

Example A.5.1

Test the /factor/ coprimeness of the polynomials

$$\bar{a}/z_1,z_2/ = z_1^2 z_2 + z_1 z_2^2 + z_1^2 + z_2^2 + 2z_1 z_2 + z_1 + z_2$$
$$\bar{b}/z_1,z_2/ = z_1^3 + z_1^2 z_2 + z_1 + z_2 \qquad /A.5.8/$$

The polynomials /A.5.8/ can be written as

$$\bar{a}/z_1,z_2/ = /z_2+1/[z_1^2+z_2/1+z_2/+z_2] = /z_1+1/[z_2^2+z_2/1+z_1/+z_1]$$
$$\bar{b}/z_1,z_2/ = /z_1^2+1//z_2+z_1/$$

Hence

$$\hat{a}/z_1,z_2/ = z_1^2 + z_1/1+z_2/ + z_2$$
$$\hat{b}/z_1,z_2/ = z_1^3 + z_1^2 z_2 + z_1 + z_2$$
$$\tilde{a}/z_1,z_2/ = z_2^2 + z_2/1+z_1/ + z_1$$
$$\tilde{b}/z_1,z_2/ = z_2+z_1$$

and

$$\hat{A} = \begin{bmatrix} 0 & 1 \\ -z_2, & -/1+z_2/ \end{bmatrix}, \hat{B} = \begin{bmatrix} 0 & 1 & 0 \\ 0 & 0 & 1 \\ -z_2 & -1 & -z_2 \end{bmatrix}, \tilde{A} = \begin{bmatrix} 0 & 1 \\ -z_1, & -/1+z_1/ \end{bmatrix}, \tilde{B} = [-z_1]$$

It is easy to check that

$$\det\left[\hat{B}^2 + /1+z_2/\hat{B} + z_2 I_3\right] = \begin{vmatrix} z_2, & 1+z_2, & 1 \\ -z_2, & -1+z_2, & 1 \\ -z_2, & -1-z_2, & 1 \end{vmatrix} = 0$$

and

$$\det\left[\tilde{B}^2 + /1+z_1/\tilde{B} + z_1 I_1\right] = z_1^2 - z_1/1+z_1/ + z_1 = 0$$

Therefore, the polynomials /A.5.8/ are not factor coprime.

Lemma A.5.1

If

$$A = \begin{bmatrix} 0 & 1 & 0 & \cdots & 0 \\ 0 & 0 & 1 & \cdots & 0 \\ \cdot & \cdot & \cdot & \cdots & \cdot \\ 0 & 0 & 0 & \cdots & 1 \\ -a_0 & -a_1 & -a_2 & \cdots & -a_{n-1} \end{bmatrix}, \quad B = \begin{bmatrix} b_0 & b_1 & \cdots & b_{n-1} \end{bmatrix} \qquad /A.5.9/$$

then

$$b_{n-1}A^{n-1} + \ldots + b_1 A + b_0 I_n = \begin{bmatrix} B \\ BA \\ \cdot \\ \cdot \\ BA^{n-1} \end{bmatrix} \qquad /A.5.10/$$

Proof

Let $H_i$ be ith row of the matrix $H = b_{n-1}A^{n-1}+\ldots+b_1 A+b_0 I_n$.
It is easy to check that $H_1 = B$. Note that $e_i = e_{i-1} A$ for $i = 1,2,\ldots,n$, where $e_i$ is the ith row of $I_n$.

Thus, we have

$$H_i = e_i H = e_{i-1} AH = e_{i-1} HA = H_{i-1} A \qquad /A.5.11/$$

/A.5.11/ follows from /A.5.11/ for $i = 2,3,\ldots,n$ and $H_1 = B$. □

Note that without loss of generality we can assume that $n_1 \geqslant m_2$ and $n_2 \geqslant m_2$ /otherwise we can change the role of $n_i$ and $m_i$, $i = 1,2$/.

From Cayley-Hamilton theorem and Lemma A.5.1 it follows that the conditions /A.5.6a/ are equivalent to the following

$$\det \begin{bmatrix} \hat{a} \\ \hat{a}\hat{B} \\ \cdot & \cdot & \cdot \\ \hat{a}\hat{B}^{m_1-1} \end{bmatrix} \neq 0 \qquad /A.5.12a/$$

and  /A.5.12a/

$$\det \begin{bmatrix} \tilde{a} \\ \tilde{a}\tilde{B} \\ \cdot & \cdot & \cdot \\ \tilde{a}\tilde{B}^{m_2-1} \end{bmatrix} \neq 0$$

where
$$\hat{a} = [\hat{a}_0, \hat{a}_1, \ldots, \hat{a}_{m_1-1}]$$

$$\tilde{a} = [\tilde{a}_0, \tilde{a}_1, \ldots, \tilde{a}_{m_2-1}]$$

Thus, we have established the following

### Theorem A.5.2

2-D primitive polynomials /A.5.2/ are /factor/ coprime if and only if the conditions /A.5.12a/ are satisied.

For example for /A.5.8/ we have $m_1 = 3$, $m_2 = 1$ and $n_1 = n_2 = 2$. Therefore, instead of the determinant

$$\det \left[ \hat{A}^3 + z_2 \hat{A}^2 + \hat{A} + z_2 I_2 \right]$$

it is sufficient to find

$$\det \begin{bmatrix} \tilde{b} \\ \tilde{b} \hat{A} \end{bmatrix} = \begin{bmatrix} z_2 & 1 \\ -z_2 & -1 \end{bmatrix}$$

Other different tests for checking the factor coprimeness of 2-D polynomials can be found in [15, 3, 4, 5].

### 5.3. Zero coprimeness of 2-D polynomial matrices

Consider two 2-D polynomial matrices $A = A/z_1, z_2/ \in F^{m \times p}[z_1, z_2]$ and $B = B/z_1, z_2/ \in F^{m \times q}[z_1, z_2]$ with $p+q \geq m \geq 1$. The matrices are zero left-coprime /ZLC/ if there exists no a pair $/z_1, z_2/$ which is a zero of all the m×m minors of the matrix

$$C = C/z_1, z_2/ = [A \mid B] \quad /A.5.13/$$

### Theorem A.5.3 [18]

A and B are ZLC if and only if there exist two polynomial matrices $X = X/z_1, z_2/ \in F^{p \times m}[z_1, z_2]$, $Y = Y/z_1, z_2 \in F^{q \times m}[z_1, z_2]$ such that

$$AX + BY = I_m \quad /A.5.14/$$

## Proof

To prove sufficiency we write /A.5.14/ in the form

$$\begin{bmatrix} A & \vdots & B \end{bmatrix} \begin{bmatrix} X \\ Y \end{bmatrix} = I_m \qquad /A.5.15/$$

From /A.5.15/ it follows that rank $\begin{bmatrix} A \vdots B \end{bmatrix} = m$ for all $/z_1,z_2/$. This implies that no $/z_1,z_2/$ is a common zero of all the mxm minors of /A.5.13/ and A and B are ZLC. To prove necessity let assume that A and B are ZLC. Let $\Delta_{i_1 i_2 \ldots i_m}/z_1,z_2/$ denote the mxm minor of /A.5.13/ formed with the given m rows and the m **distinct** columns numbered $i_1, i_2,\ldots,i_m$. From the definition of zero left coprimeness it follows that these $\dfrac{/p+q/!}{/p+q-m/!\,m!}$ polynomials are devoid of any common zeros. Thus, there exist polynomials $a_{i_1 i_2 \ldots i_m}/z_1,z_2/$ such that

$$\sum_{i_1,i_2,\ldots,i_m} a_{i_1 i_2 \ldots i_m}/z_1,z_2/\, \Delta_{i_1 i_2 \ldots i_m}/z_1,z_2/ = 1 \qquad /A.5.16/$$

Let K be any /p+q/xm real constant matrix whose mxm minors, say $M^K_{i_1,i_2 \ldots i_m}$ are all nonzero. Note that such matrix K always exists for $p+q \geq m$. Let

$$\Lambda = \text{diag}[\lambda_1, \lambda_2, \ldots, \lambda_{p+q}] \qquad /A.5.17/$$

where $\lambda_1, \lambda_2, \ldots, \lambda_{p+q}$ are independent variables
and

$$D = D/z_1,z_2,\lambda_1,\ldots,\lambda_{p+q}/ = C\Lambda K \qquad /A.5.18/$$

From the Cauchy-Binet theorem we have

$$\Delta = \Delta/z_1,z_2,\lambda_1,\ldots,\lambda_{p+q}/ =$$
$$= \sum_{i_1,i_2,\ldots,i_m} \lambda_{i_1} \lambda_{i_2} \ldots \lambda_{i_m}\, \Delta_{i_1 i_2 \ldots i_m}/z_1,z_2/ M^K_{i_1 i_2 \ldots i_m}$$

Thus, for every one of the $\dfrac{/p+q/!}{/p+q-m/!\,m!}$ m-tuples $/i_1,i_2,\ldots,i_m/$ we obtain

$$\Delta_{i_1 i_2 \ldots i_m} M^K_{i_1 i_2 \ldots i_m} = \frac{\partial^m /z_1, z_2, \lambda_1, \ldots, \lambda_{p+q}/}{\partial \lambda_{i_1} \partial \lambda_{i_2} \ldots \partial \lambda_{i_m}} \Bigg|_{\substack{\lambda_1=0 \\ \vdots \\ \lambda_{p+q}=0}} \qquad /A.5.19/$$

Let $D_a = D_a/z_1,z_2,\lambda_1,\ldots\lambda_{p+q}/$ be the mxm polynomial matrix adjoint to D. Taking into account that

$$\Delta I_m = D\, D_a = CAKD_a$$

from /A.5.19/ we obtain

$$\Delta_{i_1 i_2 \ldots i_m} I_m = C\, Z_{i_1 i_2 \ldots i_m} \qquad /A.5.20/$$

where

$$Z_{i_1 i_2 \ldots i_m} = Z_{i_1 i_2 \ldots i_m}/z_1,z_2/ = \frac{1}{M^K_{i_1 i_2 \ldots i_m}} \frac{\partial^m \Lambda K D_a}{\partial \lambda_{i_1} \partial \lambda_{i_2} \ldots \partial \lambda_{i_m}} \Bigg|_{\substack{\lambda_1=0 \\ \vdots \\ \lambda_{p+q}=0}}$$

$$/A.5.21/$$

By combining /A.5.16/ and /A.5.20/ we reach the desired result

$$C\, Z = AX + BY = I_m$$

where

$$Z = Z/z_1,z_2/ = \sum_{i_1,i_2,\ldots,i_m} a_{i_1 i_2 \ldots i_m}/z_1,z_2/\, Z_{i_1 i_2 \ldots i_m}/z_1,z_2/.$$

## Theorem A.5.4

There exist four polynomial matrices $X = X/z_1,z_2/ \in F^{p \times m}[z_1,z_2]$, $Y = Y/z_1,z_2/ \in F^{q \times m}[z_1,z_2]$, $D = D/z_1,z_2/ \in F^{t-m/xt}[z_1,z_2]$, $E = E/z_1,z_2/ \in F^{tx/t-m/}[z_1,z_2]$ such that

$$\begin{bmatrix} A & | & B \\ \hline & D & \end{bmatrix} \begin{bmatrix} X & | & \\ Y & | & E \end{bmatrix} = I_t \qquad /t = p+q/ \qquad /A.5.22/$$

if and only if the matrices A, B are ZLC.

### Proof

To prove the necessity note that any common zero $/z_1,z_2/$ of all the mxm minors of $[A | B]$ must appear as a zero of

$$\det \begin{bmatrix} A & | & B \\ \hline & D & \end{bmatrix} \qquad /A.5.23/$$

Thus, no such matrix can possess a determinant which is a nonzero constant. From /A.5.22/ it follows that /A.5.22/ should be nonzero constant. Therefore /A.5.22/ implies the zero left coprimeness of A and B. To prove the sufficiency let assume that A and B are ZLC. By theorem A.5.3 there exist X and Y such that /A.5.15/ holds. It can be easily shown that there exist two polynomial matrices $\bar{D} = \bar{D}/z_1, z_2/ \in F^{/t-m/xm}[z_1, z_2]$, $\bar{E} = \bar{E}/z_1, z_2/ \in F^{tx/t-m/} z_1, z_2$ such that

$$\det \begin{bmatrix} A & | & B \\ \hline & \bar{D} & \end{bmatrix} = a/z_2/ \in F[z_2] \qquad /A.5.24a/$$

and

$$\det \begin{bmatrix} X & | & \\ \hline Y & | & \bar{E} \end{bmatrix} = b/z_1/ \in F[z_1] \qquad /A.5.24b/$$

It is easy to check that

$$\begin{bmatrix} A & | & B \\ \hline & D & \end{bmatrix} \begin{bmatrix} X & | & \\ \hline Y & | & \bar{E} \end{bmatrix} = \begin{bmatrix} I_m & C\bar{E} \\ \bar{D}Z & \bar{D}\bar{E} \end{bmatrix} \qquad /A.5.25/$$

where

$$C = \begin{bmatrix} A & | & B \end{bmatrix} \quad \text{and} \quad Z = \begin{bmatrix} X \\ Y \end{bmatrix}$$

After performing some row and column operations, which leave determinants /A.5.24/ unchanged, we obtain

$$\begin{bmatrix} C \\ \bar{D}/I_t - ZC \end{bmatrix} \begin{bmatrix} Z & | & /I_t - ZC/\bar{E} \end{bmatrix} = \begin{bmatrix} I_m & 0 \\ 0 & \bar{D}/I_t - ZC/\bar{E} \end{bmatrix} \qquad /A.5.26/$$

since $/I_t - ZC/^2 = /I_t - ZC/$.

It follows from /A.5.26/ and /A.5.24/ that $\det \bar{D}/I_j - ZC/\bar{E} = a/z_2/b/z_1/$. By theorem A.4.2 there exist $L_1 = L_1/z_1, z_2/ \in F^{/t-m/x/t-m/}[z_1, z_2]$, $L_2 = L_2/z_1, z_2/ \in F^{/t-m/x/t-m/}[z_1, z_2]$ such that $\det L_1 = b/z_1/$, $\det L_2 = a/z_2/$ and $L_2 L_1 = \bar{D}/I_t - ZC/\bar{E}$.
It is easy to show that

$$D = L_2^{-1} \bar{D}/I_t - ZC/ \quad \text{and} \quad E = /I_t - ZC/ \bar{E} L_1^{-1}$$

are both polynomial matrices in $z_1$ and $z_2$.

Hence /A.5.26/ yields /A.5.22/. □

## 5.4. Zero coprimeness of 2-D polynomials

Consider a 2-D polynomials

$$a/z_1,z_2/ = a_n z_2^n + \ldots + a_1 z_2 + a_0$$
$$b/z_1,z_2/ = b_m z_2^m + \ldots + b_1 z_2 + b_0$$
/A.5.27/

where

$$a_i = a_i/z_1/ \in F[z_1] \quad \text{for } i = 0,1,\ldots,n$$
$$b_j = b_j/z_1/ \in F[z_1] \quad \text{for } j = 0,1,\ldots,m$$

Polynomials $a/z_1,\bar{z}_2/$ and $b/z_1,z_2/$ are zero coprime /ZC/ if and only if they have no zero $/z_1,z_2/$ in common.

From theorem A.5.3 it follows the following

### Lemma A.5.2

The polynomials /A.5.27/ are ZC if and only if there exist polynomials $x/z_1,z_2/$, $y/z_1,z_2/$ such that

$$a/z_1,z_2/x/z_1,z_2/ + b/z_1,z_2/y/z_1,z_2/ = 1 \qquad /A.5.28/$$

Let us define for /A.5.27/ the /n+m/x/n+m/ polynomial matrix in $z_1$

$$R = R/z_1/ = \begin{bmatrix} a_0 & a_1 & \cdots & a_n & 0 & 0 & \cdots & 0 \\ 0 & a_0 & \cdots & a_{n-1} & a_n & 0 & \cdots & 0 \\ \vdots & & & & & & & \vdots \\ 0 & 0 & \cdots & 0 & a_0 & a_1 & \cdots & a_n \\ b_0 & b_1 & \cdots & b_m & 0 & 0 & \cdots & 0 \\ 0 & b_0 & \cdots & b_{m-1} & b_m & 0 & \cdots & 0 \\ \vdots & & & & & & & \vdots \\ 0 & 0 & \cdots & 0 & b_0 & b_1 & \cdots & b_m \end{bmatrix} \begin{matrix} \left. \begin{matrix} \\ \\ \\ \\ \end{matrix} \right\} m \\ \\ \left. \begin{matrix} \\ \\ \\ \\ \end{matrix} \right\} n \end{matrix} \qquad /A.5.29/$$

### Theorem A.5.5 [10]

The polynomials /A.5.27/ are ZC if and only if there exists an /n+m/-row polynomial vector $p/z_1/$ such that

$$p/z_1/R = \begin{bmatrix} 1 & 0 & \cdots & 0 \end{bmatrix} \qquad /A.5.30/$$

**Proof**

First we shall show that zero coprimeness of /A.5.27/ implies /A.5.30/. From Lemma A.5.2 it follows that if /A.5.27/ are ZC then there exist polynomials $p_1/z_1,z_2/$, $p_2/z_1,z_2/$ such that

$$a/z_1,z_2/p_1/z_1,z_2/ + b/z_1,z_2/p_2/z_1,z_2/ = 1 \qquad /A.5.31/$$

Note that

$$a/z_1,z_2/p_1/z_1,z_2/ + b/z_1,z_2/p_2/z_1,z_2/ = p/z_1/RS \qquad /A.5.32/$$

where

$$p/z_1/ = [p_1/z_1/\ p_2/z_1/], \quad p_1/z_1/ \in F^m[z_1], \quad p_2/z_1/ \in F^n[z_1]$$

$$S^T = \begin{bmatrix} 1 & z_2 & \cdots & z_2^{n+m} \end{bmatrix}$$

$$p_1/z_1,z_2/ = p_1/z_1/ \begin{bmatrix} 1 \\ z_1 \\ \vdots \\ z_1^{m-1} \end{bmatrix}, \quad p_2/z_1,z_2/ = p_2/z_1 \begin{bmatrix} 1 \\ z_2 \\ \vdots \\ z_2^{n-1} \end{bmatrix}$$

Taking into account that

$$\begin{bmatrix} 1 & 0 & \cdots & 0 \end{bmatrix} S = 1$$

and /A.5.31/, /A.5.32/ we obtain

$$p/z_1/RS = \begin{bmatrix} 1 & 0 \cdots 0 \end{bmatrix} S \qquad /A.5.33/$$

The condition /A.5.30/ follows from /A.5.33/. Conversely, if /A.5.30/ holds, then from /A.5.33/ and /A.5.32/ the equation /A.5.32/ follows and by Lemma A.5.2 the polynomials /A.5.27/ are ZC. □

**Theorem A.5.6** [10]

Let the matrix R be invertible. The polynomials /A.5.27/ are ZC if and only if the first row of $R^{-1}$ is a polynomial vector.

**Proof**

If the first row of $R^{-1}$ is a polynomial vector, then $p/z_1/ = \begin{bmatrix} 1 & 0 \cdots 0 \end{bmatrix} R^{-1}$

is a polynomial vector which satisfies /A.5.30/. Thus, by theorem A.5.5 the polynomials /A.5.27/ are ZC. Conversely, if the polynomials /A.5.27/ are ZC, then by theorem A.5.5 there exists a row vector $p/z_1/$ such that /A.5.30/ holds. It follows from /A.5.30/ that the first row of $R^{-1}$ is a polynomial vector. □

If R is a unimodular matrix w.r.t $F[z_1,z_2]$, then $R^{-1}$ is also unimodular. In this particular case from theorem A.5.6 we have the following

### Corollary

If R is a unimodular matrix then the polynomials /A.5.27/ are ZC.

### Example A.5.2

Test the zero coprimeness of the polynomials

$$a/z_1,z_2/ = 1 + z_1 z_2 + z_2^2$$
$$b/z_1,z_2/ = z_2$$
/A.5.34/

In this case $n = 2$, $m = 1$ and the matrix

$$R = R/z_1/ = \begin{bmatrix} 1 & z_1 & 1 \\ 0 & 1 & 0 \\ 0 & 0 & 1 \end{bmatrix}$$
/A.5.35/

is unimodular. Thus, by the corollary the polynomials /A.5.34/ are ZC. It is easy to check that for /A.5.35/ and $p/z_1/ = [1,-z_1,-1]$ the condition /A.5.30/ holds.

## 6. Matrix fraction description

### 6.1. Extraction of greatest common divisors

Consider a 2-D rational matrix $G/z_1,z_2/ \in F^{m \times n}/z_1,z_2/$. It can be always written as

$$G/z_1,z_2/ = BA^{-1}$$
/A.6.1a/

or

$$G/z_1,z_2/ = A_1^{-1} B_1$$
/A.6.1b/

where

$$A = A/z_1,z_2/ \in F^{n\times n}[z_1,z_2], \quad B = B/z_1,z_2/ \in F^{m\times n}[z_1,z_2],$$
$$A_1 = A_1/z_1,z_2/ \in F^{m\times m}[z_1,z_2], \quad B_1 = B_1/z_1,z_2/ \in F^{m\times n}[z_1,z_2]$$

### Definition A.6.1

The right /left/ matrix fraction representation /A.6.1a/ //A.6.1.b// is called irreducible if A and B are right coprime /$A_1$ and $B_1$ are left-coprime/.

### Lemma A.6.1 [15]

Let
$$U \begin{bmatrix} A \\ \hline B \end{bmatrix} = \begin{bmatrix} R \\ \hline 0 \end{bmatrix} \qquad /A.6.2/$$

where U, A, B, R are 2-D polynomial matrices and $\det U \in F[z_2]$. If R has a primitive /left/ factorization $R = \bar{R} R^*$ in $F[z_2][z_1]$, then $R^*$ is a right common factor of A and B.

### Proof

From /A.6.2/ we have
$$\begin{bmatrix} A \\ \hline B \end{bmatrix} = U^{-1} \begin{bmatrix} R \\ \hline 0 \end{bmatrix}$$

and
$$[I \det U] \begin{bmatrix} A \\ \hline B \end{bmatrix} = [U_1 \mid U_2] \begin{bmatrix} R \\ \hline 0 \end{bmatrix} = U_1 \bar{R} R^* \qquad /A.6.3/$$

where
$$[U_1 \mid U_2] = \mathrm{Adj}\, U$$

Using Lemma A.4.1 to /A.6.3/ we obtain
$$\begin{bmatrix} A \\ \hline B \end{bmatrix} = [I \det U]^{-1} U_1 \bar{R} R^* = H R^* \qquad /A.6.4/$$

where $H = [I \det U]^{-1} U_1 \bar{R}$ is a polynomial matrix.

From /A.6.4/ it follows that $R^*$ is a right common factor od A and B. □

To find a greatest common right divisor /GCRD/ of polynomial matrices $A = A/z_1,z_2/$, $B = B/z_1,z_2/$ the following algorithm can be used.

## Algorithm A.6.1 [15]

**Step 1** Using the primitive factorization algorithm w.r.t $F[z_2][z_1]$ find $A^*$, $B^*$ and $R_o$ such that

$$\begin{bmatrix} A \\ B \end{bmatrix} = \begin{bmatrix} A^* \\ B^* \end{bmatrix} R_o$$

where $\begin{bmatrix} A^* \\ B^* \end{bmatrix}$ is primitive in $F[z_2][z_1]$.

**Step 2** Find the Hermite form w.r.t $F[z_2][z_1]$ of $\begin{bmatrix} A^* \\ B^* \end{bmatrix}$, i.e. R and U such that

$$U \begin{bmatrix} A^* \\ B^* \end{bmatrix} = \begin{bmatrix} R \\ 0 \end{bmatrix} \qquad /A.6.5/$$

where $\det U \in F[z_2]$.

**Step 3** Using the primitive factorization algorithm w.r.t $F[z_2][z_1]$ find $\bar{R}$ and $R^*$ such that

$$R = \bar{R} R^* \qquad /A.6.6/$$

$D = R^* R_o$ is the GCRD of A and B.

To prove the hypotesis we need to show that $R^*$ is a GCRD of $A^*$ and $B^*$. From Lemma A.6.1 it follows that $R^*$ is a CRD of $A^*$ and $B^*$. Thus, we need only to show that $R^*$ is divisible by any CRD of $A^*$ and $B^*$. Assuming that $D^*$ is any CRD of $A^*$ and $B^*$ we shall prove that $R^*$ and $D^*$ are related by some polynomial matrix. From the assumption we have

$$\begin{bmatrix} A^* \\ B^* \end{bmatrix} = \begin{bmatrix} \bar{A} \\ \bar{B} \end{bmatrix} D^* \qquad /A.6.7/$$

Let

$$U = \begin{bmatrix} U_1 & U_2 \\ U_3 & U_4 \end{bmatrix}$$

From /A.6.5/ and /A.6.7/ we get

$$\begin{bmatrix} R \\ 0 \end{bmatrix} = U \begin{bmatrix} A^* \\ B^* \end{bmatrix} = U \begin{bmatrix} \bar{A} \\ \bar{B} \end{bmatrix} D^* = \begin{bmatrix} CD^* \\ 0 \end{bmatrix} \qquad /A.6.8/$$

where $C = U_1 \bar{A} + U_2 \bar{B}$

and

$$R = \bar{R} R^* = C D^* \qquad /A.6.9/$$

Since $D^*$ is primitive and $\det \bar{R} \in F[z_2]$, by Lemma A.4.1 we can conclude that

$$R^* = \bar{R}^{-1} CD^* = HD^*$$

where $H = R^{-1} C$ is a polynomial matrix. □

### Example A.6.1

Find GCRD of the matrices

$$A = \begin{bmatrix} z_1 & z_1 z_2 \\ 0 & z_2 \end{bmatrix}, \qquad B = \begin{bmatrix} z_1 & 0 \\ z_1 z_2 & z_2 \end{bmatrix}$$

Using the algorithm A.6.1 we obtain

#### Step 1

$$\begin{bmatrix} A \\ B \end{bmatrix} = \begin{bmatrix} z_1 & z_1 z_2 \\ 0 & z_2 \\ z_1 & 0 \\ z_1 z_2 & z_2 \end{bmatrix} = \begin{bmatrix} 1 & z_1 \\ 0 & 1 \\ 1 & 0 \\ z_2 & 1 \end{bmatrix} \begin{bmatrix} z_1 & 0 \\ 0 & z_2 \end{bmatrix}$$

Hence

$$\begin{bmatrix} A^* \\ B^* \end{bmatrix} = \begin{bmatrix} 1 & z_1 \\ 0 & 1 \\ 1 & 0 \\ z_2 & 1 \end{bmatrix} \qquad \text{and} \qquad R_o = \begin{bmatrix} z_1 & 0 \\ 0 & z_2 \end{bmatrix}$$

#### Step 2

$$U \begin{bmatrix} A^* \\ B^* \end{bmatrix} = \begin{bmatrix} 0 & 0 & 1 & 0 \\ 0 & 1 & 0 & 0 \\ 1 & -z_1 & -1 & 0 \\ 0 & -1 & -z_2 & 1 \end{bmatrix} \begin{bmatrix} 1 & z_1 \\ 0 & 1 \\ 1 & 0 \\ z_2 & 1 \end{bmatrix} = \begin{bmatrix} 1 & 0 \\ 0 & 1 \\ 0 & 0 \\ 0 & 0 \end{bmatrix}$$

Hence
$$R = \begin{bmatrix} 1 & 0 \\ 0 & 1 \end{bmatrix}$$

**Step 3**
$$R = \bar{R}\,R^* = \begin{bmatrix} 1 & 0 \\ 0 & 1 \end{bmatrix}\begin{bmatrix} 1 & 0 \\ 0 & 1 \end{bmatrix}$$

and
$$D = R^* R_o = \begin{bmatrix} z_1 & 0 \\ 0 & z_2 \end{bmatrix}$$

Similar results can be obtained for greatest common left divisor /GCLD/.

### 6.2. Bezout identity and some other theorems

**Theorem A.6.1** /Bezout identity/ [14]

Let $N_R/z_1,z_2/$, $D_R/z_1,z_2//N_L/z_1,z_2/$, $D_L/z_1,z_2//$ be two right /left/ coprime polynomial matrices, then exists a polynomial matrix in $z_2$, say, $E_R/z_2/$ /$E_L/z_2//$ and two polynomial matrices $X_R/z_1,z_2/$, $Y_R/z_1,z_2//X_L/z_1,z_2/$, $Y_L/z_1,z_2//$ such that

$$X_R/z_1,z_2/\,D_R/z_1,z_2/ + Y_R/z_1,z_2/N_R/z_1,z_2/ = E_R/z_2/ \qquad /\text{A.6.10a}/$$

$$/N_L/z_1,z_2/Y_L/z_1,z_2/ + D_L/z_1,z_2/X_L/z_1,z_2/ = E_L/z_2// \qquad /\text{A.6.10b}/$$

**Proof**

From theorem 3.10 it follows that $N_R/z_1,z_2/$, $D_R/z_1,z_2/$ are also right 1-D coprime w.r.t $F/z_2/[z_1]$. Thus, there exist two polynomial matrices in $z_1$ with coefficients in $F/z_2/$, say, $\bar{X}_R/z_1,z_2/$, $\bar{Y}_R/z_1,z_2/$ such that

$$\bar{X}_R/z_1,z_2/D_R/z_1,z_2/ + \bar{Y}_R/z_1,z_2/N_R/z_1,z_2/ = I \qquad /\text{A.6.11}/$$

Premultiplying /A.6.11/ by suitable diagonal polynomial matrix in $z_2$ we obtain /A.6.10a/. In a similar manner we can prove /A.6.10b/. □

**Theorem A.6.2** [15]

If
$$G/z_1,z_2/ = N_R D_R^{-1} = D_L^{-1} N_L \qquad /\text{A.6.12}/$$

with $N_R = N_R/z_1,z_2/$, $D_R = D_R/z_1,z_2/$ right coprime and $N_L = N_L/z_1,z_2/$, $D_L/z_1,z_2/$ left coprime, then

$$\det D_R = \det D_L \qquad /A.6.13/$$

Proof

From /A.6.12/ it follows that $N_L D_R = D_L N_R$. Taking into account this and /A.6.10/ we obtain

$$\begin{bmatrix} X_R & Y_R \\ N_L & -D_L \end{bmatrix} \begin{bmatrix} D_R & Y_L \\ N_R & -X_L \end{bmatrix} = \begin{bmatrix} E_R & W \\ 0 & E_L \end{bmatrix} \qquad /A.6.14/$$

where
$$W = X_R Y_L - Y_R X_L.$$

Let
$$U = \begin{bmatrix} X_R & Y_R \\ N_L & -D_L \end{bmatrix} \quad \text{and} \quad V = \begin{bmatrix} D_R & Y_L \\ N_L & -X_L \end{bmatrix}$$

Thus, from /A.6.14/ we have

$$\det U \det V = \det E_R \det E_L \in F[z_2]$$

From the equation

$$U \begin{bmatrix} D_R & 0 \\ N_R & I \end{bmatrix} = \begin{bmatrix} E_R & Y_R \\ 0 & -D_L \end{bmatrix}$$

it follows that

$$\det U \det D_R = - \det E_R \det D_L \qquad /A.6.15a/$$

By symmetry in a similar manner we can also obtain the relation

$$\det \bar{U} \det D_R = - \det \bar{E}_R \det D_L \qquad /A.6.15b/$$

where $\det \bar{U}$, $\det \bar{E}_R \in F[z_1]$.

Note that /A.6.15/ imply /A.6.13/.  □

Theorem A.6.3 [15]

Let
$$P = V T^{-1} F$$

where V, T and F are 2-D polynomial matrices.

If V and T are right coprime, then P is a polynomial matrix if and only if $T^{-1}F$ is a polynomial matrix.

Proof

To prove necessity, we assume that V and T are right coprime. Then, by theorem A.6.1, there exist polynomial matrices X, Y such that

$$XT + YV = E/z_2/ \qquad /A.6.16/$$

Postmultiplying /A.6.16/ by $T^{-1}F$ we obtain

$$XF + VP = E/z_2/\, T^{-1}F \qquad /A.6.17/$$

By assumption P is a polynomial matrix, then

$$T^{-1}F = E^{-1}/z_2//XF + VP/$$

is a polynomial matrix in $z_1$ with coefficients in $F/z_2/$.

By symmetry between $z_1$ and $z_2$ we can also show that $T^{-1}F$ is a polynomial matrix in $z_2$ with coefficients in $F/z_1/$. Hence $T^{-1}F$ is a polynomial matrix in $z_1$ and $z_2$.

If $T^{-1}F$ is a polynomial matrix, then P is also a polynomial matrix as a product of two polynomial matrices. □

Theorem A.6.4 [15]

Let

$$N_R\, D_R^{-1} = D_L^{-1}\, N_L \qquad /A.6.18/$$

where $N_R$, $D_R$, $D_L$ and $N_L$ are 2-D polynomial matrices.

If $N_R, D_R$ are right coprime and $N_L$, $D_L$, $D_R$, B are both left coprime, then $D_L$, $N_L B$ are left coprime.

Proof

We shall prove the hypotesis by contradiction. Assume that $D_L$, $N_L B$ are not left coprime. Thus, $D_L^{-1}\, N_L B$ can be reduced, i.e.

$$D_L^{-1}\, N_L B = \bar{N}\, \bar{D}^{-1}$$

and

$$N_R D_R^{-1} B \bar{D} = \bar{N}$$

By theorem A.6.3 $K = D_R^{-1} B\bar{D}$ is a polynomial matrix and

$$K \bar{D}^{-1} = D_R^{-1} B \qquad /A.6.19/$$

Note that $\bar{D}$ has a determinantal degree less than the one of $D_R$. Thus, /A.6.19/ contradicts the assumption that $D_R, B$ are left coprime. □

### Theorem A.6.5 [15]

Let

$$V T^{-1} U = N D^{-1} \qquad /A.6.20/$$

where V, T, U, N and D are 2-D polynomial matrices.

If T, U are left coprime and T, V; D, N are both right coprime, then

$$\det T = \det D \qquad /A.6.21/$$

### Proof

Let $V_L$, $T_L$ be left coprime polynomial matrices such that

$$V T^{-1} = T_L^{-1} V_L \qquad /A.6.22/$$

From theorem A 6.2 it follows that $\det T = \det T_L$.
Substitution of /A.6.22/ into /A.6.20/ yields

$$V T_L^{-1} V_L U = N D^{-1}$$

By theorem A.6.4, $T_L$ and $V_L U$ are left coprime. Thus, again by theorem A.6.2 we have $\det T_L = \det D$ and /A.6.21/. □

## 7. 2-D polynomial matrix equations

### 7.1. Problem formulation

Let $R[z_1, z_2]$ denote the ring of polynomials in $z_1$ and $z_2$ with real coefficients and let $R^{k \times l}[z_1, z_2]$ be the set of kxl polynomial matrices with entries in $R[z_1, z_2]$.
Consider a 2-D polynomial matrix equation of the form

$$AX + BY = C \qquad /A.7.1/$$

where

$A = A/z_1,z_2/ \in R^{l \times p}[z_1,z_2]$, $\quad B = B/z_1,z_2/ \in R^{l \times q}[z_1,z_2]$ and
$C = C/z_1,z_2/ \in R^{l \times m}[z_1,z_2]$ are given.

By a solution to /A.7.1/ we mean any pair of $X = X/z_1,z_2/ \in R^{p \times m}[z_1,z_2]$, $Y = Y/z_1,z_2/ \in R^{q \times m}[z_1,z_2]$ satisfying /A.7.1/.

The problem can be formulated as follows. Given A, B and C, find a solution to /A.7.1/. Similarly, the problem can be formulated for the equation

$$XA + YB = C \qquad /A.7.1'/$$

where $A = A/z_1,z_2/ \in R^{p \times l}[z_1,z_2]$, $B = B/z_1,z_2/ \in R^{q \times l}[z_1,z_2]$, $C = C/z_1,z_2/ \in R^{m \times l}[z_1,z_2]$ are given and $X = X/z_1,z_2/ \in R^{m \times p}[z_1,z_2]$, $Y = Y/z_1,z_2/ \in R^{m \times q}[z_1,z_2]$ are unknown.

Note that /A.7.1'/ can be reduced to /A.7.1/ by transposition. Thus, further we shall consider only the equation /A.7.1/.

## 7.2. Necessary and sufficient conditions for the existence of a solution to the problem

Let us define the matrices $\bar{A} \in R^{r \times r}[z_1,z_2]$, $\bar{B} \in R^{r \times q}[z_1,z_2]$, $\bar{C} \in R^{r \times m}[z_1,z_2]$, $\bar{X} \in R^{r \times m}[z_1,z_2]$, $\bar{Y} \in R^{q \times m}[z_1,z_2]$ as follows

$$\bar{A} = \begin{bmatrix} A & 0 \\ 0 & 0 \end{bmatrix}, \quad \bar{B} = \begin{bmatrix} B \\ 0 \end{bmatrix}, \quad \bar{C} = \begin{bmatrix} C \\ 0 \end{bmatrix}, \quad \bar{X} = \begin{bmatrix} X \\ X' \end{bmatrix}, \quad \bar{Y} = Y \qquad /A.7.2/$$

where $r = \max/l, p+q/$ and $X' \in R^{/r-p/ \times m}[z_1,z_2]$ is arbitrary.

Using /A.7.2/ we can write the equation

$$\bar{A}\bar{X} + \bar{B}\bar{Y} = \bar{C} \qquad /A.7.3/$$

It is easy to prove the following.

### Lemma A.7.1

The equation /A.7.1/ has a solution X, Y if and only if the equation /A.7.3/ has a solution $\bar{X}$, $\bar{Y}$.

Note that there exists always a unimodular matrix $U \in R^{/r+q/ \times /r+q/}[z_1,z_2]$ /det $U \in R \setminus \{0\}$/ such that

$$[\bar{A} \mid \bar{B}] U = [G \mid 0] \qquad /A.7.4/$$

where $G \in R^{r \times r}[z_1, z_2]$.

**Lemma A.7.2**

The matrix $G$ satisfying /A.7.4/ is the geatest common left divisor /GCLD/ of $\bar{A}$ and $\bar{B}$.

**Proof**

Since
$$U = \begin{bmatrix} U_1 & U_2 \\ U_3 & U_4 \end{bmatrix}, \quad U_1 \in R^{r \times r}[z_1, z_2], \quad U_4 \in R^{q \times q}[z_1, z_2] \qquad /A.7.5/$$

is unimodular, the inverse matrix

$$U^{-1} = V = \begin{bmatrix} V_1 & V_2 \\ V_3 & V_4 \end{bmatrix}, \quad V_1 \in R^{r \times r}[z_1, z_2], \quad V_4 \in R^{q \times q}[z_1 z_2]$$

is also unimodular.

From /A.7.4/ we have

$$[A \mid B] = [G \mid 0] \begin{bmatrix} V_1 & V_2 \\ V_3 & V_4 \end{bmatrix}$$

and
$$\bar{A} = G V_1, \quad \bar{B} = G V_2$$

Thus, $G$ is common left divisor /CLD/ of $\bar{A}$ and $\bar{B}$.

From the equation
$$\bar{A} U_1 + \bar{B} U_3 = G \qquad /A.7.6/$$

it follows that $G$ is GCLD of $\bar{A}$ and $\bar{B}$. □

**Theorem A.7.1** 7

The equation /A.7.1/ has a solution if and only if GCLD od $\bar{A}$ and $\bar{B}$ is a left divisor of $\bar{C}$.

**Proof**

First we shall show that if /A.7.1/ has a solution $X_o$, $Y_o$ and $G$ is GCLD of $\bar{A}$ and $\bar{B}$, then $G$ is a left divisor of $\bar{C}$.

Substitution of $\bar{A} = G\bar{A}_o$, $\bar{B} = G\bar{B}_o$ into $\bar{A}\bar{X}_o + \bar{B}\bar{Y}_o = \bar{C}$, where $\bar{X}_o = \begin{bmatrix} X_o \\ X' \end{bmatrix}$, $\bar{Y}_o = Y_o$ yields

$$G / \bar{A}_o \bar{X}_o + \bar{B}_o \bar{Y}_o / = \bar{C}$$

Thus, G is a left divisor of $\bar{C}$.

Now, we shall show that if GCLD of $\bar{A}$ and $\bar{B}$ is a left divisor of $\bar{C}$, then /A.7.1/ has a solution. If G is GCLD of $\bar{A}$ and $\bar{B}$, then there exists a unimodular matrix /A.7.5/ such that /A.7.4/ and /A.7.6/ hold. Postmultiplying /A.7.6/ by $\bar{C}_o$ and taking into account that $\bar{C} = G\bar{C}_o$ we obtain a solution to /A.7.3/ in the form

$$\bar{X}_o = U_1 \bar{C}_o, \qquad \bar{Y}_o = U_3 \bar{C}_o$$

By Lemma A.7.1 the equation /A.7.1/ has a solution. □

### Theorem A.7.2

If $\bar{X}_o$, $\bar{Y}_o$ is a particular solution to /A.7.3/ then the general solution $\bar{X}$, $\bar{Y}$ to /A.7.3/ has the form

$$\bar{X} = \bar{X}_o - \bar{B}_1 T, \qquad \bar{Y} = \bar{Y}_o + \bar{A}_1 T \qquad /A.7.7/$$

where $\bar{A}_1$, $\bar{B}_1$ are polynomial matrices satisfying the condition

$$\bar{A}\bar{B}_1 = \bar{B}\bar{A}_1 \qquad /A.7.8/$$

and T is an arbitrary polynomial matrix of appropriate size.

### Proof

Substituting /A.7.7/ into /A.7.3/ and using /A.7.8/ we obtain

$$\bar{A}\bar{X} + \bar{B}\bar{Y} = \bar{A}\bar{X}_o + \bar{B}\bar{Y}_o + /\bar{B}\bar{A}_1 - \bar{A}\bar{B}_1/T = \bar{C}$$

since

$$\bar{A}\bar{X}_o + \bar{B}\bar{Y}_o = \bar{C}$$

From /A.7.4/ and /A.7.5/ we have

$$\bar{A}U_2 + \bar{B}U_4 = 0$$

Therefore, we can assume that $\bar{B}_1 = -U_2$, $\bar{A}_1 = U_4$ and the general solution to /A.7.3/ can be written as

$$\begin{bmatrix} \bar{X} \\ \bar{Y} \end{bmatrix} = \begin{bmatrix} U_1 & U_2 \\ U_3 & U_4 \end{bmatrix} \begin{bmatrix} \bar{C}_o \\ T \end{bmatrix} \qquad /A.7.9/$$

### Theorem A.7.3 [7]

The equation /A.7.1/ has a solution for any C if and only if A and B are zero left coprime /ZLC/.

### Proof

By theorem A.5.5 there exist two polynomial matrices $\hat{X}$ and $\hat{Y}$ such that

$$A\hat{X} + B\hat{Y} = I \qquad /A.7.10/$$

if and only if A and B are ZLC.

Postmultiplying /A.7.10/ by C we obtain

$$A\hat{X}C + B\hat{Y}C = C$$

Thus,, the equation /A.7.1/ has a solution in the form

$$X = \hat{X}C, \quad Y = \hat{Y}C \qquad /A.7.11/$$

$\square$

In the proof of theorem A.5.5 it was shown that

$$\begin{bmatrix} X \\ Y \end{bmatrix} = \sum_{i_1, i_2, \ldots, i_l} a_{i_1 i_2 \ldots i_l} Z_{i_1 i_2 \ldots i_l} \qquad /A.7.12/$$

where $a_{i_1, i_2, \ldots, i_l}$ are polynomials in $z_1, z_2$ defined by /A.5.27/ and $Z_{i_1 i_2 \ldots i_l}$ are polynomial matrices in $z_1, z_2$ defined by /A.5.32/.

From /A.7.12/ and /A.7.11/ we have a solution to /A.7.1/ in the form

$$\begin{bmatrix} X \\ Y \end{bmatrix} = \begin{bmatrix} \hat{X} \\ \hat{Y} \end{bmatrix} C = \sum_{i_1, i_2, \ldots, i_l} a_{i_1, i_2 \ldots i_l} Z_{i_1 i_2 \ldots i_l} C \qquad /A.7.13/$$

Let in /A.7.7/ be $\deg_{z_2} \bar{X}_o > \deg_{z_2} \bar{B}_1$ and $\det B_1 = b_N/z_1/z_2^N + b_{N-1}/z_1/z_2^{N-1} + \ldots + b_o/z_1/$. To find a minimal degree solution w.r.t X of /A.7.1/ let assume that $B_1$ is invertible and $b_N/z_1/ = b_N \in R \setminus \{0\}$.

Under these assumptions there exist polynomial matrices $Q_1 = Q_1/z_1,z_2/$, $R_1 = R/z_1,z_2/$ such that

$$\bar{X}_o = \bar{B}_1 Q_1 + R_1 \qquad /A.7.14/$$

with $\deg_{z_2} \bar{B}_1 > \deg_{z_2} R_1$.

Sunstitution of /A.7.14/ into /A.7.7/ yields

$$\bar{X} = R_1 - \bar{B}_1/T-Q_1/$$

For $T = Q_1$ we obtain a minimal degree solution w.r.t X of /A.7.3/, $\bar{X} = R_1$ and $\bar{Y} = \bar{Y}_o + \bar{A}_1 Q_1$ and next the desired minimal degree solution X, Y of /A.7.1/.

## Algorithms

Three algorithms for finding a solution X, Y to /A.7.1/ will be presented.

### Algorithm A.7.1

Using elementary column operations carry out the reduction

$$\begin{bmatrix} A & B & C \\ I_1 & 0 & 0 \\ 0 & I_q & 0 \end{bmatrix} \longrightarrow \begin{bmatrix} A & B & 0 \\ I_1 & 0 & U_1 \\ 0 & I_q & U_2 \end{bmatrix} \qquad /A.7.15/$$

If can be easily shown that a solution to /A.7.1/ is given by $X = -U_1$ and $Y = -U_2$.

**Remark** Note that not for all A, B and C the reduction /A.7. / can be carried out.

### Example A.7.1

Find a solution to /A.7.1/ with

$$A = \begin{bmatrix} z_1 & z_2 \\ z_1^2 & z_2^2 \\ z_1^2 & z_2^2 \end{bmatrix}, \quad B = \begin{bmatrix} 0 \\ 1 \end{bmatrix}, \quad C = \begin{bmatrix} z_1, & z_1 z_2 + z_2 \\ z_1^2+1, & z_1^2 z_2 + z_2^2 \end{bmatrix} \qquad /A.7.16/$$

Using elementary column operations we carry out the reduction

$$\begin{bmatrix} A & B & C \\ I_1 & 0 & 0 \\ 0 & I_q & 0 \end{bmatrix} = \begin{bmatrix} z_1 & z_2 & 0 & z_1 & z_1z_2+z_2 \\ z_1^2 & z_2^2 & 1 & z_1^2+1 & z_1^2z_2+z_2^2 \\ 1 & 0 & 0 & 0 & 0 \\ 0 & 1 & 0 & 0 & 0 \\ 0 & 0 & 1 & 0 & 0 \end{bmatrix} \rightarrow \begin{bmatrix} z_1 & z_2 & 0 & 0 & 0 \\ z_1^2 & z_2^2 & 1 & 0 & 0 \\ 1 & 0 & 0 & -1 & -z_2 \\ 0 & 1 & 0 & 0 & -1 \\ 0 & 0 & 1 & -1 & 0 \end{bmatrix}$$

Hence the desired solution is

$$X = \begin{bmatrix} 1 & z_2 \\ 0 & 1 \end{bmatrix}, \qquad Y = \begin{bmatrix} 1 & 0 \end{bmatrix}$$

From the proofs of theorem A.5.5 and theorem A.7.3 we have the following

**Algorithm A.7.2** [12]

**Step 1** Find the minors $\Delta_{i_1 i_2 \ldots i_l} /z_1, z_2/$ of the matrix

$$D = \begin{bmatrix} A & B \end{bmatrix}$$

**Step 2** Find polynomials $a_{i_1 i_2 \ldots i_l} /z_1, z_2/$ which satisfy the equation

$$\sum_{i_1, i_2, \ldots, i_l} a_{i_1 i_2 \ldots i_l} /z_1, z_2/ \Delta_{i_1 i_2 \ldots i_l} /z_1, z_2/ = 1$$

**Step 3** Find a matrix $K \in R^{/p+q/ \times 1}$ whose all minors $M^K_{i_1 i_2 \ldots i_l}$ are nonzero.

**Step 4** Find $F = D \Lambda K$ and its adjoint matrix $F_a$ where $\Lambda = \text{diag}[\lambda_1, \lambda_2, \ldots, \lambda_{p+q}]$.

**Step 5** Find $E = \Lambda K F_a$.

**Step 6** Find the matrices

$$Z_{i_1 i_2 \ldots i_l} = \frac{1}{M^K_{i_1 i_2 \ldots i_l}} \frac{\partial^l E}{\partial \lambda_{i_1} \partial \lambda_{i_2} \ldots \partial \lambda_{i_l}} \Big|_{\substack{\lambda_1 = 0 \\ \ldots \\ \lambda_{p+q} = 0}}$$

**Step 7** Find the desired solution X, Y using

$$\begin{bmatrix} X \\ Y \end{bmatrix} = \sum_{i_1,i_2,\ldots,i_l} a_{i_1 i_2 \ldots i_l} Z_{i_1 i_2 \ldots i_l} C \qquad /A.7.17/$$

**Remark:** The algorithm can be used if A and B are ZLC.

**Example A.7.2**

Find a solution to /A.7.1/ with

$$A = \begin{bmatrix} z_1 & 0 \\ 1 & z_2 \end{bmatrix}, \quad B = \begin{bmatrix} -1 \\ z_2 \end{bmatrix}, \quad C = \begin{bmatrix} z_1 z_2 & , & -z_2 \\ z_2 - z_1 & , & 1 \end{bmatrix}$$

It is to check that in this case the matrices A and B are ZLC and therefore the algorithm A.7.2 can be used.

**Step 1** The minors $\Delta_{i_1 i_2}$ of the matrix

$$D = \begin{bmatrix} A & \vdots & B \end{bmatrix} = \begin{bmatrix} z_1 & 0 & -1 \\ 1 & z_2 & z_2 \end{bmatrix}$$

are

$$\Delta_{12} = \begin{vmatrix} z_1 & 0 \\ 1 & z_2 \end{vmatrix} = z_1 z_2 \;,\quad \Delta_{13} = \begin{vmatrix} z_1 & -1 \\ 1 & z_2 \end{vmatrix} = z_1 z_2 + 1 \;,$$

$$\Delta_{23} = \begin{vmatrix} 0 & -1 \\ z_2 & z_2 \end{vmatrix} = z_2$$

**Step 2** It is easy to check that the polynomials

$$a_{12} = -1, \quad a_{13} = 1, \quad a_{23} = 0 \qquad /A.7.18a/$$

or

$$a_{12} = -2, \quad a_{13} = 1, \quad a_{23} = z_1 \qquad /A.7.18b/$$

satisfy the equation

$$a_{12}\Delta_{12} + a_{13}\Delta_{13} + a_{23}\Delta_{23} = 1$$

**Step 3** The matrix

$$K = \begin{bmatrix} 1 & 0 \\ 0 & 1 \\ 1 & 1 \end{bmatrix}$$

has all nonzero minors $M^K_{i_1 i_2}$

**Step 4**

$$F = D\Lambda K = \begin{bmatrix} z_1 & 0 & -1 \\ 1 & z_2 & z_2 \end{bmatrix} \begin{bmatrix} \lambda_1 & 0 & 0 \\ 0 & \lambda_2 & 0 \\ 0 & 0 & \lambda_3 \end{bmatrix} \begin{bmatrix} 1 & 0 \\ 0 & 1 \\ 1 & 1 \end{bmatrix} =$$

$$= \begin{bmatrix} z_1\lambda_1 - \lambda_3, & -\lambda_3 \\ \lambda_1 + z_2\lambda_2, & z_2(\lambda_2 + \lambda_3) \end{bmatrix}$$

and

$$F_a = \text{Adj} F = \begin{bmatrix} z_2(\lambda_2 + \lambda_3), & \lambda_3 \\ -\lambda_1 - z_2\lambda_3, & z_1\lambda_1 - \lambda_3 \end{bmatrix}$$

**Step 5**

$$E = \Lambda K F_a = \begin{bmatrix} z_2\lambda_1(\lambda_2 + \lambda_3), & \lambda_1\lambda_3 \\ -\lambda_1\lambda_2 - z_2\lambda_2\lambda_3, & z_1\lambda_1\lambda_2 - \lambda_2\lambda_3 \\ \lambda_3(z_2\lambda_2 - \lambda_1), & z_1\lambda_1\lambda_3 \end{bmatrix}$$

**Step 6**

$$Z_{12} = \frac{1}{M^K_{12}} \left. \frac{\partial^2 E}{\partial \lambda_1 \partial \lambda_2} \right|_{\substack{\lambda_1=0 \\ \lambda_2=0 \\ \lambda_3=0}} = \begin{bmatrix} z_2 & 0 \\ -1 & z_1 \\ 0 & 0 \end{bmatrix}$$

$$Z_{13} = \frac{1}{M^K_{13}} \left. \frac{\partial^2 E}{\partial \lambda_1 \partial \lambda_3} \right|_{\substack{\lambda_1=0 \\ \lambda_2=0 \\ \lambda_3=0}} = \begin{bmatrix} z_2 & 1 \\ 0 & 0 \\ -1 & z_1 \end{bmatrix}$$

$$Z_{23} = \frac{1}{M^K_{23}} \left. \frac{\partial^2 E}{\partial \lambda_2 \partial \lambda_3} \right|_{\substack{\lambda_1=0 \\ \lambda_2=0 \\ \lambda_3=0}} = \begin{bmatrix} 0 & 0 \\ z_2 & 1 \\ -z_2 & 0 \end{bmatrix}$$

Step 7  Using /A.7.17/ we obtain

for the polynomials /A.7.18a/

$$\begin{bmatrix} X \\ Y \end{bmatrix} = /a_{12}Z_{12} + a_{13}Z_{13} + a_{23}Z_{23}/C = \begin{bmatrix} z_2-z_1, & 1 \\ z_1^2, & -/z_1+z_2/ \\ \hline -z_1^2, & z_1+z_2 \end{bmatrix}$$

and for the polynomials /A.7.18b/

$$\begin{bmatrix} X \\ Y \end{bmatrix} = \begin{bmatrix} z_2-z_1-z_1z_2^2, & 1+z_2^2 \\ z_1z_2+z_1^2/1+z_2^2/, & -z_1/1+z_2^2/-2z_2 \\ \hline -z_1^2/1+z_2^2/, & z_1/1+z_2^2/+z_2 \end{bmatrix}$$

From considerations presented in A.7.2 the following algorithm follows.

## Algorithm A.7.3 [7]

Step 1  Find $r = \max/1,p+q/$ and $\bar{A}$, $\bar{B}$, $\bar{C}$ defined by /A.7.2/.

Step 2  Using elementary column operations carry out the reduction

$$\begin{bmatrix} A & B \\ I_r & 0 \\ 0 & I_q \end{bmatrix} \rightarrow \begin{bmatrix} G & 0 \\ U_1 & U_2 \\ U_3 & U_4 \end{bmatrix}$$

and find G, $U_1$, $U_2$, $U_3$ and $U_4$.

Step 3  Using /A.7.9/ find $\bar{X}$, $\bar{Y}$ and next X, Y from /A.7.2/.

## Example A.7.3

Solve /A.7.1/ for /A.7.16/ using the algorithm A.7.3.

Step 1  In this case $n = 2$, $l = m = p = 2$, $g = 1$, $r = \max/1,p+q/ = 3$ and

$$\bar{A} = \begin{bmatrix} z_1 & z_2 & 0 \\ z_1^2 & z_2^2 & 0 \\ 0 & 0 & 0 \end{bmatrix}, \quad \bar{B} = \begin{bmatrix} 0 \\ 1 \\ 0 \end{bmatrix}, \quad \bar{C} = \begin{bmatrix} z_1 & , & z_1z_2+z_2 \\ z_1^2+1 & , & z_1^2z_2+z_2^2 \\ 0 & & 0 \end{bmatrix}$$

**Step 2** Using elementary column operations we carry out the reduction

$$\begin{bmatrix} A & B \\ I_r & 0 \\ 0 & I_q \end{bmatrix} = \left[\begin{array}{ccc|c} z_1 & z_2 & 0 & 0 \\ z_1^2 & z_2^2 & 0 & 1 \\ 0 & 0 & 0 & 0 \\ \hline 1 & 0 & 0 & 0 \\ 0 & 1 & 0 & 0 \\ 0 & 0 & 1 & 0 \\ 0 & 0 & 0 & 1 \end{array}\right] \rightarrow \left[\begin{array}{ccc|c} z_1 & z_2 & 0 & 0 \\ z_1^2 & z_2^2 & 1 & 0 \\ 0 & 0 & 0 & 0 \\ \hline 1 & 0 & 0 & 0 \\ 0 & 1 & 0 & 0 \\ 0 & 0 & 0 & 1 \\ 0 & 0 & 1 & 0 \end{array}\right]$$

Hence

$$G = \begin{bmatrix} z_1 & z_2 & 0 \\ z_1^2 & z_2^2 & 1 \\ 0 & 0 & 0 \end{bmatrix}, \quad U = \begin{bmatrix} U_1 & U_2 \\ U_3 & U_4 \end{bmatrix} = \left[\begin{array}{ccc|c} 1 & 0 & 0 & 0 \\ 0 & 1 & 0 & 0 \\ 0 & 0 & 0 & 1 \\ 0 & 0 & 1 & 0 \end{array}\right]$$

**Step 3** It is easy to verify that

$$\bar{C}_o = \begin{bmatrix} 1-z_2t_1 & , & z_2t_2 \\ z_1t_1 & , & 1+z_1/1-t_2/ \\ 1+z_1z_2/z_1-z_2/t_1 & , & z_1z_2/z_1-z_2//1-t_2/ \end{bmatrix}$$

satisfies the equation

$$\begin{bmatrix} z_1 & z_2 & 0 \\ z_1^2 & z_2^2 & 1 \\ 0 & 0 & 0 \end{bmatrix} \bar{C}_o = \begin{bmatrix} z_1 & , & z_1z_2+z_2 \\ z_1^2+1 & , & z_1^2z_2+z_2^2 \\ 0 & , & 0 \end{bmatrix}$$

for arbitrary polynomials $t_1$ and $t_2$

__Step 4__ Using /A.7.9/ we obtain

$$\begin{bmatrix} \bar{X} \\ \bar{Y} \end{bmatrix} = U \begin{bmatrix} \bar{C}_o \\ T \end{bmatrix} = \left[ \begin{array}{ccc} 1-z_2 t_1 & , & z_2 t_2 \\ z_1 t_1 & , & 1+z_1/1-t_2/ \\ t_3 & , & t_4 \\ \hline 1+z_1 z_2/z_1-z_2/t_1 & , & z_1 z_2/z_1-z_2//1-t_2/ \end{array} \right]$$

where $t_3, t_4$ are arbitrary polynomials in $z_1$ and $z_2$.

The desired solution

$$X = \begin{bmatrix} 1-z_2 t_1, & z_2 t_2 \\ z_1 t_1, & 1+z_1/1-t_2/ \end{bmatrix}, \quad Y = \begin{bmatrix} 1+z_1 z_2/z_1-z_2/t_1, & z_1 z_2/z_1-z_2//1-t_2/ \end{bmatrix}$$

depends on two arbitrary polynomials $t_1$ and $t_2$.

Other algorithms for finding a solution to /A.7.1/ are given in [6, 9, 17, 16].

# REFERENCES

[1] J.Białecki: New necessary and sufficient condition for factor-coprimeness of two dimensional polynomials. Proc. IV Polish-English Seminar on "Real Time Process Control", Jabłonna, May 30-June 2, 1983, pp. 20-26

[2] N.K.Bose: A criterion to determine if two multivariable polynomials are relatively prime. Proc. IEEE, vol. 60, No.1, January 1972, pp. 134-135

[3] B.Eichstaedt: A polynomial matrix approach to computation of l.c.m. and g.c.d of polynomials over a factorial ring with an application to the case of 2-D polynomials. Proc. IV Polish-English Seminar on "Real Time Process Control", Jabłonna, May 30- June 2, 1983, pp. 65-74

[4] M.Gajowniczek: Evaluation of the greatest common divisor of 2-D polynomials. Proc. IV Polish-English Seminar on "Real Time Process Control", Jabłonna, May 30-June 2, 1983, pp. 84-88

[5] E.I.Jury, B.D.O.Anderson: Generalized Bezontian and Sylvester matrices in multivariable linear control. IEEE Trans. Autom. Control, vol. AC-21, August 1976, pp. 551-556

[6] T.Kaczorek: Algorithm for solving 2-D polynomial matrix equations. Bull.Acad.Polon.Sér.sci techn., vol.31, No. 1-12, 1983, pp. 51-57

[7] T.Kaczorek: A new method of analysis and synthesis for n-D dynamical systems. Proc. 6th Int. Congress of Cybernetics and Systems. Sept. 10-14, 1984, Paris /in press/

[8] T.Kaczorek: Extension of Sylvester's theorem to two-dimensional systems. Bull.Acad.Polon.Sér.sci.techn., vol. 30, No.1-2, 1982, pp. 53-58

[9] T.Kaczorek: New algorithms of solving 2-D polynomial equations. Bull. Acad.Polon. Sér. sci.techn., vol. 30, No. 5-6, 1982, pp. 77-83

[10] T.Kaczorek: New tests for zero coprimeness of 2-D polynomials. Prace Naukowe Politechniki Warszawskiej ELEKTRYKA, No. 74, 1983, pp. 15-21

[11] T.Kaczorek: Polynomial matrix equations in two indeterminants. Bull Acad.Polon.Sér. sci techn., vol. 30, No. 1-3, 1982, pp. 39-44

[12] T.Kaczorek: Solving of 2-D polynomial matrix equations. Proc. 3rd Int. Conf. on Functional Differential Systems and Related Topics. Błażejewko 1984, Poland /in press/

[13] J.Klamka: Function of 2-D matrix. Foundations of Control Engineering, vol. 9, 1984 /in press/

[14] E.B.Lee, S.H.Żak: Smith forms over $R[z_1,z_2]$. IEEE Trans.Autom. Control, vol. AC-28, No.1, January 1983, pp. 115-118

[15] M.Morf, B.C.Lévy, S.Y.Kung: New results in 2-D systems theory, part I: 2-D polynomial matrices, factorization and coprimeness. Proc. of IEEE, vol. 65, No.6, June 1977, pp.861-872

[16] M.Šebek: 2-D Exact model matching. IEEE Trans. Autom.Control, vol AC-28, No.2, February 1983, pp. 215-217

[17] M.Šebek: 2-D polynomial equations. Kybernetika, vol. 19, No.3, 1983, pp. 212-224

[18] D.Youla, G.Gnavi: Notes of n-Dimensional System Theory. IEEE Trans.Circ. and Systems. vol.CAS-26.No.2,Feb.1979,pp.105-111.